DNAPL
Site Evaluation

Robert M. Cohen
James W. Mercer

GeoTrans, Inc.
46050 Manekin Plaza, Suite 100
Sterling, Virginia 21066

John Matthews
EPA Project Officer

Robert S. Kerr Environmental Research Laboratory
Office of Research and Development
U.S. Environmental Protection Agency
Ada, Oklahoma 74820

CRC Press
Taylor & Francis Group
Boca Raton London New York

CRC Press is an imprint of the
Taylor & Francis Group, an **informa** business

NOTICE

The information in this document has been funded wholly or in part by the United States Environmental Protection Agency under contract 68-C8-0058 to Dynamac Corporation. This report has been subjected to the Agency's peer and administrative review and has been approved for publication as an EPA document. Mention of trade names or commercial products does not constitute endorsement or recommendation for use.

All research projects making conclusions or recommendations based on environmentally related measurements and funded by the Environmental Protection Agency are required to participate in the Agency Quality Assurance Program. This project did not involve environmentally related measurements and did not involve a Quality Assurance Project Plan.

FOREWORD

EPA is charged by Congress to protect the Nation's land, air and water systems. Under a mandate of national environmental laws focused on air and water quality, solid waste management and the control of toxic substances, pesticides, noise and radiation, the Agency strives to formulate and implement actions which lead to a compatible balance between human activities and the ability of natural systems to support and nurture life.

The Robert S. Kerr Environmental Research Laboratory is the Agency's center of expertise for investigation of the soil and subsurface environment. Personnel at the laboratory are responsible for management of research programs to: (a) determine the fate, transport and transformation rates of pollutants in the soil, the unsaturated and the saturated zones of the subsurface environment; (b) define the processes to be used in characterizing the soil and subsurface environment as a receptor of pollutants; (c) develop techniques for predicting the effect of pollutants on ground water, soil, and indigenous organisms; and (d) define and demonstrate the applicability and limitations of using natural processes, indigenous to soil and subsurface environment, for the protection of this resource.

Dense Nonaqueous Phase Liquids (DNAPLs), such as some chlorinated solvents, wood preservative wastes, coal tar wastes, and pesticides, are immiscible fluids with densities greater than water. As a result of widespread production, transportation, utilization, and disposal of hazardous DNAPLs, particularly since 1940, there are numerous DNAPL contamination sites in the United States. The potential for serious long-term contamination of groundwater by DNAPL chemicals is high at many sites due to their toxicity, limited solubility (but much higher than drinking water limits), and significant migration potential in soil gas, groundwater, and/or as a separate phase.

The goal of the EPA's Subsurface Cleanup and Mobilization Processes (SCAMP) research program is to improve the effectiveness of remedial activities at sites with subsurface contamination from DNAPLs. This report was prepared as part of the SCAMP research program and is designed to guide and assist investigators involved in the planning, implementation, and evaluation of DNAPL site characterization studies.

Clinton W. Hall

Clinton W. Hall
Director
Robert S. Kerr Environmental Research Laboratory

iv

CONTENTS

CONTENTS

CONTENTS

CONTENTS

CONTENTS

LIST OF TABLES

LIST OF TABLES

LIST OF TABLES

LIST OF TABLES

LIST OF FIGURES

LIST OF FIGURES

LIST OF FIGURES

LIST OF FIGURES

LIST OF FIGURES

LIST OF FIGURES

LIST OF FIGURES

LIST OF FIGURES

ACKNOWLEDGEMENTS

Several associates at GeoTrans, Inc. helped to prepare this document. Robin Parker authored much of Chapter 10. Anthony Bryda contributed to Chapter 9. Barry Lester performed the analytical modeling of soil gas transport described in Chapter 5.3.20. Tim Rogers helped research historic chemical production discussed in Chapter 3. James Mitchell and Brenda Cole prepared many of the figures contained herein. We are also indebted to Steve Schmelling, Chuck Newell, and Tom Sale for their reviews, and to the many investigators whose work forms the basis of this report.

1 EXECUTIVE SUMMARY

The potential for serious long-term contamination of groundwater by DNAPL chemicals is high at many sites due to their toxicity, limited solubility (but much higher than drinking water standards), and significant migration potential in soil gas, groundwater, and/or as a separate phase liquid. DNAPL chemicals, particularly, chlorinated solvents, are among the most prevalent contaminants identified in groundwater supplies and at contamination sites.

Remedial activities at a contaminated site need to account for the possible presence of DNAPL. If remediation is implemented at a DNAPL site, yet does not consider the DNAPL, the remedy will underestimate the time and effort required to achieve remediation goals. Thus, adequate site characterization is required to understand contaminant behavior and to make remedial decisions.

Based on the information presented, the main findings of this document include the following.

(1) The major types of DNAPLs are halogenated solvents, coal tar and creosote, PCB oils, and miscellaneous or mixed DNAPLs. Of these types, the most extensive subsurface contamination is associated with halogenated (primarily chlorinated) solvents due to their widespread use and properties (high density, low viscosity, significant solubility, and high toxicity).

(2) The physical and chemical properties of subsurface DNAPLs can vary considerably from that of pure DNAPL compounds due to: the presence of complex chemical mixtures; the effects of in-situ weathering; and, the fact that much DNAPL waste consists of off-specification materials, production process residues, and spent materials.

(3) DNAPL chemicals migrate in the subsurface as volatiles in soil gas, dissolved in groundwater, and as a mobile, separate phase liquid. This migration is governed by transport principles and the following chemical and media specific properties: saturation, interfacial tension, wettability, capillary pressure, residual saturation, relative permeability, solubility, vapor pressure, volatilization, density, and viscosity.

(4) DNAPL chemical migration is controlled by the interaction of these properties and principles with site-specific hydrogeologic and DNAPL release conditions. Based on this information, a conceptual model may be developed concerning the behavior of DNAPL in the subsurface. Various quantitative methods can be employed to examine DNAPL chemical transport within the framework provided by a site conceptual model. Conceptual models are used to guide site characterization and remedial activities.

(5) Subsurface DNAPL is acted upon by three distinct forces due to: (1) gravity (sometimes referred to as buoyancy), (2) capillary pressure, and (3) hydrodynamic pressure (also known as the hydraulic or viscous force). Each force may have a different principal direction of pressure and the subsurface movement of immiscible fluid is determined by the resolution of these forces.

(6) Gravity promotes the downward migration of DNAPL. The fluid pressure exerted at the base of a DNAPL body due to gravity is proportional to the DNAPL body height, the density difference between DNAPL and water in the saturated zone, and the absolute DNAPL density in the vadose zone.

(7) Capillary pressure resists the migration of nonwetting DNAPL from larger to smaller openings in water-saturated porous media. It is directly proportional to the interfacial liquid tension and the cosine of the DNAPL contact angle, and is inversely proportional to pore radius. Fine-grained layers with small pore radii, therefore, can act as capillary barriers to DNAPL migration. Alternatively, fractures, root holes, and coarse-grained strata with relatively large openings provide preferential pathways for nonwetting DNAPL migration. Capillary pressure effects cause lateral spreading of DNAPL above capillary barriers and also act to immobilize DNAPL at residual saturation and in stratigraphic traps. This trapped DNAPL is a long-term source of groundwater contamination and thereby hinders attempts to restore groundwater quality.

(8) The hydrodynamic force due to hydraulic gradient can promote or resist DNAPL migration and is usually minor compared to gravity and capillary pressures. The control on DNAPL movement exerted by the hydrodynamic force increases with: (a) decreasing gravitational pressure due to reduced DNAPL density and thickness; (b) decreasing capillary pressure due to the presence of coarse media, low interfacial tension, and a relatively high contact angle; and (c) increasing hydraulic gradient.

Mobile DNAPL can migrate along capillary barriers (such as bedding planes) in a direction opposite to the hydraulic gradient. This complicates site characterization.

(9) DNAPL presence and transport potential at contamination sites needs to be characterized because: (a) the behavior of subsurface DNAPL cannot be adequately defined by investigating miscible contaminant transport due to differences in properties and principles that govern DNAPL and solute transport; (b) DNAPL can persist for decades or centuries as a significant source of groundwater and soil vapor contamination; and (c) without adequate precautions or understanding of DNAPL presence and behavior, site characterization activities may result in expansion of the DNAPL contamination and increased remedial costs.

(10) Specific objectives of DNAPL site evaluation may include: (a) estimation of the quantities and types of DNAPLs released and present in the subsurface; (b) delineation of DNAPL release source areas; (c) determination of the subsurface DNAPL zone; (d) determination of site stratigraphy; (e) determination of immiscible fluid properties; (f) determination of fluid-media properties; and (g) determination of the nature, extent, migration rate, and fate of contaminants. The overall objectives of DNAPL site evaluation are to facilitate adequate assessments of site risks and remedies, and to minimize the potential for inducing unwanted DNAPL migration during remedial activities.

(11) Delineation of subsurface geologic conditions is critical to site evaluation because DNAPL movement can be largely controlled by the capillary properties of subsurface media. It is particularly important to determine, if practicable, the spatial distribution of fine-grained capillary barriers and preferential DNAPL pathways (e.g., fractures and coarse-grained strata).

(12) Site characterization should be a continuous, iterative process, whereby each phase of investigation and remediation is used to refine the conceptual model of the site.

(13) During the initial phase, a conceptual model of chemical presence, transport, and fate is formulated based on available site information and an understanding of the processes that control chemical distribution. The potential presence of DNAPL at a site should be considered in the initial phase of site characterization planning. Determining DNAPL presence should be a high priority at the onset of site investigation to guide the selection of site characterization methods. Knowledge or suspicion of DNAPL presence requires that special precautions be taken during field work to minimize the potential for inducing unwanted DNAPL migration.

(14) Assessment of the potential for DNAPL contamination based on historical site use information involves careful examination of: (a) land use since site development; (b) business operations and processes; (c) types and volumes of chemicals used and generated; and, (d) the storage, handling, transport, distribution, dispersal, and disposal of these chemicals and operation residues. Pertinent information can be obtained from corporate records, government records, historical society documents, interviews with key personnel, and historic aerial photographs.

(15) DNAPL presence can also be: (a) determined directly by visual examination of subsurface samples; (b) inferred by interpretation of chemical analyses of subsurface samples; and/or (c) suspected based on interpretation of anomalous chemical distribution and hydrogeologic data. However, due to limited and complex distributions of DNAPL at some sites, its occurrence may be difficult to detect, leading to inadequate site assessments and remedial designs.

(16) Under ideal conditions, DNAPL presence can be identified by direct visual examination of soil, rock, and fluid samples. Direct visual detection may be difficult, however, where DNAPL is colorless, present in low saturation, or distributed heterogeneously. Direct visual detection can be enhanced by the use of hydrophobic dye to colorize NAPL in water and soil samples during a shake test, ultraviolet fluorescence analysis (for fluorescent DNAPLs), and/or centrifugation to separate fluid phases. For volatile NAPLs, analysis of organic vapors emitted from soil samples can be used to screen samples for further examination.

(17) Indirect methods for assessing the presence of DNAPL in the subsurface rely on comparing

measured chemical concentrations to effective solubility limits for groundwater and to calculated equilibrium partitioning concentrations for soil and groundwater. Where present as a separate phase, DNAPL compounds are generally detected at <10% of their aqueous solubility limit in groundwater due to the effects of non-uniform groundwater flow, variable DNAPL distribution, the mixing of groundwater in a well, and the reduced effective solubility of individual compounds in a multi-liquid NAPL mixture. Typically, dissolved contaminant concentrations >1% of the aqueous solubility are highly suggestive of NAPL presence. Concentrations less than 1% of the solubility limit, however, are not necessarily indicative of NAPL absence. In soil, contaminant concentrations in the percent range are generally indicative of NAPL presence. However, NAPL may also be present at much lower concentrations. The presence of subsurface DNAPL can also be inferred from anomalous contaminant distributions, such as higher dissolved concentrations with depth beneath a shallow release area or higher concentrations upgradient hydraulically from a release area.

(18) Noninvasive methods can often be used during the early phases of field work to optimize the cost-effectiveness of a DNAPL site characterization program. Specifically, surface geophysical surveys, soil gas analysis, and air photointerpretation can facilitate characterization of contaminant source areas, geologic controls on contaminant movement (e.g., stratigraphy and utilities), and the extent of subsurface contamination. Conceptual model refinements derived using these methods reduce the risk of spreading contaminants during subsequent invasive field work.

(19) The value of surface geophysics at most DNAPL sites will be to aid characterization of waste disposal areas, stratigraphic conditions, and potential routes of contaminant migration. The use of surface geophysical surveys for direct detection of NAPL is currently limited by a lack of: (a) demonstrable methods, (b) documented successes, and (c) environmental geophysicists trained in these techniques.

(20) Soil gas analysis can be an effective screening tool for detecting volatile organic compounds in the vadose zone. Consideration should be given to its use during the early phases of site investigation to assist delineation of volatile DNAPL in the vadose zone, contaminant source areas, contaminated shallow groundwater, and contaminated soil gas; and, thereby, guide subsequent invasive field work.

(21) Aerial photographs should be acquired during the initial phases of site characterization study to facilitate analysis of waste disposal practices and locations, drainage patterns, geologic conditions, signs of vegetative stress, and other factors relevant to contamination site assessment. Additionally, aerial photograph fracture trace analysis should be considered at sites where bedrock contamination is a concern.

(22) Following development of the site conceptual model based on available information and noninvasive field methods, invasive techniques will generally be required to advance site characterization and enable the conduct of risk and remedy assessments. Various means of subsurface exploration are utilized to directly observe and measure subsurface materials and conditions. Generally, these invasive activities include: (a) drilling and test pit excavation to characterize subsurface solids and liquids, and (b) monitor well installation to sample fluids, and to conduct fluid level surveys, hydraulic tests, and borehole geophysical surveys.

(23) Invasive methods will generally be used to: (a) delineate DNAPL source (entry) areas; (b) define the stratigraphic, lithologic, structural, and/or hydraulic controls on the movement and distribution of DNAPL, contaminated groundwater, and contaminated soil gas; (c) characterize fluid and fluid-media properties that affect DNAPL migration and the feasibility of alternative remedies; (d) estimate or determine the nature and extent of contamination, and the rates and directions of contaminant transport; (e) evaluate exposure pathways; and, (f) design monitoring and remedial systems.

(24) The risk of enlarging the zone of chemical contamination by use of invasive methods is an important consideration that must be evaluated during site characterization. Drilling, well installation, and pumping activities typically present the greatest risk of promoting DNAPL migration during site investigation. Precautions should be taken to minimize these risks. DNAPL transport caused by characterization activities may: (a)

heighten the risk to receptors; (b) increase the difficulty and cost of remediation; and (c) generate misleading data, leading to the development of a flawed conceptual model, and flawed assessments of risk and remedy.

(25) Drilling methods have a high potential for promoting downward DNAPL migration. For example, DNAPL trapped in structural or stratigraphic lows can be mobilized by site characterization activities (e.g., drilling through a DNAPL pool). To minimize drilling risks, DNAPL investigators should: (a) avoid unnecessary drilling within the DNAPL zone; (b) minimize the time during which a boring is open; (c) minimize the length of hole which is open at any time; (d) use telescoped casing drilling techniques to isolate shallow contaminated zones from deeper zones; (e) utilize knowledge of site stratigraphy and chemical distribution, and carefully examine subsurface materials brought to the surface as drilling proceeds, to avoid drilling through a barrier layer beneath DNAPL; (f) consider using a dense drilling mud to prevent DNAPL from sinking down the borehole during drilling; and, (g) select optimum well materials and grouting methods based on consideration of site-specific chemical compatibility.

(26) Monitor wells are installed to characterize immiscible fluid distributions, flow directions and rates, groundwater quality, and media hydraulic properties. Pertinent data are acquired by conducting fluid thickness and elevation surveys, hydraulic tests, and borehole geophysical surveys. The locations and design of monitor wells are selected based on consideration of the site conceptual model and specific data collection objectives. Inadequate well design can increase the potential for causing vertical DNAPL migration and misinterpretation of fluid elevation and thickness measurements.

(27) As knowledge of the site increases and becomes more complex, the conceptual model may take the form of either a numerical or analytical model. Data collection continues until the conceptual model is proven sufficiently.

(28) At some sites, it will be very difficult and/or impractical to determine the subsurface DNAPL distribution and DNAPL transport pathways in a detailed manner. Subsurface characterization is particularly complicated by the presence of heterogeneous strata, fractured media, and complex DNAPL releases. The significance of incomplete site characterization should be analyzed during the conduct of risk and remedy assessments.

(29) During implementation of a remedy, the subsurface system is stressed. This provides an opportunity to monitor and not only learn about the effectiveness of the remediation, but to learn more about the subsurface. Therefore, remediation (especially pilot studies) should be considered part of site characterization, yielding data that may allow improvements to be made in the conduct of the remediation effort.

(30) Site characterization, data analysis, and conceptual model refinement are iterative activities which should satisfy the characterization objectives needed to facilitate risk and remedy assessments. As emphasized throughout this manual, site characterization includes both data collection and data interpretation. For sites involving DNAPL, challenges are presented to both activities, but perhaps data interpretation is most challenging. For that reason, in addition to site characterization techniques, data interpretation should be emphasized during all phases of site characterization.

There is no practical cookbook approach to DNAPL site investigation or data analysis. Each site presents variations of contaminant transport conditions and issues. Although there are no certain answers to many of the DNAPL site evaluation issues, this manual provides a framework for their evaluation.

2 INTRODUCTION

Dense nonaqueous phase liquids (DNAPLs), such as some chlorinated solvents, creosote based wood-treating oils, coal tar wastes, and pesticides, are immiscible fluids with a density greater than water. As a result of widespread production, transportation, utilization, and disposal of hazardous DNAPLs, particularly since 1940, there are numerous DNAPL contamination sites in North America and Europe. The potential for serious long-term contamination of groundwater by some DNAPL chemicals at many sites is high due to their toxicity, limited solubility (but much higher than drinking water limits), and significant migration potential in soil gas, groundwater, and/or as a separate phase (Figure 2-1). DNAPL chemicals, especially chlorinated solvents, are among the most prevalent groundwater contaminants identified in groundwater supplies and at waste disposal sites.

The subsurface movement of DNAPL is controlled substantially by the nature of the release, the DNAPL density, interfacial tension, and viscosity, porous media capillary properties, and, usually to a lesser extent, hydraulic forces. Below the water table, non-wetting DNAPL migrates preferentially through permeable pathways such as soil and rock fractures, root holes, and sand layers that provide relatively little capillary resistance to flow. Visual detection of DNAPL in soil and groundwater samples may be difficult where the DNAPL is transparent, present in low saturation, or distributed heterogeneously. These factors confound characterization of the movement and distribution of DNAPL even at sites with relatively homogenous soil and a known, uniform DNAPL source. The difficulty of site characterization is further compounded by fractured bedrock, heterogeneous strata, multiple DNAPL mixtures and releases, etc.

Obtaining a detailed delineation of subsurface DNAPL, therefore, can be very costly and may be impractical using conventional site investigation techniques. Furthermore, the risk of causing DNAPL migration by drilling or other actions may be substantial and should be considered prior to commencing field work. Although DNAPL can greatly complicate site characterization, failure to adequately define its presence, fate, and transport can result in misguided investigation and remedial efforts. Large savings and environmental benefits can be realized by conducting studies and implementing remedies in a cost-effective manner. Cost-effective DNAPL site management requires an understanding of DNAPL properties and migration processes, and of the methods available to investigate and interpret the transport and fate of DNAPL in the subsurface.

Lighter-than-water NAPLs (LNAPLs) which do not sink through the saturated zone, such as petroleum products, are also present and cause groundwater contamination at numerous sites. Although many of the same principles and concerns apply at both LNAPL and DNAPL sites, LNAPL site characterization is not specifically addressed in this document.

2.1 MANUAL OBJECTIVES

This manual is designed to guide investigators involved in the planning and implementation of characterization studies at sites suspected of having subsurface contamination by DNAPLs. Specifically, the document is intended to:

- Summarize the current state of knowledge for characterizing DNAPL-contaminated sites;

- Develop a framework for planning and implementing DNAPL site characterization activities;

- Provide a detailed discussion of the types of data, tools, and methods that can be used to identify, characterize, and monitor DNAPL sites, and an analysis of their utility, limitations, risks, availability, and cost;

- Identify and illustrate methods, including the development of conceptual models, to interpret contaminant fate and transport at DNAPL sites based on the data collected;

- Assess new and developing site characterization methodologies that may be valuable and identify additional research needs; and,

- Review the scope of the DNAPL contamination problem, the properties of DNAPLs and media, and DNAPL transport processes to provide context for understanding DNAPL site characterization.

The primary goal of this manual is to help site managers minimize the risks and maximize the cost-effectiveness of site investigation/remediation by providing the best information available to describe and evaluate activities that can be used to determine the presence, fate, and transport of subsurface DNAPL contamination.

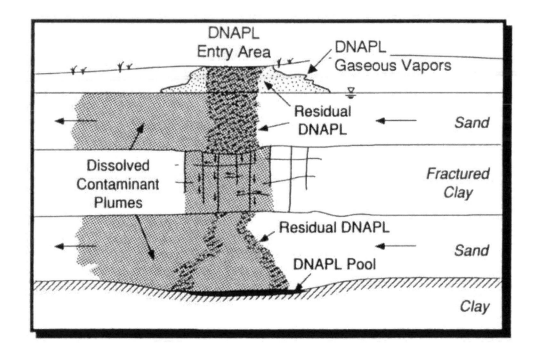

Figure 2-1.   DNAPL chemicals are distributed in several phases: dissolved in groundwater, adsorbed
to soils, volatilized in soil gas, and as residual and mobile immiscible fluids (modified
from Huling and Weaver, 1991; WCGR, 1991).

2.2 DERIVATION OF MANUAL

DNAPL contamination has been identified by USEPA at numerous hazardous waste sites and is suspected to exist at many others. Growing recognition of the significance of DNAPL contamination has prompted the USEPA's Robert S. Kerr Environmental Research Laboratory to sponsor research and technology transfer concerning DNAPL issues. Pertinent DNAPL documents published recently by the Kerr Laboratory include: *Laboratory Investigation of Residual Liquid Organics from Spills, Leaks, and the Disposal of Hazardous Wastes in Groundwater* (EPA/600/6-90/004), *Dense Nonaqueous Phase Liquids -- Ground Water Issue Paper* (EPA/540/4-91-002), *Dense Nonaqueous Phase Liquids -- A Workshop Summary, Dallas Texas, April 17-18, 1991* (EPA/600/R-92/030) and *Estimating Potential for Occurrence of DNAPL at Superfund Sites* (EPA Quick Reference Fact Sheet, 8/91).

An objective of the DNAPL Workshop Meeting held in Dallas, Texas during April 1991 was to develop a document covering DNAPL fundamentals, site characterization, remedial alternatives, and promising topics for further investigation. Although the document that evolved from the Workshop summarizes the main observations and conclusions of participants regarding DNAPL site investigation and remediation concerns, it is not a detailed site characterization manual.

2.3 DNAPL SITE INVESTIGATION ISSUES

Numerous site characterization guidance documents have been developed by USEPA and others during the past decade (Table 2-1). Few of these manuals focus on characterization issues specific to DNAPL sites. The primary additional investigation concerns at DNAPL sites relate to: (a) the risk of inducing DNAPL migration by drilling, pumping, or other field activities; (b) the use of special sampling and measurement methods to assess DNAPL presence and migration potential; and, (c) development of a cost-effective characterization strategy that accounts for DNAPL chemical transport processes, the risk of inducing DNAPL movement during field work, and the data required to select and implement a realistic remedy.

The potential presence of DNAPL at a contamination site dictates consideration of the following issues/questions during scoping, conduct, and interpretation of a site characterization study.

- What questions should be asked prior to initiating a field investigation to determine if a site is a potential DNAPL site? What do preexisting site data indicate regarding the presence and distribution of DNAPL?

- If DNAPL contamination is possible, how should the site be sampled? What should the goal of sampling be and how does this affect the sampling strategy?

- What are the risks of sampling and what technical considerations should be made in order to balance those risks against the risks of not sampling? How do these risks and technical considerations vary with different geologic environments and DNAPLs?

- Given some knowledge of the type of DNAPL presence and site hydrogeology, can a preliminary conceptual model of DNAPL chemical migration be formulated? If so, how can the site investigation be optimized based on the development of a conceptual model?

- How do we determine where to place and install monitor wells? What materials should be used for well construction and other equipment which contacts DNAPL?

- Is it possible to isolate zones of DNAPL contamination through use of telescoped well construction (e.g. double-cased wells) or other techniques?

- What methods are available for characterizing DNAPL sites? How accessible, reliable, and costly are the various methods?

- How should noninvasive techniques, such as soil gas analysis and surface geophysical surveys, be used to characterize DNAPL sites?

- How can the extent, volume, and mobility of subsurface DNAPL be estimated?

- What data are necessary to facilitate evaluation of alternative remedial options at DNAPL sites? What data are necessary to conduct risk assessments? What data are necessary to meet regulatory requirements?

Although there are no certain answers to many of these questions, this manual provides a framework for their evaluation.

Table 2-1. Contamination Site Investigation Guidance Documents.

AASHTO, 1988. Manual on subsurface investigations, American Association of State Highway and Transportation Officials, 391 pp.

Aller, L., T.W. Bennett, G. Hackett, R.J. Petty, J.H. Lehr, H. Sedoris, D.M. Nielsen, and J.E. Denne, 1989. Handbook of Suggested Practices for the Design and Installation of Ground-Water Monitoring Wells, USEPA-600/4-89/034, National Water Well Association, Dublin, Ohio, 398 pp.

American Petroleum Institute, 1989. A guide to the assessment and remediation of underground petroleum releases, API Publication 1628, Washington, D.C., 81 pp.

Barcelona, M.J., J.P. Gibb, J.A. Helfrich, and E.E. Garske, 1985. Practical guide for ground-water sampling, USEPA/600/2-85-104, 169 pp.

Barcelona, M.J., J.P. Gibb, and R.A. Miller, 1983. A guide to the selection of materials for monitoring well construction and ground-water sampling, Illinois State Water Survey Contract Report 327 to USEPA R.S. Kerr Environmental Research Laboratory, EPA Contract CR-809966-01, 78 pp.

Barcelona, M.J., J.F. Keely, W.A. Pettyjohn, and A. Wehrmann, 1987. Handbook ground water, USEPA/625/6-87/016, 212 pp.

Berg, E.L., 1982. Handbook for sampling and sample preservation of water and wastewater, USEPA/600/4-82-029.

Claasen, H.C., 1982. Guidelines and techniques for obtaining water samples that accurately represent the water chemistry of an aquifer, USGS Open-File Report 82-1024, 49 pp.

dePastrovich, T.L., Y. Baradat, R. Barthell, A. Chiarelli, and D.R. Fussell, 1979. Protection of groundwater from oil pollution, CONCAWE (Conservation of Clean Air and Water - Europe), The Hague, 61 pp.

Devitt, D.A., R.B. Evans, W.A. Jury, and T.H. Starks, 1987. Soil gas sensing for detection and mapping of volatile organics, USEPA/600/8-87/036, 266 pp.

Dunlap, W.J., J.F. McNabb, M.R. Scalf, and R.L. Cosby, 1977. Sampling for organic chemicals and microorganisms in the subsurface, USEPA/600/2-77/176 (NTIS PB272679).

Electric Power Research Institute, 1985. Preliminary results on chemical changes in groundwater samples due to sampling devices, EPRI EA-4118 Interim Report, Palo Alto, California.

Electric Power Research Institute, 1989. Techniques to develop data for hydrogeochemical models, EPRI EN-6637, Palo Alto, California.

Fenn, D., E. Cocozza, J. Isbister, O. Braids, B. Yard, and P. Roux, 1977. Procedures manual for ground water monitoring at solid waste disposal facilities, USEPA SW-611, 269 pp.

Ford, P.J., D.J. Turina, and D.E. Seely, 1984. Characterization of hazardous waste sites -- A methods manual, Volume II Available sampling methods, USEPA-600/4-84-076 (NTIS PB85-521596).

Fussell, D.R., H. Godjen, P. Hayward, R.H. Lilie, A. Marco, and C. Panisi, 1981. Revised inland oil spill clean-up manual, CONCAWE (Conservation of Clean Air and Water - Europe), The Hague, Report No. 7/81, 150 pp.

Gas Research Institute, 1987. Management of manufactured gas plant sites, GRI-87/0260, Chicago, Illinois.

GeoTrans, Inc., 1983. RCRA permit writer's manual ground-water protection, USEPA Contract No. 68-01-6464, 263 pp.

GeoTrans, Inc., 1989. Groundwater monitoring manual for the electric utility industry, Edison Electric Institute, Washington, D.C.

Klute, A., ed., 1986. Methods of Soil Analysis, Soil Science Society of America, Inc., Madison, WI, 1188 pp.

Mercer, J.W., D.C. Skipp, and D. Giffin, 1990. Basics of pump-and-treat ground-water remediation technology, USEPA-600/8-90/003.

Nielsen, D.M. (ed.), 1991. Practical Handbook of Ground Water Monitoring, Lewis Publishers, Chelsea, Michigan, 717 pp.

NJ Department of Environmental Protection, 1988. Field sampling procedures manual, Hazardous Waste Program, NJDEP, Trenton, New Jersey.

Rehm, B.W., T.R. Stolzenburg, and D.G. Nichols, 1985. Field measurement methods for hydrogeologic investigations: A critical review of the literature, Electric Power Research Institute Report EPRI EA-4301, Palo Alto, California.

Sara, M.N., 1989. Site assessment manual, Waste Management of North America, Inc., Oak Brook, Illinois.

Scalf, M.R., J.F. McNabb, W.J. Dunlap, R.L. Cosby, and J.S. Fryberger, 1981. Manual of Ground-Water Quality Sampling Procedures, National Water Well Association, Worthington, Ohio, 93 pp.

Simmons, M.S., 1991. Hazardous Waste Measurements, Lewis Publishers, Chelsea, Michigan, 315 pp.

Sisk, S.W., 1981. NEIC manual for groundwater/subsurface investigations at hazardous waste sites, USEPA-330/9-81-002 (NTIS PB82-103755).

Skridulis, J., 1984. Comparison of guidelines for monitoring well design, installation and sampling practices, NUS Corporation Report to USEPA, Contract No. 68-01-6699.

Summers, K.V., and S.A. Gherini, 1987. Sampling guidelines for groundwater quality, Electric Power Research Institute report EPRI EA-4952, Palo Alto, California.

USEPA, 1986. RCRA ground-water monitoring technical enforcement guidance document, OSWER-9550.1.

USEPA, 1987. A compendium of Superfund field operations methods, USEPA/540/P-87/001.

USEPA, 1987. Handbook -- Ground Water, EPA/625/6-87/016, 212 pp.

USEPA, 1988. Guidance for conducting remedial investigations and feasibility studies under CERCLA, USEPA/540-G-89/004.

USEPA, 1988. Field screening methods catalog - User's guide, USEPA/540/2-88/005.

USEPA, 1989. Interim final RCRA facility investigation (RFI) guidance, USEPA/530/SW-89/031.

USEPA, 1991. Seminar publication -- Site characterization of subsurface remediation, EPA/625/4-91/026, 259 pp.

USGS, 1977. National handbook of recommended methods for water-data acquisition.

Wilson, L.G., 1980. Monitoring in the vadose zone: A review of technical elements and methods, USEPA-600/7-80-134, 168 pp.

2.4 DNAPL SITE INVESTIGATION PRACTICE

Remedial investigations were conducted at numerous DNAPL sites in the United States during the 1980s pursuant to requirements of the Comprehensive Environmental Response, Compensation, and Liability Act (CERCLA) and the Resource Conservation and Recovery Act (RCRA). Unfortunately, dissemination of procedures and strategies for DNAPL site investigation has been slow. This is due to prior limited recognition of the DNAPL problem and the fact that many DNAPL sites are the subject of litigation which has stifled publication of investigation results.

Within the past few years, however, recognition of the significance of DNAPL contamination has grown considerably. This is evidenced by and due to events which include publication of an English translation of Friedrich Schwille's DNAPL experiments entitled *Dense Chlorinated Solvents in Porous and Fractured Media* in 1988, increasing focus on DNAPL issues during the annual conference on *Petroleum Hydrocarbons and Organic Chemicals in Ground Water* sponsored by the National Ground Water Association and the American Petroleum Institute, presentation of DNAPL contamination short courses by the Waterloo Centre for Groundwater Research, organization of the *Conference on Subsurface Contamination by Immiscible Fluids* by the International Association of Hydrologists in Calgary during April 1990, and increased attention given to DNAPL issues by USEPA and technical journals (such as *Ground Water, Ground Water Monitoring Review, Water Resources Research*, and the *Journal of Contaminant Hydrology*).

2.5 MANUAL ORGANIZATION

The organization of this document is outlined in the DNAPL site characterization flow chart given in Figure 2-2. Discussions of the scope of the DNAPL contamination problem (Chapter 3), the properties of DNAPLs and media (Chapter 4), and DNAPL transport processes (Chapter 5) are provided as a foundation for understanding DNAPL site characterization methods and strategies. Objectives and strategies of DNAPL site characterization are described in Chapter 6. The identification of DNAPL sites based on historic information and preexisting data is addressed in Chapter 7. This is followed by an examination of: noninvasive characterization methods (i.e., soil gas analysis and surface geophysical techniques) in Chapter 8, invasive methods (i.e., test pits and drilling) in Chapter 9, and laboratory methods for characterizing fluid and media properties in Chapter 10. Several case histories illustrating investigation findings and problems specific to DNAPL sites are provided in Chapter 11. Priority research needs are discussed briefly in Chapter 12; and references for the entire document are listed in Chapter 13. Chemical properties of DNAPL chemicals, a listing of parameters and conversion factors, and a glossary of terms related to DNAPL site contamination are given in Appendices A, B, and C, respectively. The Executive Summary (Chapter 1) outlines the main findings of this document.

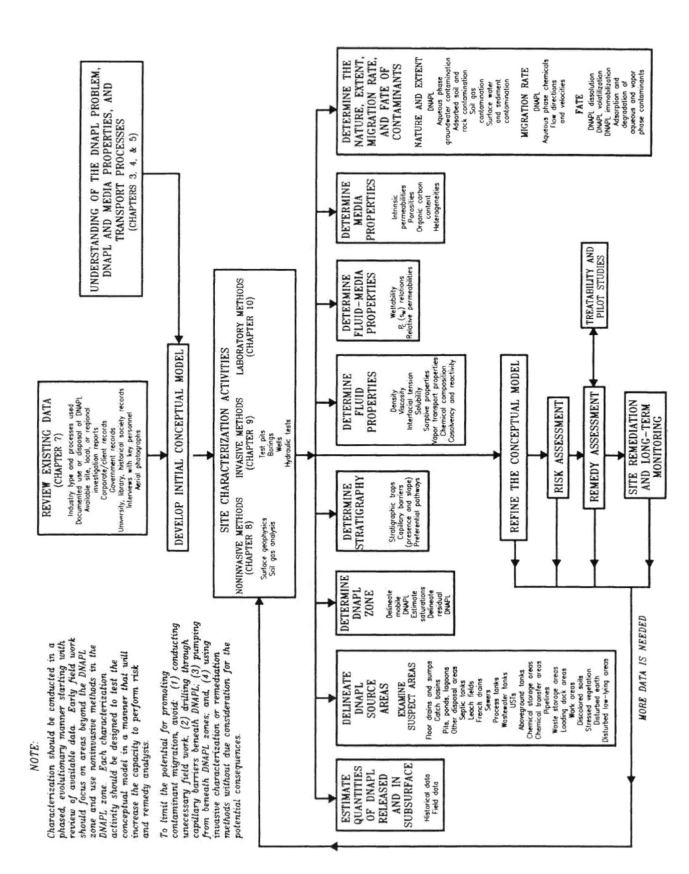

Figure 2-2. DNAPL site characterization flow chart.

3 DNAPL TYPES AND SCOPE OF PROBLEM

3.1 INTRODUCTION

The potential for widespread contamination of groundwater by DNAPLs is substantial because of the extensive production, transport, utilization, and disposal of large volumes of DNAPL chemicals during the 20th Century. Some DNAPL wastes are older, such as coal tar generated as early as the 1820s at manufactured gas sites in the eastern U.S. There are literally thousands of sites in North America where DNAPLs may have been released to the subsurface in varying quantities (NRC, 1990, p.24). At some of these sites, tons of DNAPL were released over time and DNAPL presence is obvious in subsurface materials. At most sites, however, the limited volume of DNAPL present hinders direct identification even though it is sufficient to provide a source for significant groundwater contamination. As a result, DNAPL chemicals are frequently detected at contamination sites even where DNAPL presence has not been determined.

The most prevalent DNAPL types are outlined in Table 3-1 with summary information on DNAPL density, viscosity, production, and usage, and selected contamination site references. The major DNAPL types include: halogenated solvents, coal tar and creosote, PCB oils, and miscellaneous or mixed DNAPLs. Of these types, the most extensive subsurface contamination is associated with halogenated (primarily chlorinated) solvents, either alone or within mixed DNAPL sites, due to their widespread use and properties (high density, low viscosity, significant solubility, and high toxicity). As shown in Figure 3-1, pure chlorinated solvents are generally much more mobile than creosote/coal tar and PCB oil mixtures due to their relatively high density:viscosity ratios.

Although the physical properties of pure DNAPL chemicals are well-defined, the physical properties of subsurface DNAPLs can vary considerably from that of the pure DNAPL compounds due to:

- the presence of complex chemical mixtures,

- the effects of in-situ weathering (dissolution, volatilization, and degradation of the less persistent DNAPL fractions), and,

- the fact that much DNAPL waste consists of off-specification materials, production process residues, and spent materials.

Physical and chemical properties of numerous DNAPL chemicals are provided in Appendix A.

3.2 HALOGENATED SOLVENTS

Halogenated solvents, particularly chlorinated hydrocarbons, and brominated and fluorinated hydrocarbons to a much lesser extent, are DNAPL chemicals encountered at contamination sites. These halocarbons are produced by replacing one or more hydrogen atoms with chlorine (or another halogen) in petrochemical precursors such as methane, ethane, ethene, propane, and benzene. Many bromocarbons and fluorocarbons are manufactured by reacting chlorinated hydrocarbon intermediates (such as chloroform or carbon tetrachloride) with bromine and fluorine compounds, respectively. DNAPL halocarbons at ambient environmental conditions include: chlorination products of methane (methylene chloride, chloroform, carbon tetrachloride), ethane (1,1-dichloroethane, 1,2-dichloroethane, 1,1,1-trichloroethane, 1,1,2,2-tetrachloroethane), ethene (1,1-dichloroethene, 1,2-dichloroethene isomers, trichloroethene, tetrachloroethene), propane (1,2-dichloropropane, 1,3-dichloropropene isomers), and benzene (chlorobenzene, 1,2-dichlorobenzene, 1,4-dichlorobenzene); fluorination products of methane and ethane such as 1,1,2-trichlorofluormethane (Freon-11) and 1,1,2-trichlorotrifluorethane (Freon-113); and, bromination products of methane (bromochloromethane, dibromochloromethane, dibromodifluoromethane, bromoform), ethane (bromoethane, 1,1,2,2-tetrabromoethane), ethene (ethylene dibromide), and propane (1,2-dibromo-3-chloropropane).

Although most chlorinated solvents were first synthesized during the 1800s, large-scale production generally began around the middle of the 1900s primarily for use as solvents and chemical intermediates. Annual production rates at five-year intervals between 1920 and 1990 for selected chlorinated solvents and other DNAPL chemicals are shown in Table 3-2 and Figure 3-2. An estimated 29 billion lbs of chlorinated hydrocarbons were produced in the U.S. during 1990 (U.S. International Trade Commission, 1991). The uses of six common chlorinated solvents are illustrated in Figure 3-3.

Fluorocarbons were discovered in the search for improved refrigerants by General Motors in 1930. In the U.S.,

Table 3-1. DNAPL Types: Sources, Use, Properties, Contamination, and References.

DNAPL Type, Derivation, Specific Gravity and Absolute Viscosity	Uses and Contamination Sources	Selected Statistics on Extent of DNAPL Production, Usage, and Contamination	Selected References
Halogenated Hydrocarbon Solvents (such as dichloroethene, trichloroethene, tetrachloroethene, dichloroethane, trichloroethane, methylene chloride, chloroform, carbon tetrachloride, bromoform, ethylene dibromide, chlorobenzene, and chlorotoluene). Halogenated hydrocarbon solvents are produced by replacing one or more hydrogen atoms with chlorine (or another halogen) in petrochemical precursors such as methane, ethylene, and benzene. Specific gravities at 20° C. of the pure halogenated solvents listed above range from approximately 1.08 (chlorotoluene) to 2.89 (bromoform). Absolute viscosities at 20° C. of the pure compounds listed above range from approximately 0.36 cp (1,1-dichloroethene) to 2.04 cp (bromoform).	Chemical manufacturing Solvent manufacturing, reprocessing, and/or packaging Vapor degreasing operations Commercial dry cleaning operations Electronic equipment manufacturing Dry plasma etching of semiconductor chips Computer parts and products manufacturing Metal parts/products manufacturing Aircraft and automotive manufacturing, maintenance and repair operations Machine shops and metal works Tool-and-die plants Musical instrument manufacturing Photographic film manufacturing and processing Dye and paint manufacturing Pharmaceutical manufacturing Plastics manufacturing Flame retardant materials manufacturing Refrigerants manufacturing Military equipment manufacturing and maintenance Printing presses and publishing operations Septic tank cleaners Textile processing, dying, and finishing operations Solvent and carrier fluid formulations in rubber coatings, solvent soaps, printing inks, adhesives and glues, sealants, polishes, lubricants, and silicones Insecticide and herbicide production Waste disposal sites	Halogenated (particularly chlorinated) solvents have been widely used in manufacturing and cleaning industries since about the 1940s. Production and use of these solvents has generally proliferated with time, but demand for some solvents has diminished due to environmental, cost, and/or other factors. U.S. production estimates for selected chlorinated solvents in 1990 include: 484,000,000 lbs of chloroform; 461,000,000 lbs of methylene chloride; 802,000,000 lbs of 1,1,1-trichloroethane; 372,000,000 lbs of tetrachloroethene; and 237,000,000 lbs of chlorobenzene (U.S. International Trade Commission, 1991). An estimated 620,400,000 lbs of waste solvents were produced by degreasing operations in 1974. Halogenated solvents are among the contaminants most frequently detected at subsurface contamination sites. They are associated with waste disposal and chemical releases from numerous industries and operations. Due to their extensive release, high density, low viscosity, and toxic nature, chlorinated solvents present the most severe DNAPL contamination problem.	Chlorinated solvent contamination at the Gloucester Landfill in Ottawa, Ontario (Jackson et al., 1985; Jackson and Patterson, 1989; Jackson et al., 1990; and LeSage et al., 1990) Solvent contamination at the IBM Dayton facility in South Brunswick, N.J. (Althoff et al., 1981; CH2M-Hill, 1989; Stipp, 1991) Solvent contamination of fractured shale at Oak Ridge, Tennessee (Kueper et al., 1991) VOC contamination at the General Mills site in Minneapolis, Minnesota (CH2M-Hill, 1989) Chlorinated solvent contamination of production wells in Birmingham, U.K. (Rivett et al., 1990) Release of 1,2-dichloroethane due to a train derailment in British Columbia (Dakin and Holmes, 1987) Other References: Begor et al. (1989), CH2M-Hill (1989), Feenstra and Cherry (1988), Mackay and Cherry (1989), Holmes and Cambell (1990), Robertson (1992), Schaumburg (1990)

Table 3-1. DNAPL Types: Sources, Use, Properties, Contamination, and References.

DNAPL Type, Derivation, Specific Gravity and Absolute Viscosity	Uses and Contamination Sources	Selected Statistics on Extent of DNAPL Production, Usage, and Contamination	Selected References
Coal Tar and Creosote Derived from the destructive distillation of coal in coke ovens and retorts, coal tar is composed of thousands of hydrocarbons dominated by PAHs (polycyclic aromatic hydrocarbons with a substantial content of naphthalene compounds) that are mixed with lesser amounts of tar acids such as phenol and cresol; N-, S-, and O-heterocyclic aromatic compounds; and <5% BTEX (benzene, toluene, ethylbenzene, and xylenes). Creosote consists of various coal tar distillates (primarily the 200–400° C. fractions) blended to meet product standards. It is estimated to contain approximately 85% PAHs, 10% phenolic compounds, and 5% N-, S-, and O- heterocyclics. For wood-treatment applications, creosote may be applied undiluted or mixed with coal tar or a petroleum oil (such as diesel fuel) in ratios that range from 80:20 to 50:50, creosote:carrier. The specific gravity of creosote is typically between 1.01 and 1.05, but ranges up to 1.14 in certain blends. The specific gravity of coal tar ranges between 1.01 and 1.18. The viscosity of creosote and coal tar is generally much greater than that of water, typically in the range of 10 to 70 cp.	Wood treating plants Coal tar distillation plants Steel industry coking operations Manufactured gas (coal gasification) plants Roofing tars Road tars	There are an estimated 700 active and abandoned wood-treatment plants in the U.S. In 1978, 188 of 631 wood-treatment plants in the U.S., operating predominantly along the eastern seaboard, in the southeast, and in the Pacific northwest, were reported using creosote and/or coal tar (McGinnis, 1989). Manufactured gas plants produced "town" gas for lighting and heating primarily between 1850 and 1950. More than 900 coal gasification plants were operational in the U.S. in 1920 (Rhodes, 1979). There were 64 coal tar producers and 24 coal tar distillation plants producing creosote in the U.S. in 1972 (USDA, 1980). In 1986, an estimated 1000 million lbs of creosote were utilized by 415-550 creosoting operations in the U.S. (Mueller et al., 1989). In 1990, an estimated 596 million liters of crude coal tar and 297 million liters of creosote oil were produced in the U.S. (U.S. International Trade Commission, 1991). Large quantities of DNAPL were released at some wood treatment and manufactured gas plant sites due to the longterm use and/or generation of large quantities of coal tar and creosote chemicals and the waste disposal practices at these sites. DNAPL contamination at wood treatment sites derives from leaking tanks and pipelines, dripping treated lumber, leaking holding ponds, etc. In 1988, approximately 40 creosote/coal tar wood treatment sites were on the National Priority List of CERCLA sites (USEPA, 1988).	Union Pacific Railroad Tie Plant in Laramie, Wyoming (Sale and Piontek, 1988; Sale et al., 1988; and Sale et al., 1989) Abandoned creosote waste site in Conroe, Texas (Bedient et al., 1984) American Creosote Works site in Pensacola, Florida (Troutman et al., 1984; Mattraw and Franks, 1984; Goerlitz et al., 1985; USGS, 1985; and, Franks, 1987) Reilly Tar site in St. Louis Park, Minnesota (Erlich et al., 1982; and, Hult and Schoenberg, 1984) Manufactured gas plant site in Stroudsburg, Pennsylvania (Villaume et al., 1983; and, Villaume, 1985) Manufactured gas plant site in Wallingford, Connecticut (Conway et al., 1985; and, Quinn et al., 1985) Other references: Austin (1984), Baechler and MacFarlane (1990), Belanger et al. (1990), Edison Electric Institute (1984), Feenstra and Cherry (1990), Ghiorse et al. (1990), GRI (1987), Konasewich et al. (1990), Litherland and Anderson (1990), McGinnis et al. (1991), Mueller et al. (1989), Murarka (1990), Raven and Beck (1990), Rosenfeld and Plumb (1991), USDA (1980), USEPA (1989d), Villaume (1984)

Table 3-1. DNAPL Types: Sources, Use, Properties, Contamination, and References.

DNAPL Type, Derivation, Specific Gravity and Absolute Viscosity	Uses and Contamination Sources	Selected Statistics on Extent of DNAPL Production, Usage, and Contamination	Selected References
PCBs (and mixtures of PCBs and organic solvents such as chlorinated benzenes and mineral oil)	Transformer/capacitor oil production, reprocessing, and disposal facilities	In the U.S., the only large producer of PCBs was Monsanto Chemical Co., which sold them between 1929 and 1977 under the Aroclor trademark. PCB production peaked in 1970 when more than 85 million lbs were produced in the U.S. by Monsanto of which 57% was Aroclor 1242 (HEW, 1972). Due to environmental concerns, Monsanto ceased production of Aroclor 1260 in 1971; restricted the sale of other PCBs to totally enclosed systems (electrical transformers, capacitors, and electromagnets) in 1972; and, ceased all production and sale of PCBs in 1977. In 1979, USEPA issued final rules under the 1976 Toxic Substances Control Act restricting the manufacture, processing, use, and distribution of PCBs to specifically exempted and authorized activities.	Transformer oil contamination site in Regina, Saskatchewan (Roberts et al., 1982; Schwartz et al., 1982; and, Anderson and Pankow, 1986; Atwater, 1984)
PCBs (Polychlorinated Biphenyls) are extremely stable, nonflammable, dense, and viscous liquids that have been primarily used as insulators in electrical transformers and capacitors. They are formed by substituting chlorine atoms for hydrogen atoms on a biphenyl (double benzene ring) molecule. Commercial PCBs are a series of technical mixtures, consisting of many isomers and compounds. Monsanto Chemical Co. sold PCBs using the trade name Aroclor●. Each Aroclor is identified by a four-digit number such as 1254. The first two digits, 12, indicate the twelve carbons in the biphenyl double ring. The last two digits indicate the weight percent of chlorine in the PCB product mix. Aroclor 1016 which contains approximately 41% chlorine by weight, however, was not named using this convention.	Between 1929 and 1971, PCBs were sold for use as dielectric fluids in electrical transformers and capacitors, in oil-filled switches, electromagnets, voltage regulators, heat transfer media, fire retardants, hydraulic fluids on machines that handled hot metals to reduce fire hazards, lubricants, cutting oils, plasticizers, wax extenders, carbonless copy paper, paints, inks, adhesives, vacuum pumps, gas-transmission turbines, and dedusting agents.	In 1977, it was estimated that of the 1.25 billion lbs of PCBs sold in the U.S. since 1929, 750 million lbs (60%) were still in use, 290 million lbs (23%) were in landfill and dumps, 150 million lbs (12%) had been otherwise released to the environment, and only 55 million lbs (5%) had been destroyed by incineration or degraded in the environment (Lavigne, 1990). The 1242, 1254, and 1260 Aroclors comprise approximately 80% of the PCBs produced in the U.S. by Monsanto.	Contamination at a PCB storage and transfer station in Smithville, Ontario (Feenstra, 1989; Mclehwain et al, 1989) Other references: Addison (1983), Alford-Stevens (1986), Derks (1990), Feenstra (1989), Griffin and Chian (1980), HEW (1972), Hutzinger et al. (1974), Lavigne (1990), Mclehwain et al. (1989), Miller (1982), Moein et al. (1976), Monsanto (undated), Monsanto (1988), Moore and Walker (1991), NRCC (1980), USEPA (1983), USEPA (1990c), Wagner (1991)
	In 1972, Monsanto restricted sales of PCBs to applications involving only closed electrical systems (transformers, capacitors, and electromagnets).	Approximately 200,000 transformers containing askarel (a generic name for PCB fluid) were in use in the U.S. Half of these were estimated to be still in service in 1990 (Derks, 1990). Approximately 17% of CERCLA sites involve PCB contamination (Haley et al., 1990). The potential for DNAPL migration is greatest at sites where PCBs were produced, reprocessed, and/or disposed in quantity.	
Specific gravities at 25° C. and viscosities at 38° C. of several PCBs are: PCB — Sp.G. — Viscosity Aroclor 1221 — 1.18 — 12 cp Aroclor 1232 — 1.27 — 37 cp Aroclor 1016 — 1.37 — 55 cp Aroclor 1242 — 1.38 — 63 cp Aroclor 1248 — 1.44 — 150 cp Aroclor 1254 — 1.53 — 1900 cp Aroclor 1260 — 1.62 — sticky resin Note that physical properties of PCB products are modified by mixture with chlorobenzenes, mineral oil, or other solvents.			
DNAPL mixtures and uncommon DNAPLs including pesticides and herbicides	Chemical industry facilities Waste handling, reprocessing, and disposal sites	Many industrial waste disposal sites contain complex DNAPL mixtures derived from off-specification materials and process residues.	Chlorinated organic chemical contamination at hazardous waste sites in Niagara Falls, N.Y. (Cohen et al., 1987; Faust et al., 1990; and, Pinder et al, 1990) Motco CERCLA site in LaMarque, Texas (Connor et al., 1989; and, Newell et al., 1991)

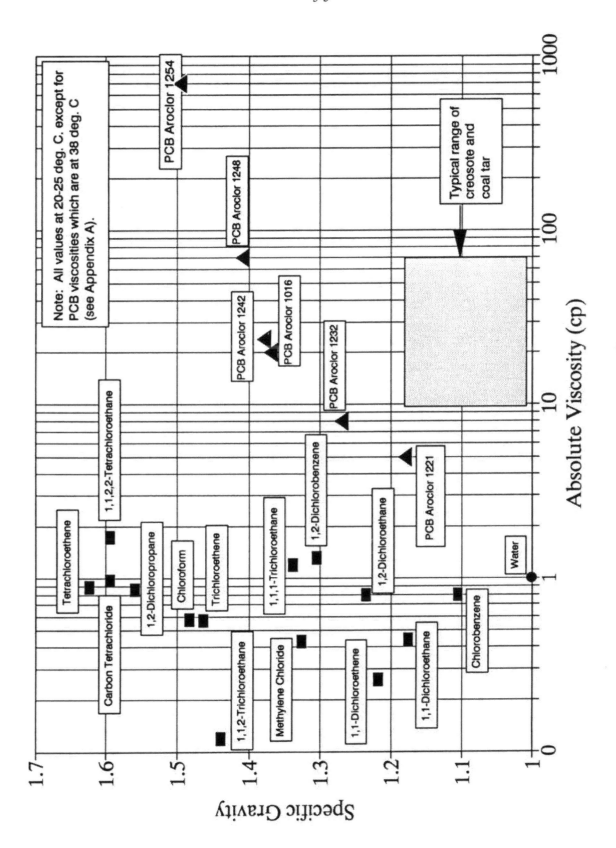

Figure 3-1. Specific gravity versus absolute viscosity for some DNAPLs. DNAPL mobility increases with increasing density:viscosity ratios.

Table 3-2. U.S. Production of Selected DNAPL Chemicals in lbs (U.S. International Trade Commission data, no entry means no data available).

Chemical Compound	1920	1925	1939	1935	1940	1945	1950	1955	1960	1965	1970	1975	1980	1985	1990
Aniline	3.92E+07	1.33E+07	2.64E+07	3.26E+07	5.57E+07	8.72E+07	9.80E+07	1.32E+08	1.20E+08	1.96E+08	3.98E+08	4.07E+08	6.59E+08	7.16E+08	9.88E+08
o-Anisidine									1.30E+06	1.59E+06	1.85E+06				
Benzyl chloride	1.25E+06	5.73E+05			2.38E+06		1.04E+07	1.22E+07	2.14E+07	6.20E+07	7.51E+07				
Bromoethane						4.03E+05		9.22E+06	1.27E+07	1.43E+07	2.10E+07	3.60E+07			
Carbon disulfide						3.55E+08	4.26E+08	5.66E+08	5.23E+08	7.57E+08	7.21E+08	4.79E+08	3.77E+08		
Carbon tetrachloride				5.58E+07	1.01E+08	1.93E+08	2.17E+08	2.87E+08	3.72E+08	5.94E+08	1.01E+09	9.06E+08	7.10E+08	6.46E+08	4.13E+08
Chlorobenzene	4.83E+06	8.69E+06				2.38E+08	3.83E+08	4.36E+08	6.05E+08	5.46E+08	4.85E+08	3.06E+08	2.83E+08		2.37E+08
Chloroform				1.92E+06		9.22E+06	2.03E+07	4.04E+07	7.64E+07	1.53E+08	2.40E+08	2.62E+08	3.53E+08	2.75E+08	4.84E+08
Chloropicrin					3.08E+06							5.70E+06	5.42E+06	1.09E+07	
1,2-Dibromo-3-chloro-propane									3.08E+06	3.43E+06					
Dibutyl phthalate						4.57E+07	1.98E+07	2.39E+07	1.89E+07	2.00E+07	2.29E+07	1.23E+07	1.23E+07	2.17E+07	1.74E+07
1,2-Dichlorobenzene				2.90E+06	5.85E+06	1.18E+07	1.74E+07	2.56E+07	2.47E+07	4.11E+07	6.62E+07	5.47E+07	5.47E+07	2.17E+07	4.87E+07
1,2-Dichloroethane							3.05E+08	5.10E+08	1.27E+09	2.46E+09	7.46E+09	7.98E+09	1.11E+10	1.21E+10	1.38E+10
1,2-Dichloropropane										6.11E+07		8.42E+07	7.70E+07		
Diethyl phthalate						9.70E+06	1.61E+07	1.58E+07	1.68E+07	1.80E+07	2.06E+07	1.17E+07	2.09E+07	1.72E+07	
Dimethyl phthalate						1.88E+07	3.77E+06	3.95E+06	3.39E+06	4.41E+06	8.12E+06	6.77E+06	7.04E+06	7.65E+06	1.25E+07
Ethylene dibromide											2.97E+08	2.75E+08			
Methylene chloride							3.97E+07	7.40E+07	1.13E+08	2.11E+08	4.23E+08	4.97E+08	5.64E+08	4.67E+08	4.61E+08
Nitrobenzene	5.32E+07		3.92E+07	4.82E+07	6.91E+07	1.16E+08	3.97E+07	1.76E+08	1.62E+08	2.80E+08	5.48E+08	4.14E+08	6.12E+08	9.13E+08	
2-Nitrotoluene	2.17E+06	3.34E+06			1.10E+01										
Parathion								5.17E+06	7.43E+06	1.66E+07	1.53E+07				
Tetrachloroethene								1.78E+08	2.09E+08	4.29E+08	7.07E+08	6.79E+08	7.65E+08	6.78E+08	3.72E+08
1,2,4-Trichlorobenzene								1.52E+07			9.34E+06				
1,1,1-Trichloroethane											3.66E+08	4.59E+08	6.92E+08	8.69E+08	8.02E+08
1,1,2-Trichlorofluoro-methane									7.24E+07	1.70E+08	2.44E+08	2.70E+08	1.58E+08	1.76E+08	1.34E+08

U.S. Production of Selected DNAPLs
1920–1990

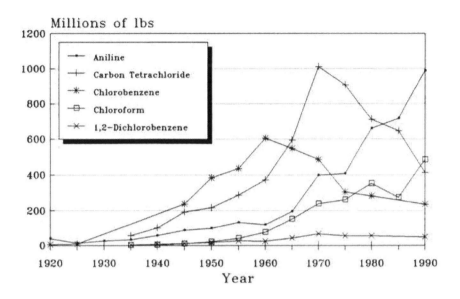

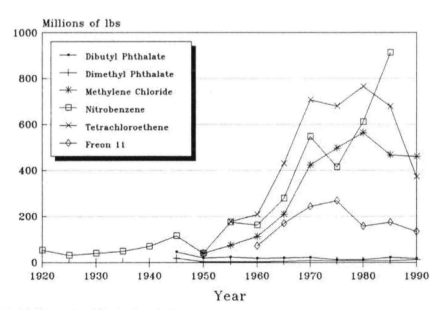

Ref: US International Trade Commission

Figure 3-2. U.S. production of selected DNAPLs in millions of lbs per year between 1920 and 1990.

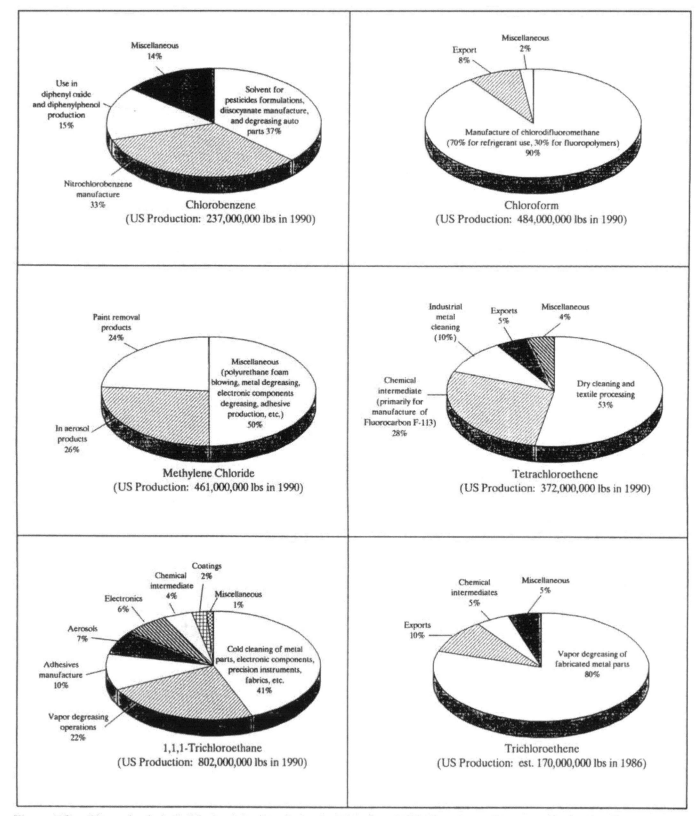

Figure 3-3. Uses of selected chlorinated solvents in the U.S. circa 1986 (data from Chemical Marketing Reporter and U.S. International Trade Commission).

fluorocarbon refrigerants have been produced primarily by duPont under the trade name Freon. In 1990, approximately 918 million lbs of fluorinated hydrocarbons were produced in the U.S. (U.S. International Trade Commission, 1991). Of the 793 million lbs of fluorocarbons produced in 1980: 46% were used as refrigerants, 20% as foam blowing agents, 16% as solvents, 7% as fluoropolymers (such as teflon), and <1% as aerosol propellants (Austin, 1984). Prior to 1974, when concerns arose regarding atmospheric ozone depletion, aerosol propellants were the main end use (52%) of fluorocarbons.

In 1980, approximately 60% of domestic bromine output was used to manufacture ethylene dibromide (EDB) for use in engine fuel antiknock fluids to prevent lead oxide deposition (Austin, 1984). Use of EDB for this purpose, however, has diminished with the phaseout of leaded gasoline. Brominated hydrocarbon DNAPLs are used as fire retardants and fire extinguishing agents, and in a variety of other products.

Chlorinated solvents are frequently detected in groundwater supplies and at disposal sites. For example, as shown in Table 3-3, chlorinated solvents account for ten of the twenty organic contaminants detected most frequently at contamination sites. The halogenated solvents present an extremely high contamination potential due to their extensive production and use, relatively high mobility as a separate phase (high density:viscosity ratio), significant solubility, and high toxicity. This is reflected in the physical properties, production, and drinking water standards data for several common halogenated solvents provided in Table 3-4.

Subsurface contamination derived from halogenated solvents are associated with industries that produce and/or use these DNAPLs and waste disposal sites. Their subsurface presence is caused by: leakage from tanks, pipelines, drums, and other containers; spillage during filling operations; and intentional discharge to landfills, pits, ponds, sewers, etc. Halogenated solvents are frequently present in mixed DNAPL sites.

3.3 COAL TAR AND CREOSOTE

Coal tar and creosote are complex chemical mixture DNAPLs derived from the destructive distillation of coal in coke ovens and retorts. These oily DNAPLs are generally translucent brown to black, and are characterized by specific gravities that range between 1.01

and 1.20, viscosities much higher than water (typically 10 to 70 cp), and the distinctive odor of naphthalene (moth balls).

Historically, coal tar has been produced by coal tar distillation plants and as a byproduct of manufactured gas plant and steel industry coking operations. Manufactured gas plants began producing illuminating or "town" gas for lighting and heating, and by-products for chemical production, in several eastern cities circa 1820. More than 900 gasification plants were operational in the U.S. by 1920 (Rhodes, 1979). Substantial manufactured gas production continued until about 1950 when natural gas became widely available via pipeline. Currently, approximately 98% of coal-tar produced in the U.S. is a by-product of blast furnace coke production (Austin, 1984). In 1990, U.S. production of crude coal tar was 158 million gallons (U.S. International Trade Commission, 1991).

During the coking process, coal is heated to between 450° and 900° C for approximately 16 hours and transformed to coke and vapors by pyrolysis. The evolved coal vapors are then condensed to produce water and approximately 8 to 9 gallons of liquid tar per ton of coal upon cooling. With a specific gravity of between about 1.15 and 1.20, coal tar sinks and is separated for processing. This coal tar is then distilled fractionally to yield approximately: (1) 5% light oil (up to 200° C), (2) 17% chemical or middle oil (200°-250°C), (3) 7% heavy oil (250°-300° C), (4) 9% anthracene oil (300°-350°C), and (5) 62% pitch (Edison Electric Institute, 1984).

Creosote consists of various coal tar distillates (primarily the 200°C to 400°C fraction) which are blended to meet American Wood-Preservers' Association (AWPA) product standards. These creosote blends are then used alone or diluted with coal tar, petroleum, or, to a very limited extent, pentachlorophenol. Existing AWPA product specifications and physical properties are given in Table 3-5. As shown, the specific gravity of creosote products without petroleum oil ranges from 1.07 to 1.13 (at 38° C to 15.5° C for water). Dilution with petroleum reduces the specific gravity of creosote-petroleum solutions to approximately 1.01 to 1.05.

Commercial utilization of creosote/coal tar for pressure treating lumber in the United States commenced circa 1870 with construction of a plant in Mississippi (USDA, 1980). Numerous pressure-treating plants were constructed between 1870 and 1925 to meet the growing demand for treated railroad cross-ties and bridge timbers.

Table 3-3. Frequency of Detection of Most Common Organic Contaminants at Hazardous Waste Sites
(USEPA, 10/17/91; and Plumb and Pitchford, 1985).

RANKING BASED ON NUMBER OF SITES AT WHICH ORGANIC CONTAMINANT WAS DETECTED IN ANY MEDIUM (USEPA, 10/17/91)	ORGANIC CONTAMINANT	DNAPL CHEMICAL?	PERCENTAGE OF 1300 SITES AT WHICH CONTAMINANT WAS DETECTED IN ANY MEDIUM (USEPA, 10/17/91)	RANKING BASED ON NUMBER OF SITES AT WHICH ORGANIC CONTAMINANT WAS DETECTED IN GROUNDWATER (PLUMB AND PITCHFORD, 1985)	PERCENTAGE OF 183 SITES AT WHICH CONTAMINANT WAS DETECTED IN GROUNDWATER (PLUMB AND PITCHFORD, 1985)
1	Toluene	No	60.5	2	31.15
2	Trichloroethene	Yes	57.3	1	34.43
3	Methylene Chloride	Yes	54.7	3	31.15
4	Benzene	No	53.2	7	27.32
5	Tetrachloroethene	Yes	51.8	4	31.15
6	Ethylbenzene	No	47.5	11	25.14
7	1,1,1-Trichloroethane	Yes	47.1	9	26.78
8	Chloroform	Yes	45.4	10	25.14
9	Xylenes	No	44.3	--	--
10	bis(2ethylhexyl) phthalate	No	41.8	6	28.42
11	Acetone	No	40.0	20	12.02
12	1,1-Dichloroethane	Yes	39.7	5	28.42
13	Phenol	No	39.4	14	19.13
14	trans-1,2-Dichloroethene	Yes	38.4	8	27.32
15	Napthalene	No	35.5	18	12.57
16	1,1-Dichloroethene	Yes	33.2	13	20.22
17	1,2-Dichloroethane	Yes	32.7	12	21.31
18	Vinyl Chloride	No	32.1	15	16.39
19	2-Butanone	No	31.8	--	--
20	Chlorobenzene	Yes	31.4	16	16.39
23	Dibutyl Phthalate	Yes	30.3	17	15.30
40	Chloroethane	No	18.1	19	12.57

Table 3-4. Production, physical property, and RCRA groundwater action level data for selected DNAPL chemical.

DNAPL	U.S. Production (lbs/yr)	Specific Gravity	Absolute Viscosity (cp)	Solubility (mg/L)	Safe Drinking Water Act Maximum Concentration Limits (MCLs) (mg/L)	Ratio of Solubility Limit to RCRA Action Level
Carbon tetrachloride	4.13E+08	1.594	0.97	800	0.005	160,000
Chlorobenzene	2.37E+08	1.106	0.80	500	0.1	5,000
Chloroform	4.84E+08	1.483	0.58	8000	--	--
1,2-Dichloroethane	1.38E+10	1.235	0.80	8690	0.005	1,738,000
Methylene chloride	4.61E+08	1.327	0.43	20,000	0.005	4,000,000
Tetrachloroethene	3.72E+08	1.623	0.89	150	0.005	30,000
1,1,1-Trichloroethane	8.02E+08	1.339	0.83	1360	0.2	6,800
Trichloroethene	1.65E+08	1.464	0.57	1100	0.005	220,000
1,1,2-Trichloro-fluoromethane	1.34E+08	1.487	0.42	1100	--	--
Water		1.00	1.12			

Notes: Production data for 1990 (U.S. International Trade Commission), except for trichloroethene which is for 1986 (Halogenated Solvents Industry Alliance).
Physical property data at 20-25° C is excerpted from Appendix A.
All MCLs are final (Federal Register, 1/30/91, Vol.56, No.20), except for methylene chloride which was proposed in 7/90.

Table 3-5. American Wood-Preservers' Association standards for creosote and coal tar products.

SPECIFICATIONS	NEW MATERIAL		OLD MATERIAL	
	Not less than	Not more than	Not less than	Not more than
	P1/P13-91 STANDARD FOR COAL TAR CREOSOTE FOR LAND, FRESHWATER, AND COASTAL WATER USE			
% Water by volume	--	1.5	--	3.0
% Matter insoluble in xylene by weight	--	3.5	--	4.5
% Coke residue by weight	--	9.0	--	10.0
Specific gravity at 38° C compared to water at 15.5° C: Whole Creosote	1.08	1.13	1.08	1.13
Fraction 235-315° C	1.025	--	1.025	--
Fraction 315-355° C	1.085	--	1.085	--
Distillation -- The distillate % by weight on a water-free basis shall be within the following limits: Up to 210° C	--	5	--	5
Up to 235° C	--	25	--	25
Up to 315° C	32	--	32	--
Up to 355° C	52	--	52	--
	P2-90 STANDARD FOR CREOSOTE SOLUTIONS			
% Water by volume	--	1.5	--	3.0
% Matter insoluble in xylene by weight	--	3.5	--	4.5
% Coke residue by weight	--	9.0	--	10.0
Specific gravity at 38° C compared to water at 15.5° C: Whole Creosote	1.08	1.13	1.08	1.13
Fraction 235-315° C	1.025	--	1.025	--
Fraction 315-355° C	1.085	--	1.085	--
Distillation -- The distillate % by weight on a water-free basis shall be within the following limits: Up to 210° C	--	5	--	5
Up to 235° C	--	25	--	25
Up to 315° C	32	--	32	--
Up to 355° C	52	--	52	--

Notes:

1. The products must be derived entirely from the carbonization of bituminous coal.
2. Creosote-petroleum oil solutions shall consist solely of specified proportions of coal tar creosote which meets AWPA Standard P1 and of petroleum oil which meets AWPA Standard P4. No creosote-petroleum oil solution shall contain less than 50% by volume of such creosote or more than 50% by volume of such petroleum oil.
3. The P4-86 Standard for petroleum oil for blending with creosote specifies that petroleum oil for blending with creosote have a specific gravity at 60°F/60°F not less than 0.96; not more than 1% water and sediment; a flash point not less than 175° F; and a kinematic viscosity between 4.2 and 10.2 cSt at 210° F.

The industry continued to expand thereafter due to the demand for treated poles by developing utility companies. Prior to the introduction of pentachlorophenol-petroleum mixtures for wood preservation in the 1930s, creosote/coal tar was the only potent wood preservative available.

According to U.S. International Trade Commission records, the production of creosote oil in the United States has decreased from 145 million gallons in 1953 to 63 million gallons in 1990. Although creosote has lost market share since the 1950s to inorganic arsenical and pentachlorophenol preservatives, it remains the dominant wood preservative in the U.S., particularly for treating railroad cross-ties, switch ties, and pilings.

Creosote solutions were used at approximately 188 of 631 wood-treating plants operating in the U.S. during 1978 (USDA, 1980). Most of these plants are located in the east/southeast and west/northwest wood-growing belts as shown in Figure 3-4. Creosote and coal tar are supplied to wood-treating operations by coal tar distillation plants. In 1972, there were 24 coal tar distillation plants producing creosote in the U.S. (Table 3-6).

During 1976, of an estimated 1.18 billion lbs of creosote and coal tar used to treat approximately 47% of all commercially preserved wood in the U.S.: 415 million lbs (35%) were used as straight creosote; 626 million lbs (53%) were used in creosote/coal tar solutions (averaging 63% creosote and 37% coal tar); 137 million lbs (12%) were used in creosote-petroleum solutions typically with 50 to 70% creosote; and 2 million lbs (<0.2%) were used in creosote-pentachlorophenol solutions (USDA, 1980). Creosote-petroleum mixtures are utilized primarily in the Central and Western U.S. for treating railroad ties.

In addition to wood-preservation, coal tar is used for road, roofing, and water-proofing solutions. Considerable use of coal tar is also made for fuels.

Creosote and coal tar are complex mixtures containing more than 250 individual compounds. Creosote is estimated to contain 85% polycyclic aromatic hydrocarbons (PAHs), 10% phenolic compounds, and 5% N-, S-, and O- heterocyclic compounds. The composition of creosote and coal tar are quite similar, although coal tar generally includes a light oil component (<5% of the total) consisting of monocyclic aromatic compounds such as benzene, toluene, ethylbenzene, and xylene (BTEX). Chemical composition data for coal tar and creosote are given in Table 3-7.

Creosote and coal tar contamination of the subsurface are associated with wood-treating plants (Figure 3-4), former manufactured gas plants, coal tar distillation plants, and steel industry coking plants. In 1988, there were 55 wood-preserving contamination sites on the National Priority List of CERCLA sites. Creosote and/or coal tar are a source of groundwater and soil contamination at approximately forty of these sites (Table 3-8).

Prior to the 1970s, liquid wastes (including creosote and coal tar) from wood-treating plants were typically discharged to ponds, sumps, and/or streams. Many plants had small (1 to 4 acres) unlined ponds to trap the DNAPL wastes as effluent discharged to streams or public water treatment facilities (McGinnis, 1989). As a result, many wood-treating sites have large volumes of DNAPL-contaminated soils in the vicinity of former discharge ponds. Soil contamination at wood-treating plants is also prevalent in the wood treating, track, and storage areas due to preservative drippage from wood as it is being moved and stored, and around preservative tanks and pipelines due to spillage and leaks. Consistent with the composition of creosote and coal tar, PAHs (in addition to BTEX compounds) are common contaminants detected in groundwater at wood-treating sites (Rosenfeld and Plumb, 1991).

Before development of the coal tar distillation industry in about 1887, most coal tar derived from manufactured gas plants was apparently disposed at or near the plant site (Villaume et al., 1983). Due to their long period of operation and voluminous coal tar generation, substantial DNAPL contamination is associated with many former manufactured gas plants. GRI (1987) recently reviewed available site investigation reports for 33 former manufactured gas plants. Heavily contaminated soils or sludges were found at and near coal tar ponds, coal tar holding tanks, pipeline and tank spill and leak sites, and, in DNAPL stratigraphic traps.

DNAPLs similar to coal tar and creosote were produced by manufactured gas plants that used crude oil rather than coal. Additionally, refinery (petroleum) coke units process heavy end hydrocarbon to produce coke and Bunker C oil with a typical density of 1.01 g/cm^3.

3.4 POLYCHLORINATED BIPHENYLS (PCBs)

PCBs are extremely stable, nonflammable, dense, and viscous liquids that are formed by substituting chlorine atoms for hydrogen atoms on a biphenyl (double benzene

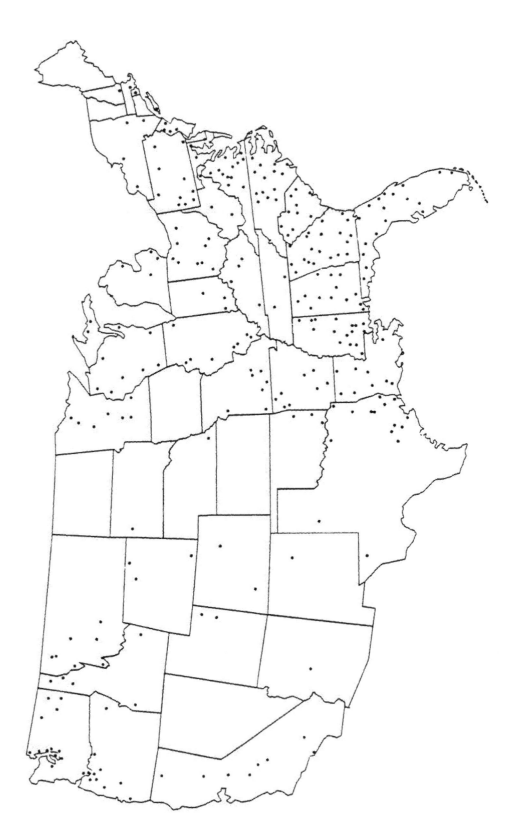

Figure 3-4.   Locations of wood-treating plants in the United States (modified from McGinnis, 1989).

Table 3-6. Creosote production in the United States in 1972 by plant[a].

Plant	Estimated Plant Capacity (million lbs/yr)	Estimated Annual Production (million lbs)
Allied Chemicals Corporation		250-350
Detroit, Michigan	100-200	
Ensely, Alabama	100-200	
Ironton, Ohio	100-200	
Koppers Company, Inc.		350-450
Cicero (Chicago), Illinois	100-200	
Follansbee, West Virginia	100-200	
Fontana, California	200-300	
Houston, Texas	10-20	
Portland, Oregon	10-20	
Kearny (Seaboard), New Jersey	10-20	
St. Paul, Minnesota	10-20	
Swedeland, Pennsylvania	10-20	
Woodward, Alabama	100-200	
Youngstown, Ohio	100-200	
Reilly Tar and Chemical Corp.		50-100
Cleveland, Ohio	10-20	
Granite City, Illinois	10-20	
Ironton, (Provo), Utah	10-20	
Lone Star, Texas	10-20	
Chattanooga, Tennessee	10-20	
USS Chemicals		250-350
Clairton, Pennsylvania	100-300	
Fairfield, Alabama	100-200	
Gary, Indiana	100-200	
The Western Tar Products Corp.		20-40
Memphis, Tennessee	10-20	
Terre Haute, Indiana	10-20	
Witco Chemical Corporation		10-20
Point Comfort, Texas	10-20	----
TOTAL ANNUAL PRODUCTION (1972)		1,150

[a]USDA (1980)

Table 3-7. Composition of creosote and coal tar, solubility of pure coal tar compounds, proposed RCRA groundwater action levels, and prevalence in groundwater at wood-treating sites.

CREOSOTE/COAL TAR FRACTION and Compounds	CREOSOTE 1=Commercial creosote (USEPA, 1990d) 2=U.S. creosote (USDA, 1980) 3=German creosote (USDA, 1980) 4=Creosote (Mueller et al, 1989) (Note: Relative weight percents of dominant compounds in each fraction given for creosote no. 4.)				COAL TAR 1 = Coal tar (GRI, 1987) 6 = British coal tar (USDA,1980) 7 = U.S. coal tar (USDA, 1980)			Aqueous Solubility of Pure Compound (mg/L) (Montgomery and Welkom, 1990; and Mueller et al., 1989)	RCRA Groundwater Action Levels Proposed 7/5/91 (mg/L)	% Detects in GW Samples from 5 Wood-treating Sites (Rosenfeld and Plumb, 1991)	Average Concentration in Groundwater Samples at 5 Wood-treating Sites (mg/L) (Rosenfeld and Plumb, 1991)
	1	2	3	4	5	6	7				
VOLATILE AROMATICS					5%						
Benzene					0.1%	0.25%	0.12%	1780	0.005	22%	0.033
Ethylbenzene						0.02%	0.02%	152	0.7	19%	0.039
Toluene					0.2%	0.22%	0.25%	515	2	20%	0.048
Xylene					1.0%	0.19%	0.14%	200	10	18%	0.094
Styrene						0.04%	0.02%	300	0.005		
ACID EXTRACTABLES				10%	5%						
Phenol				20	0.7%	0.57%	0.61%	82,000	20	12%	1.537
Cresols				30	1.1%	0.10%	0.97%	24,000	2		
Pentachlorophenol				10				20	0.2		
Xylenols				35	0.2%	0.48%	0.36%	5000	0.02	13%	1.219
2,4-Dimethylphenol				5							
2,3,5-Trimethylphenol											
BASE/NEUTRALS				85%	28%						
Naphthalene	17.0%		7.3%	13	10.9%	8.9%	8.8%	32	0.1	35%	3.312
Methylnaphthalenes	10.0%	2.1%	4.2%	21	2.4%	2.0%	1.9%	25		27%	0.563
Dimethylnaphthalenes	1.9%	0.8%		8	3.3%			2			
Biphenyl				8				7			
Acenaphthene	7.8%	9.0%	4.1%	4	1.3%	0.96%	1.06%	3	2	38%	0.805
Fluorene	6.0%	10.0%	9.6%	8	1.6%	0.88%	0.84%	2	0.002	34%	0.661
Phenanthrene	19.4%	21.0%	12.6%	13	4.0%	6.30%	2.66%	1	0.002	29%	1.825
Anthracene	2.5%	2.0%		13	1.1%	1.00%	0.75%	0.07		21%	0.425
Fluoranthene	11.8%	10.0%	6.8%	4				0.3	1	22%	1.025
Pyrene	8.4%	8.5%	5.0%	2				0.1	1	22%	0.666
Chrysene	4.2%	3.0%	2.8%	2				0.002	0.0002	13%	0.249

Table 3-7. Composition of creosote and coal tar, solubility of pure coal tar compounds, proposed RCRA groundwater action levels, and prevalence in groundwater at wood-treating sites.

CREOSOTE/COAL TAR FRACTION and Compounds	CREOSOTE 1=Commercial creosote (USEPA, 1990d) 2=U.S. creosote (USDA, 1990) 3=German creosote (USDA, 1980) 4=Creosote (Mueller et al., 1989) (Note: Relative weight percents of dominant compounds in each fraction given for creosote no. 4.)				COAL TAR 1 = Coal tar (GRI, 1987) 6 = British coal tar (USDA,1980) 7 = U.S. coal tar (USDA, 1980)			Aqueous Solubility of Pure Compound (mg/L) (Montgomery and Welkom, 1990; and Mueller et al., 1989)	RCRA Groundwater Action Levels Proposed 7/5/91 (mg/L)	% Detects in GW Samples from 5 Wood-treating Sites (Rosenfeld and Plumb, 1991)	Average Concentration in Groundwater Samples at 5 Wood-treating Sites (mg/L) (Rosenfeld and Plumb, 1991)
	1	2	3	4	5	6	7				
Anthraquinone				1							
2,3-Benzo[b]fluorene				1				0.002	0.00005		
Methylanthracene		4.0%		1				0.04			
Benzo[a]pyrene				1				0.003	0.0002	8%	0.057
Diphenyldimethyinaphth.			3.2%								
Diphenyloxide			3.4%			1.5%					
N,S,O-HETEROCYCLICS				5%							
Quinoline				10				6700			
Isoquinoline				10				4500			
Carbazole	5.1%	2.0%		10	1.1%	1.33%	0.60%	1			
2,4-Dimethylpyridine				10							
Benzo[b]thiophene				10				130			
Dibenzothiophene				10				2			
Dibenzofuran				10				10		28%	0.332
PITCH					62%	59.8%	63.5%				
TOTAL	94.1%	75.4%	59.3%		91%	84.5%	82.6%				

Table 3-8. Creosote and coal tar wood preserving sites on the Superfund list (modified from USEPA, 1989d).

SITE	LOCATION	CONTAMINATED MEDIA
Hocomonco Pond	Westborough, MA	GW, SW, SO
Southern Maryland Wood Treating	Hollywood, MD	GW, SW, SO
L.A. Clark & Sons	Spotsylvania City, VA	GW, SW, SO
Atlantic Wood Industries, Inc.	Portsmouth, VA	SO, GW?, SW?
Rentoldl, Inc. (VA Wood Preserving Division)	Richmond, VA	GW, SW, SO
American Creosote, Pensacola Plt	Pensacola, FL	GW, SO
Brown Wood Preserving	Live Oak, FL	SO
American Creosote, Jackson Plant	Jackson, TN	GW, SW, SO
Cape Fear Wood Preserving	Fayetteville, NC	GW, SW, SO
Koppers Co., Inc. Florence Plant	Florence, SC	GW, SW, SO
Reilly Tar, St. Louis Park Plant	St. Louis Park, MN	GW, SO
Reilly Tar & Chemical, Dover Plant	Dover, OH	GW, SO
MacGillis & Gibbs/Bell Lumber	New Brighton, MN	GW, SO
Boise-Cascade-Onan/Medtronics	Fridley, MN	GW, SO
Burlington Northern (Brainerd)	Brainerd/Baxter, MN	GW, SO
Joslyn Manufacturing & Supply Co.	Brooklyn Center, MN	GW, SO
Moss-American (Kerr-McGee Oil Co.)	Milwaukee, WI	GW, SW, SO
Galesburg/Koppers Co.	Galesburg, IL	GW, SO, SW?
Mid-South Wood Products	Mena, AR	GW, SO
Texarkana Wood Preserving Co.	Texarkana, TX	GW, SO
United Creosoting Co.	Conroe, TX	GW, SO
Bayou Bonfouca	Slidell, LA	GW, SW, SO
Midland Products	Ola/Birta, AR	GW, SO
Koppers Co., Inc., Texarkana Plant	Texarkana, TX	GW, SO
South Cavalcade Street	Houston, TX	GW, SO
North Cavalcade Street	Houston, TX	GW, SO
Arkwood	Omaha, AR	GW, SO
Baxter/Union Pacific Tie Treating	Laramie, WY	GW, SW, SO
Broderick Wood Products	Denver, CO	GW, SO
Montana Pole and Treating	Butte, MT	GW, SO, SO
Idaho Pole Co.	Bozeman, MT	GW, SW, SO
Burlington Northern, Somers Plant	Somers, MT	GW, SW, SO
Libby Groundwater (Champion International)	Libby, MT	GW, SW, SO
Koppers Co., Inc., Oroville Plant	Oroville, CA	GW, SW, SO
Southern California Edison (Visalia)	Visalia, CA	GW, SO
J.H. Baxter	Weed, CA	GW, SW, SO
Marley Cooling Tower Co.	Stocktom, CA	GW, SO
Wyckoff Co./Eagle Harbor	Bainbridge Island, WA	GW, SO, SD
American Crossarm & Conduit Co.	Chehalis, WA	SO

Notes: GW = Groundwater, SW = Surface water or lagoon, SO = Soil or lagoon sediments, SD = River sediments

ring) molecule. In the U.S., the only large producer of PCBs was Monsanto Chemical Co., which sold them between 1929 and 1977 under the Aroclor trademark for use primarily as dielectric fluids in electrical transformers and capacitors. PCBs were also sold for use in oil-filled switches, electromagnets, voltage regulators, heat transfer media, fire retardants, hydraulic fluids, lubricants, plasticizers, carbonless copy paper, dedusting agents, etc. (Table 3-1).

Production of PCBs peaked in 1970 when more than 85 million lbs were produced in the U.S. by Monsanto (HEW, 1972). Due to environmental concerns, Monsanto ceased production of Aroclor 1260 in 1971; restricted the sale of other PCBs to totally enclosed applications (transformers, capacitors, and electromagnets) in 1972; and, ceased all production and sale of PCBs in 1977. In 1979, USEPA issued final rules under the 1976 Toxic Substances Control Act restricting the manufacture, processing, use, and distribution of PCBs to specifically exempted and authorized activities. The pattern of PCB production and use in the U.S. between 1965 and 1977 is illustrated in Figure 3-5.

Commercial PCBs are a series of technical mixtures, consisting of many isomers and compounds. Each Aroclor is identified by a four-digit number such as 1254. The first two digits, 12, indicate the twelve carbons in the biphenyl double ring. The last two digits indicate the weight percent chlorine in the PCB mixture, such as 54% chlorine in Aroclor 1254. Aroclor 1016, which contains approximately 41% chlorine, however, was not named using this convention. Properties and approximate molecular compositions of the Aroclor formulations are given in Table 3-9. As shown, the Aroclors become more dense and viscous and less soluble with increasing chlorine content. The lower chlorinated formulations (Aroclors 1016 to 1248) are colorless mobile oils. Aroclor 1254 is a viscous yellow liquid, and Aroclor 1260 is a black sticky resin.

Aroclors 1242, 1254, and 1260 comprise approximately 80% of the PCBs produced by Monsanto (Feenstra, 1989); and Aroclor 1254 accounted for an estimated 57% of the more than 85 million lbs of PCBs manufactured by Monsanto during peak production in 1970 (HEW, 1972).

PCBs were frequently mixed with carrier fluids prior to use. For example, PCBs were typically diluted with 0 to 70% carrier fluid, usually chlorobenzenes or mineral oil, in askarel. Askarel is a generic name for fire-resistant dielectric fluids that were used in about 97% of the estimated 200,000 electrical transformers put in use prior to 1979 in the U.S. (Derks, 1990; Wagner, 1991; Monsanto, 1988). Askarel transformers containing 3 to 3000 gallons of PCB oil are generally employed in hazardous locations where flammability is a concern (Wagner, 1991). The mix of Aroclor and carrier fluid type and content, therefore, determines the physical properties of the PCB fluid, including its density, viscosity, solubility, and volatility. Some petroleum oil - PCB mixtures form LNAPLs.

In 1977, it was estimated that of the 1.25 billion lbs of PCBs sold in the U.S. since 1929, 750 million lbs (60%) were still in use, 290 million lbs (23%) were in landfills and dumps, 150 million lbs (12%) had been otherwise released to the environment, and only 55 million lbs (5%) had been destroyed by incineration or degraded in the environment (Lavigne, 1990). An estimated 100,000 electrical transformers containing askarel were estimated to be still in service as of 1990 (Derks, 1990).

Approximately 15% of CERCLA sites involve PCB contamination (Haley et al., 1990; and, USEPA, 1990c). Due to their widespread use and persistence, PCBs are often detected in the environment at very low concentrations. The potential for DNAPL migration is greatest at sites where PCBs were produced, utilized in manufacturing processes, stored, reprocessed, and/or disposed in quantity. The extent of PCB contamination as a DNAPL problem is unclear.

3.5 MISCELLANEOUS AND MIXED DNAPL SITES

Miscellaneous DNAPLs refer to dense, immiscible fluids that are not categorized as halogenated solvents, coal tar, creosote, or PCBs. These include some herbicides and pesticides, phthalate plasticizers, and various exotic compounds (Appendix A).

Mixed DNAPL sites refer to landfills, lagoons, chemical waste handling or reprocessing sites, and other facilities where various organic chemicals were released to the environment and DNAPL mixtures are present. Many mixed DNAPL sites derive from the disposal of off-specification products and process residues in landfills and lagoons by chemical manufacturers. Typically, these mixed DNAPL sites include a significant component of chlorinated solvents.

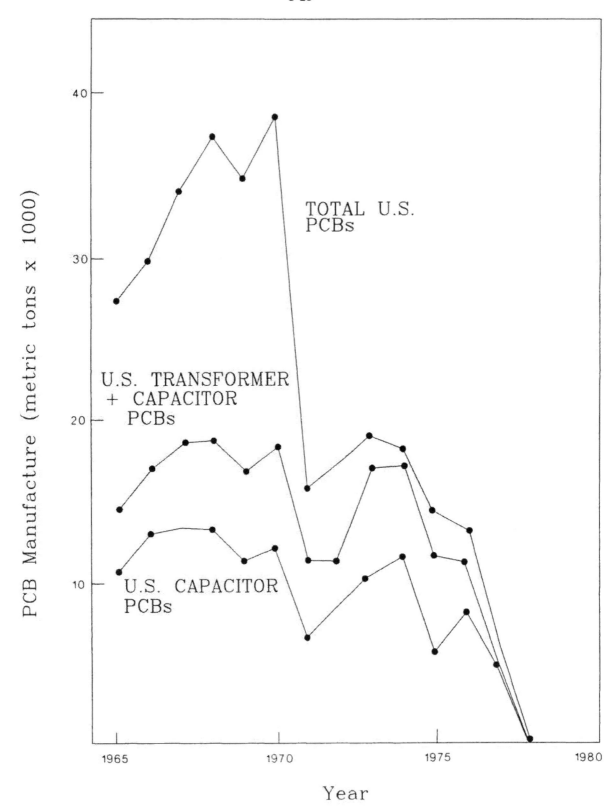

Figure 3-5. Pattern of U.S. PCB production and use between 1965 and 1977 (modified with permission from ACS. 1983).

Table 3-9.   Approximate molecular composition (%) and selected physical properties of Aroclor PCBs (from Moore and Walker, 1991; Monsanto, 1988; and Montgomery and Welkom, 1990).

COMPOSITION AND PROPERTIES	AROCLOR						
	1221	1232	1016	1242	1248	1254	1260
Biphenyl	11.0	6.0	Tr	--	--	--	--
Monochlorobiphenyl	51.0	26.0	1.0	1.0	--	--	--
Dichlorobiphenyl	32.0	29.0	20.0	17.0	1.0	--	--
Trichlorobiphenyl	4.0	24.0	57.0	40.0	23.0	--	--
Tetrachlorobiphenyl	2.0	15.0	21.0	32.0	50.0	16.0	--
Pentachlorobiphenyl	0.5	0.5	1.0	10.0	20.0	60.0	12.0
Hexachlorobiphenyl	--	--	Tr	0.5	1.0	23.0	46.0
Heptachlorobiphenyl	--	--	--	--	--	1.0	36.0
Octachlorobiphenyl	--	--	--	--	--	--	6.0
Specific Gravity (@ 25/15.5° C)	1.18	1.27	1.37	1.38	1.41	1.50	1.56
Absolute Viscosity (cp @ 38° C)	5	8	20	24	70	700	resin
Solubility (µg/L @ 25° C)	200	--	240	240	54	12	2.7
Vapor Pressure (mm @ 25° C)	0.0067	0.0046	0.0004	0.0004	0.0004	0.00008	0.00004
Log K_{ow}	2.8	3.2	4.4	4.1	6.1	6.5	6.9

Tr = trace (< 0.01%)

4 PROPERTIES OF FLUID AND MEDIA

DNAPL chemicals in the subsurface migrate as volatiles in soil gas, dissolved in groundwater, and as a mobile, separate phase. This migration is governed by several factors and principles, some of which differ from those controlling miscible contaminant transport. In this chapter, which is adapted from Mercer and Cohen (1990), properties of fluid and media associated with DNAPL flow are described. The influence of these properties on DNAPL transport at contamination sites is examined further in Chapter 5 and methods to measure fluid and media properties are documented in Chapter 10.

4.1 SATURATION

The saturation of a fluid is the volume fraction of the total void volume occupied by that fluid. Saturations range from zero to one and the saturations of all fluids sum to one. Saturation is important because it is used to define the volumetric distribution of DNAPL, and because other properties, such as relative permeability and capillary pressure, are functions of saturation.

Photographs showing variable NAPL distribution and saturation states observed during laboratory experiments are presented by Schwille (1988) and Wilson et al. (1990). In the saturated zone, nonwetting DNAPL flows selectively through the coarser, more permeable portions of heterogeneous media, averting the finer-grained zones which provide greater capillary resistance to entry. As a result, mobile DNAPL is present as globules (blobs) connected along fractures, macropores, and the larger pore openings. Water occupies the smaller pores and tends to be retained as a film between the nonwetting NAPL globules and media solids. At residual saturation, DNAPL occurs as disconnected singlet and multi-pore globules within the larger pore spaces. The fluid distribution is more complex in the vadose zone where NAPL migration occurs by displacing soil gas and sometimes water from pore throats and bodies. A variety of saturation states for different subsurface conditions are illustrated in Figure 4-1.

Measuring saturation presents a costly sampling and analysis problem. Numerous samples will generally be needed to assess the distribution of subsurface DNAPL; and sampling, sample preservation, and analytical methods must be carefully selected and implemented to produce reliable results. Ferrand et al. (1989) present a dual-gamma (^{137}Cs-^{241}Am) technique for laboratory determination of three-fluid saturation profiles in porous media. They provide porosity and saturation profiles for sands containing air, water, and trichloroethene or tetrachloroethene. Cary et al. (1991) demonstrated a procedure to extract NAPL from soil by shaking a soil-water suspension with a strip of hydrophobic, porous polyethylene in a glass jar for several hours. Estimates of NAPL content were determined by weighing the strips before and after the extraction procedure. Saturation can also be estimated gravimetrically in conjunction with various organic solvent extraction techniques and specialized distillations (Cary et al., 1989a, b), some of which are described in Chapter 10. Chemical analyses can be made to estimate total contaminant concentration and NAPL saturation. Finally, drilling and sampling, pumping tests, and borehole geophysical logging can be used to facilitate field estimates of saturation. These estimates are qualitative, subject to considerable error and uncertainty, and the methods used are largely undocumented.

4.2 INTERFACIAL TENSION

The characteristics of DNAPL movement are largely derived from interfacial tensions which exist at the interface between immiscible fluids (NAPL, air, and water). Interfacial tension is related directly to the capillary pressure across a water-NAPL interface and is a factor controlling wettability. Liquid interfacial tension develops due to the difference between the greater mutual attraction of like molecules within each fluid and the lesser attraction of dissimilar molecules across the immiscible fluid interface (Schowalter, 1979). This unbalanced force draws molecules along the interface inward, resulting in a tendency for contraction of the fluid-fluid interface to attain a minimum interfacial area (Wilson et al., 1990). Interfacial tension has been likened to a contractile skin on the NAPL surface which resists stretching of the interfacial surface (WCGR, 1991). The fluid on the concave side of the interface is at higher pressure than the fluid across the interface and the pressure difference is proportional to the degree of interface curvature. As a result of interfacial tension, nonwetting DNAPLs tend to form globules in water and water-saturated media.

The interfacial tension between a liquid and its own vapor is called surface tension. The value of the liquid interfacial tension is always less than the greater of the surface tensions for the pure liquids. This results from the mutual attraction of unlike molecules at the immiscible liquid interface.

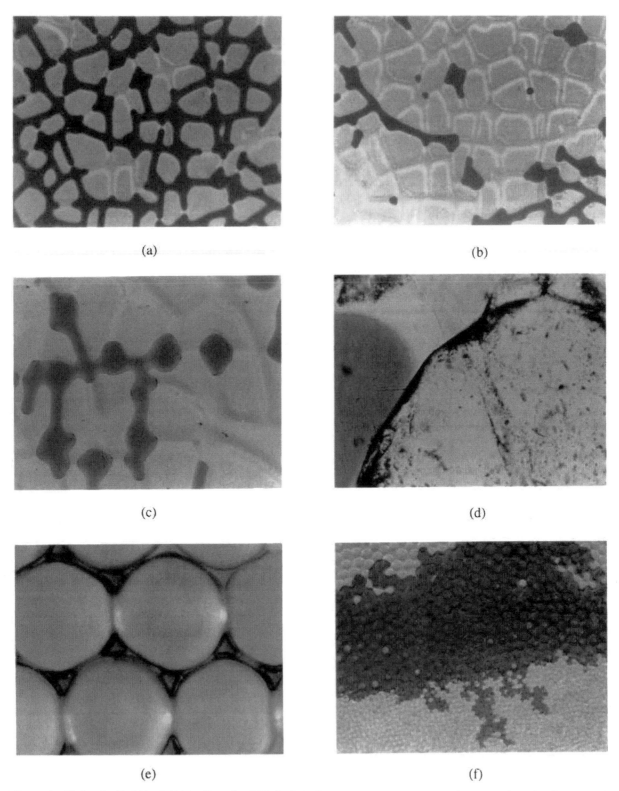

(a)

(b)

(c)

(d)

(e)

(f)

Figure 4-1. Various immiscible fluid distributions -- Dark NAPL (Soltrol) and water in a homogeneous micromodel after (a) the displacement of water by NAPL and then (b) the displacement of NAPL by water (with NAPL at residual saturation). A more complex blob in the micromodel is shown in (c). In (d), a pore body is filled with the dark non-wetting phase fluid; and is separated from the wetting phase by a light intermediate wetting fluid. The distribution of (dark) tetrachloroethene (PCE) retained in large pore spaces between moist glass beads after being dripped in from above is shown in (e). In (f), PCE infiltrating into water-saturated glass beads pooled in coarse beads overlying finer beads and PCE fingers extend into the underlying finer bead layer. Figures (a) to (d) are from Wilson et al. (1990); Figures (e) and (f) are from Schwille (1988).

Interfacial tension is measured in units of energy per unit area (or force per unit length) typically using the capillary rise or Du Nouy tensiometer methods (Chapter 10). It decreases with increasing temperature (approximately 0.1 dynes/cm/°F for crude oil-water systems) and may be affected by pH, surface-active agents, and gas in solution (Schowalter, 1979). Interfacial tensions range from zero, for completely miscible liquids, to 72 dynes/cm, the surface tension of water at 25° C (Lyman et al., 1982). Values of interfacial and surface tensions for DNAPL chemicals, however, generally range between 15 and 50 dynes/cm as shown in Appendix A.

4.3 WETTABILITY

Wettability refers to the preferential spreading of one fluid over solid surfaces in a two-fluid system; it depends on interfacial tension. Whereas the wetting fluid (usually water in a DNAPL-water system) will tend to coat solid surfaces and occupy smaller openings in porous media, the nonwetting fluid will tend to be constricted to the larger openings (i.e., fractures and relatively large pore bodies). Anderson (1986a, 1986b, 1986c, 1987a, 1987b, and 1987c) prepared an extensive literature review on wettability, its measurement, and effects on relative permeability, capillary pressure, residual NAPL saturation, and enhanced NAPL recovery.

The simplest measure of wettability is the contact angle at the fluid-solid interface (Figure 4-2). In a DNAPL-water system, if the adhesive forces between the water and solid phases exceed the cohesive forces within the water as well as the adhesive forces between the DNAPL and solid phases, then the solid-water contact angle, ϕ, measured into the water (in degrees) will be acute indicating that water, rather than DNAPL, preferentially wets the medium (Wilson et al., 1990). The porous medium is considered water-wet if ϕ is less than approximately 70°, NAPL-wet if ϕ is greater than 110°, and neutral if ϕ is between 70° and 110° (Anderson, 1986a). Contact angle measurements should be interpreted as qualitative indicators of wettability because they do not account for media heterogeneity, roughness, and pore geometry (Huling and Weaver, 1991; Wilson et al., 1990). Methods for measuring contact angles and the bulk wettability of soil and rock samples are described in Chapter 10 and by Honarpour et al. (1986), Gould (1964), Anderson (1986b), and Wilson et al. (1990).

Except for mercury, liquids (NAPL or water), rather than air, preferentially wet solid surfaces in the vadose zone.

Wettability relations in immiscible fluid systems are affected by several factors including media mineralogy, water chemistry, NAPL chemistry, the presence of organic matter or surfactants, and media saturation history. With the exceptions of organic matter (such as coal, humus, and peat), graphite, sulfur, talc and talc-like silicates, and many sulfides, most natural porous media are strongly water-wet if not contaminated by NAPL (Anderson, 1986a). Although water is typically the wetting fluid in NAPL-water systems and has been considered a perfect wetting agent in certain petroleum reservoirs (Berg, 1975; Corey, 1986; Schowalter, 1979; Smith, 1966), other researchers have documented that petroleum reservoirs, particularly dolomite and limestone, may be partially or preferentially wet by oil (Nutting, 1934; Benner and Bartell, 1941; Treiber et al., 1972; Salathiel, 1973; Leach et al., 1962; Craig, 1971; Anderson, 1986a).

NAPL wetting usually increases due to adsorption and/or deposition on mineral surfaces of organic matter and surfactants derived from NAPL or water (Honarpour et al., 1986; Thomas, 1982; Treiber et al., 1972; JBF Scientific Corp, 1981; Schowalter, 1979). For example, Villaume et al. (1983) reported that coal tar at a former manufactured gas plant site in Pennsylvania preferentially wet quartz surfaces, possibly as a result of the presence of surfactants in the coal tar. NAPL wetting has been shown to increase with aging during contact angle studies (Craig, 1971; JBF Scientific Corp., 1981), presumably due to mineral surface chemistry modifications resulting from NAPL presence. Equilibrium contact angle measurements for NAPLs containing surfactants may require aging samples for hundreds or thousands of hours (Anderson, 1986b). Another factor to consider is that a hysteresis effect is frequently observed during contact angle studies in which the contact angle is less when NAPL advances over an initially water-saturated medium than when NAPL is receding from a NAPL-contaminated medium (Villaume, 1985).

Given the heterogeneous nature of subsurface media and the factors that influence wettability, some investigators have concluded that the wetting of porous media by NAPL can be heterogeneous, or fractional, rather than uniform (Honarpour et al., 1986; Anderson, 1986a). Unfortunately, few wettability studies have been conducted on DNAPLs. Results of contact angle experiments using several DNAPLs and various substrates are provided in Table 4-1 (Arthur D. Little, Inc., 1981).

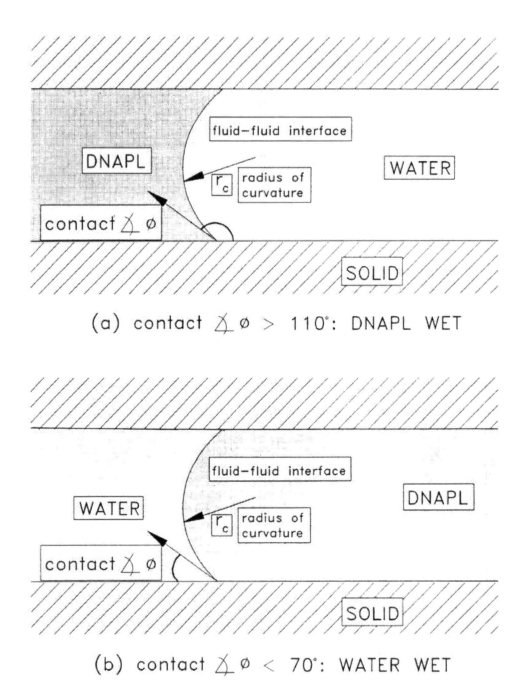

Figure 4-2. Contact angle (measured into water) relations in (a) DNAPL-wet and (b) water-wet saturated systems (modified from Wilson et al., 1990). Most saturated media are preferentially wet by water (see Chapter 4.3).

Table 4-1. Results of contact angle experiments conducted using DNAPLs by Arthur D. Little, Inc., 1981 (from Mercer and Cohen, 1990).

DNAPL	Substrate	Medium	Contact Angle (°)
Tetrachloroethene	clay	APL	23-48
Tetrachloroethene	clay	air	153-168
1,2,4-Trichlorobenzene	clay	APL	28-38
1,2,4-Trichlorobenzene	clay	air	153
Hexachlorobutadiene	clay	water	32-48
Hexachlorocyclopentadiene	clay	water	32-41
2,6-Dichlorotoluene	clay	water	30-38
4-Chlorobenzotrifluoride	clay	water	30-52
Carbon Tetrachloride	clay	water	27-31
Chlorobenzene	clay	water	27-34
Chloroform	clay	water	29-31
S-Area DNAPL	clay	APL	21-54
S-Area DNAPL	clay	water	20-37
S-Area DNAPL	clay	air	170-171
S-Area DNAPL	fine sand and silt	water	30-40
S-Area DNAPL	clayey till (30-40% clay)	water	20-37
S-Area DNAPL	Ottawa fine to coarse sand	water	33-50
Tetrachloroethene	Ottawa fine to coarse sand	water	33-45
Tetrachloroethene	Lockport Dolomite	water	16-21
Tetrachloroethene	Lockport Dolomite	air	171
S-Area DNAPL	Lockport Dolomite	water	16-19
S-Area DNAPL	Lockport Dolomite	air	164-169
S-Area DNAPL	NAPL-contaminated fine sand	APL	45-105
S-Area DNAPL	soils with vegetative matter	water	50-122
S-Area DNAPL	paper	water	31
S-Area DNAPL	wood	water	34-37
S-Area DNAPL	cotton cloth	water	31-33
S-Area DNAPL	stainless steel	water	131-154
S-Area DNAPL	clay	water (SA)	25-54
S-Area DNAPL with solvents	clay	water	15-45

Notes: Adsorbed S-Area (OCC Chemical Corporation site in Niagara Falls, NY) chemicals were detected on some of the clay samples. APL refers to aqueous phase liquids (water containing dissolved chemicals). S-Area DNAPL is composed primarily of tetrachlorobenzene, trichlorobenzenes, tetrachloroethene, hexachlorocyclopentadiene, and octachlorocyclopentene. SA refers to surface-active agents (Tide® and Alconox®) which were added to the water.

4.4 CAPILLARY PRESSURE

Capillary pressure causes porous media to draw in the wetting fluid and repel the nonwetting fluid (Bear, 1972). This is due to the dominant adhesive force between the wetting fluid and media solid surfaces. As a result of contact angle, a meniscus exists at the interface between two immiscible fluids in a pore with a radius of curvature that is proportional to the pore radius (Wilson et al., 1990). The pressure drop across this curved interface is known as capillary pressure which is equal to the difference between the nonwetting fluid pressure and the wetting fluid pressure (Figure 4-3). For a water-NAPL system with water being the wetting phase, capillary pressure, P_c, is defined as:

$$P_c = P_N - P_w \qquad (4-1)$$

where P_N is the NAPL pressure and P_w is the water pressure.

Capillary pressure is a function of interfacial tension, σ, contact angle, ϕ, and pore size (Bear, 1979):

$$P_c = (2 \, \sigma \cos \phi) \, / \, r \qquad (4-2)$$

where r is the radius of the water-filled pore resisting NAPL entry and σ is the interfacial tension between NAPL and water with the subscripts dropped. Consistent sets of units for these and other parameters are given in Appendix B. Equation 4-2 is valid only for immiscible fluid interfaces that form subsections of a sphere. Capillary pressure increases as r and ϕ decrease and as σ increases.

The capillary pressure that must be overcome for a nonwetting NAPL to enter the largest pores (which offer the least capillary pressure resistance) in a water-saturated medium is known as the threshold or displacement entry pressure. Because capillary forces can restrict the migration of NAPL into water-saturated media, fine-grained layers with small r can be capillary barriers. That is, before NAPL can penetrate a water-saturated porous medium, the NAPL pressure head must exceed the resistance of the capillary forces (e.g., Schwille, 1988). The thickness or height of a NAPL column required to develop sufficient NAPL pressure head to exceed capillary force resistance is known as the critical NAPL thickness (or height), z_n.

Capillary pressure effects explains much of the distribution and behavior of subsurface DNAPL (see

Chapter 5). DNAPL penetration of the vadose zone is influenced by the distribution of water content and pore openings; and enhanced by dry conditions and inclined, relatively permeable pathways such as those provided by fractures, root holes, and dipping bedding plane laminations. Upon reaching the capillary fringe above the water table, sinking DNAPL will tend to be obstructed and spread laterally until sufficient DNAPL thickness has accumulated to exceed the threshold entry pressure at the capillary fringe (Schwille, 1988; Cary et al., 1989b; de Pastrovich et al., 1979, and Wilson et al., 1990). Similarly, in the saturated zone, DNAPL will tend to spread laterally over fine-grained capillary barriers and sink through fractures and coarser media where possible as depicted in Figure 4-4 (Kueper and Frind, 1991a; Kueper and McWhorter, 1991; Schwille, 1988). Laboratory experiments and contamination site findings clearly demonstrate that even small-scale differences in pore size distributions can control the path of DNAPL migration.

Several relationships between the different forces affecting DNAPL migration (gravity, capillary pressure, and hydraulic gradients) for a variety of assumed conditions are given in Table 4-2. The equations in Table 4-2 approximate capillary pressure because they do not strictly account for the complicated pore geometries and distributions in natural porous media. Nevertheless, these equations provide a means to examine the conditions of subsurface DNAPL movement. Example applications of several of these relationships are provided in Chapter 5.3. Equations for estimating threshold entry pressures and critical DNAPL heights which must be exceeded for DNAPL penetration of water-saturated media are introduced below.

The threshold entry pressure can be estimated as an equivalent head of water by

$$h_c = (2 \, \sigma \cos \phi) \, / \, (r \, \rho_w \, g) \qquad (4-3)$$

where h_c is the capillary rise of the wetting fluid (water), ρ_w is the density of water, and, g is acceleration due to gravity, for conditions of hydrostatic equilibrium where water above, within, and below the DNAPL body is connected hydraulically and there exists zero capillary pressure at the top of the DNAPL body. This latter condition will be present when the top of the DNAPL body was last under imbibition conditions such as where residual DNAPL is trapped at the trailing edge of sinking DNAPL body (Kueper and McWhorter, 1991). Similarly, the critical DNAPL thickness, z_n, required for DNAPL

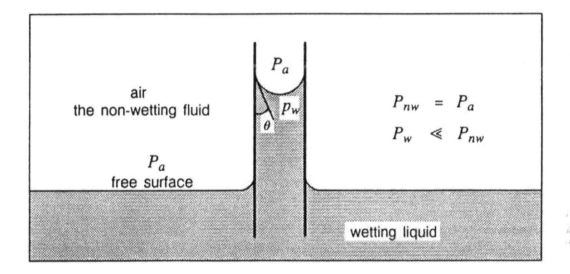

Figure 4-3.   Capillarity is exemplified by the pressure difference between the wetting fluid (water) and the non-wetting fluid (air) at the interface within a tube that causes the wetting fluid to rise above the level of the free surface (from Wilson et al., 1990); where P_{nw} is the pressure in the non-wetting phase and P_w is the pressure in the wetting phase, P_a is atmospheric pressure.

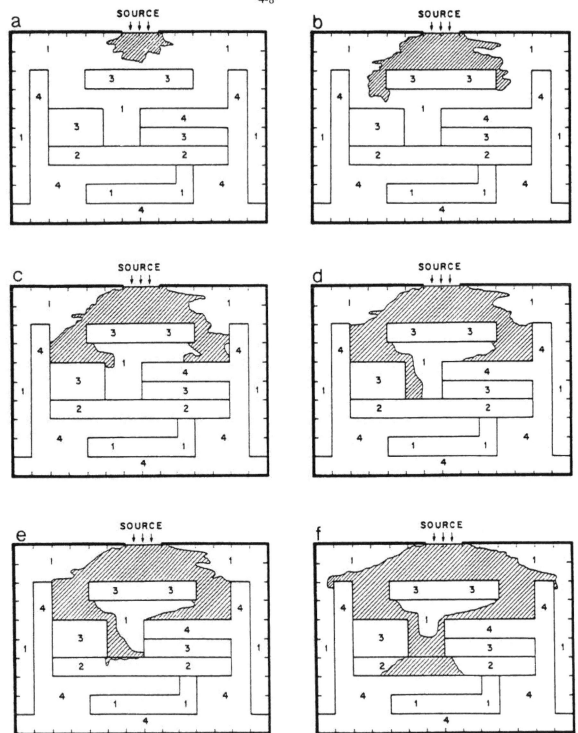

Figure 4-4.   Observed infiltration of tetrachloroethene into water-saturated parallel-plate cell containing heterogeneous sand lenses after (a) 34 s, (b) 126 s, (c) 184 s, (d) 220 s, (e) 225 s, and (f) 313 s (from Kueper and Frind, 1991a). Sands are: 1 is #16 Silica sand ($k = 5.04E\text{-}10$ m^2, $P_d = 3.77$ cm water); 2 is #25 Ottawa sand ($k = 2.05E\text{-}10$ m^2, $P_d = 4.43$ cm water); 3 is #50 Ottawa sand ($k = 5.26E\text{-}11$ m^2; $P_d = 13.5$ cm water); and 4 is #70 Silica sand ($k = 8.19E\text{-}12$ m^2; $P_d = 33.1$ cm water).

Table 4-2. Relationships between capillary pressure, gravity, and hydraulic forces useful for estimating conditions of DNAPL movement (from Kueper and McWhorter, 1991; WCGR, 1991; and Mercer and Cohen, 1990).

Condition	Equation
(a) Capillary pressure exerted on the surface of a nonwetting NAPL sphere	$P_c = P_{NAPL} - P_w = (2\sigma\cos\phi) \, / \, r$
(b) Capillary pressure exerted on the surface of NAPL in a fracture plane where b is the fracture aperture	$P_c = P_{NAPL} - P_w = (2\sigma\cos\phi) \, / \, b$
Hydrostatic Conditions	
(c) Critical height of DNAPL required for downward entry of DNAPL through the capillary fringe (the top of the saturated zone)	$z_n = (2\sigma\cos\phi) \, / \, (r \, g \, \rho_n)$
(d) Critical height of DNAPL required for downward entry of DNAPL into the water-saturated base of a lagoon where DNAPL is pooled beneath water; or, below the water table, for entry of DNAPL into a layer with smaller pore openings (assuming top of DNAPL body last existed under imbibition conditions)	$z_n = (2\sigma\cos\phi) \, / \, [rg(\rho_n - \rho_w)]$
(e) Critical height of DNAPL required for entry of DNAPL into a water-saturated fracture at the base of a lagoon where DNAPL is pooled beneath water; or, below the water table, for entry of DNAPL into a water-saturated fracture having an aperture, b, smaller than the host medium pore radii; or, below the water table, for entry of DNAPL into a water-saturated fracture segment having an aperture smaller than that of the overlying host fracture segment (assuming top of DNAPL body last existed under imbibition conditions)	$z_n = (2\sigma\cos\phi) \, / \, [bg(\rho_n - \rho_w)]$
(f) Critical height of DNAPL required below the water table, for entry of DNAPL into a layer with smaller pore openings where the top of the DNAPL body is under drainage conditions	$z_n = [P_{c(fine)} - P_{c(coarse)}] \, / \, [g(\rho_n - \rho_w)]$
(g) The stable DNAPL pool length, L_n, that can exist below the water table following initial DNAPL migration where θ is the dip angle of the base of the host medium and L_n is measured parallel to the host medium base slope	$L_n = (2\sigma \cos \phi) \, / \, [rg(\rho_n - \rho_w) \sin \theta]$
(h) The stable DNAPL pool length, L_n, within a fracture that can exist below the water table following initial DNAPL migration where θ is the dip angle of the fracture, b is the maximum fracture aperture at the leading edge of the DNAPL pool, and L_n is measured parallel to the fracture slope	$L_n = (2\sigma \cos \phi) \, / \, [bg(\rho_n - \rho_w) \sin \theta]$
Hydrodynamic Conditions	
(i) Neglecting capillary pressure effects, the critical upward hydraulic gradient, i_c, required across a DNAPL body of height z_n to prevent downward DNAPL migration in a uniform porous medium	$i_c = \Delta h / \Delta z_n = (\rho_n - \rho_w) \, / \, \rho_w$
(j) Neglecting capillary pressure effects, the minimum hydraulic head difference between the bottom and top of a DNAPL body of height z_n to prevent downward DNAPL migration in a uniform porous medium	$\Delta h = i_c \, \Delta z_n = z_n (\rho_n - \rho_w) \, / \, \rho_w$
(k) Neglecting capillary pressure effects, the critical hydraulic gradient, i_c, required to prevent the downward movement of DNAPL along the top of a dipping (angle = θ) capillary barrier (i.e., in sloping fractures, bedding planes, or within a sloping coarse layer above a fine grained layer) with i_c measured parallel to the slope	$i_c = [(\rho_n - \rho_w) \sin \theta] \, / \, \rho_w$
(l) The critical horizontal hydraulic gradient, i_c, which must exist across a DNAPL pool of length L beneath the water table to overcome capillary resistance and mobilize DNAPL in the pool (to calculate i_c for a pool of DNAPL in a horizontal fracture, replace r with the fracture aperture, b)	$i_c = (2\sigma \cos \phi) \, / \, (r\rho_w gL)$
(m) The critical upward hydraulic gradient, i_c, required to arrest the downward migration of DNAPL through an aquitard of thickness, Δz, where ΔP_c is the capillary pressure of DNAPL pooled at the top of the aquitard minus the threshold entry (displacement) pressure of the aquitard	$i_c = \Delta h / \Delta z = [(\rho_n - \rho_w)/\rho_w] + [\Delta P_c/(\rho_w g \Delta z)]$

penetration into water-saturated pores with radii, r, can be estimated by

$$z_n = (2 \sigma \cos \phi) / [r g (\rho_n - \rho_w)] \qquad (4-4)$$

where ρ_n is the DNAPL density.

Equation 4-3 is solved for a range of σ, ϕ, and r values in Figure 4-5. The critical DNAPL thickness, z_n, can be estimated using Equation 4-4, or variations thereof, for several conditions, including where

- DNAPL is pooled beneath water above the water-saturated base of a waste lagoon (Table 4-2d),

- sinking DNAPL encounters the top of the saturated zone (Table 4-2c),

- below the water table, sinking DNAPL encounters a layer with smaller pore openings (Table 4-2d),

- below the water table, sinking DNAPL accumulates in a porous medium above a fractured medium having fractures apertures that are smaller than the overlying medium pore radii (Table 4-2e), or,

- DNAPL sinking through fractures encounters fractures with smaller apertures (Table 4-2e).

For the case of DNAPL entry to the saturated zone (Table 4-2c), the effective density difference in Equation 4-4 equals ρ_n (because $\rho_{air} \approx 0$). For fractured media (Table 4-2e), h_c and z_n can be estimated by substituting a fracture aperture value, b, for r in Equations 4-3 and 4-4, respectively. In actuality, the fracture openings are irregular and the entry pressure will generally be intermediate between those calculated using the fracture aperture and radius (Kueper and McWhorter, 1991). The effects of capillary pressure on DNAPL movement in fractured media are further discussed by Kueper and McWhorter (1991).

If the top of a DNAPL body is under drainage conditions (DNAPL is invading the overlying water), then the capillary pressure at the top of the DNAPL column will equal the threshold entry pressure of the host medium (Kueper and McWhorter, 1991). As a result, capillary pressure is exerted above and below the DNAPL body and the z_n value required for downward DNAPL penetration of a finer layer from an overlying coarser layer can be estimated by

$$z_n = [P_{d(fine)} - P_{d(coarse)}] / [g (\rho_n - \rho_w)] \qquad (4-5)$$

as given in Table 4-2f where P_d is the threshold entry pressure (Kueper and McWhorter, 1991).

Under hydrostatic conditions, the potential for DNAPL penetration of progressively finer pore openings increases proportionally to the overlying DNAPL column thickness and the DNAPL-water density difference. Once DNAPL enters a vertical fracture, it will readily invade finer and finer fractures with depth due to the increase in DNAPL column height with fracture depth.

Where DNAPL encounters a coarser underlying medium, capillary pressure will work to squeeze the DNAPL into the larger openings. This effect can be demonstrated by upward rather than downward DNAPL movement after placing DNAPL (for example, at $s_n = 0.80$ and $s_w = 0.20$) in a silt layer beneath a very coarse sand with a much lower DNAPL saturation. Capillary pressure will cause a portion of the DNAPL in the silt to rise into the coarser medium against the force of gravity. Therefore, unless exhausted by residual saturation or prevented by hydraulic pressure, downward DNAPL movement will occur readily in homogeneous media or from finer to coarser layers.

In natural media, pore spaces are extremely complex and irregular, and their geometry cannot be described analytically (Bear, 1972). Due to their variability, the threshold entry pressure for each pore will be different. Macropore (e.g., fractures and worm tubes) radii can sometimes be measured directly. The effective mean pore radius in a porous medium will typically be much smaller than the mean grain radius. Several researchers have developed equations that can be used to estimate mean effective pore radius. Based on laboratory data, Hubbert (1953) defined capillary pressure and pore radius, r, in terms of mean grain diameter, d, such that

$$r \approx d/8 \qquad (4-6)$$

and used this relationship to estimate threshold entry (displacement) pressures in sediments with various grain sizes for an oil-water system as given in Table 4-3.

Leverett (1941) and others have suggested a semi-empirical method to evaluate the relationship between capillary pressure and medium properties in which mean pore radius, r, can be estimated as

$$r \approx (k/n)^{0.5} \qquad (4-7)$$

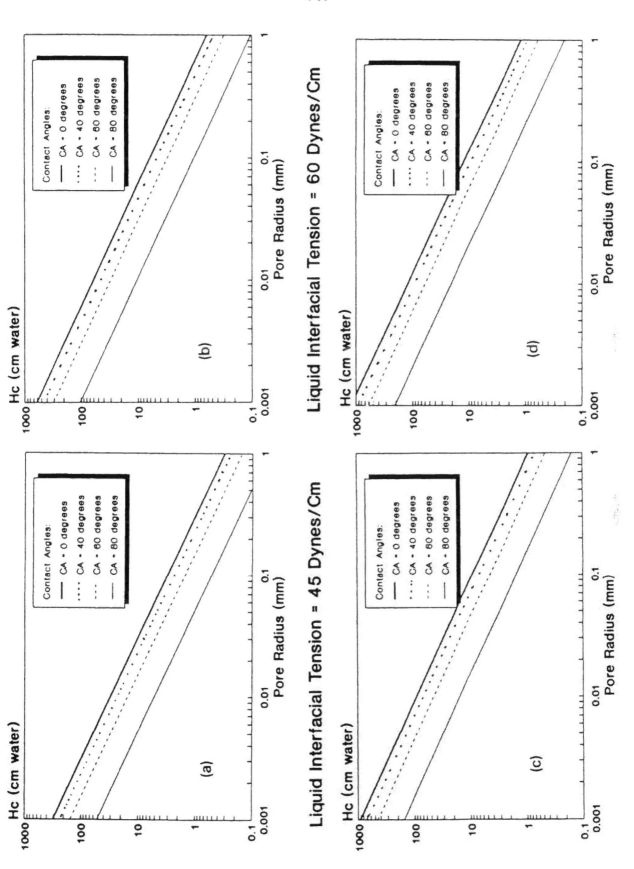

Figure 4-5. Capillary pressure as a function of liquid interfacial tension and contact angle (CA) (from Mercer and Cohen, 1990).

Table 4-3. Threshold entry (displacement) pressures in sediments with various grain sizes based on a ($\sigma \cos \phi$) value of 25 dynes/cm (0.025 N/m) and Equation 4-5 (Hubbert, 1953).

Medium	Mean Grain Size Diameter (mm)	Threshold Entry Capillary Pressure (Pa)	Equivalent Capillary Rise of Water (cm)	Equivalent Capillary Rise of Water (ft)
Clay	< 0.0039	> 100,000	>1000	>33
Silt	0.0039 - 0.063	6300 - 100,000	65 - 1000	2.1 - 33
Sand	0.063 - 2.0	200 - 6300	2.1 - 65	0.06 - 2.1
Coarse Sand	2.0 - 4.0	100 - 200	1.0 - 2.1	0.03 - 0.06

where k and n are the intrinsic permeability and porosity of the medium, respectively (Bear, 1972).

The threshold entry pressure must be measured to obtain more accurate data. This is typically done using a laboratory test cell to determine the height of DNAPL required to initiate drainage from a water-saturated soil or rock sample during measurement of the capillary pressure-saturation, $P_c(s_w)$, relationship. Methods for measuring threshold entry pressure and $Pc(s_w)$ relations are described in Chapter 10.

Equation 4-2 implies that the wetting phase will be progressively displaced from larger-to-smaller pores by the nonwetting phase with increases in nonwetting NAPL pressure head and that the $P_c(s_w)$ relationship is driven by pore size distribution and fluid interfacial tensions (Parker, 1989). Laboratory experiments demonstrate that capillary pressure can be represented as a function of saturation (e.g., Thomas, 1982). Tetrachloroethene-water drainage capillary pressure-saturation curves determined for seven sands of varying hydraulic conductivity are plotted in Figure 4-6 (Kueper and Frind, 1991b). As shown, the capillary pressure curve is typically L-shaped with a low threshold entry pressure for coarser-grained, higher permeability materials and a higher threshold entry pressure for finer-grained, lower permeability materials. DNAPL saturation increases with capillary pressure because higher capillary pressures are required to displace water from incrementally smaller pore openings.

If the $P_c(s_w)$ curve has been determined for a particular DNAPL and soil at a contamination site, then the potential for DNAPL to invade a finer underlying layer can be assessed by: (1) determining the saturation of soil samples taken immediately above the finer layer; (2) estimating the corresponding capillary pressure using the $P_c(s_w)$ curve; and (3) comparing this capillary pressure to an estimated or measured value of threshold entry pressure (P_d) for the finer layer (Kueper and McWhorter, 1991). In less permeable host media, lower DNAPL saturations will be required to exceed the threshold entry pressure requirement of an underlying capillary barrier (Figure 4-6).

To facilitate modeling analyses, laboratory $P_c(s_w)$ measurements are often fitted by nonlinear regression using an empirical parametric model such as the Brooks-Corey (1964) and van Genuchten (1980) models (see also Luckner et al., 1989; Parker, 1989). For example, the Brooks-Corey $P_c(s_w)$ relationship is

$$s_e = (P_c / P_d)^{-\lambda} \qquad (4-8)$$

where $s_e = (s_w - s_{wr})/(1 - s_{wr})$, s_w is the water saturation, s_{wr} is the residual water saturation, P_d is the threshold entry pressure of the medium corresponding to nonwetting phase entry, and λ is a pore-size distribution index. Based on the assumptions of Leverett (1941), three-phase system behavior may also be predicted (Parker, 1989; Parker et al., 1987).

Different immiscible fluid pairs produce different $P_c(s_w)$ curves in the same medium. Lenhard and Parker (1987b) measured $P_c(s_w)$ relations for several NAPLs (i.e., benzene, o-xylene, p-cymene, and benzyl alcohol) in a sandy porous medium. As part of this study, they evaluated a scaling procedure (Parker et al., 1987) applied to $P_c(s_w)$ relations of two-phase air-water, air-NAPL and NAPL-water porous media systems. Relatively good fits were obtained by matching the experimental data to multifluid versions of the Brooks-Corey (1964) and van Genuchten (1980) retention functions. Lenhard and Parker (1987b) concluded that $P_c(s_w)$ curves for any two-phase fluid system in a porous medium can be predicted by scaling the $P_c(s_w)$ relationship for a single two-phase system based on interfacial tension data as suggested by several prior investigators and illustrated in Figure 4-7. The capillary pressure, P_{ca}, measured for a particular NAPL$_a$, saturation, and soil can be scaled using liquid interfacial tension measurements to estimate the capillary pressure, P_{cb}, for NAPL$_b$ for the same saturation and soil using the relation:

$$P_{ca}/\sigma_{a-w} = P_{cb}/\sigma_{b-w} \qquad (4-9)$$

where the subscripts a, b, and w refer to NAPL$_a$, NAPL$_b$, and water, respectively (WCGR, 1991).

Similarly, $P_c(s_w)$ relations can be scaled for curves generated using different soils (Leverett, 1941; Kueper and Frind, 1991b) by

$$P_{cD} = [(P_{ca}/\sigma_a)/(k_a/n_a)]^\alpha = [(P_{cb}/\sigma_b)/(k_b/n_b)]^\alpha \quad (4-10)$$

where, for a specific saturation, P_{cD} is a dimensionless capillary pressure, P_{ca} is the capillary pressure in the NAPL$_a$-water-medium$_a$ system, σ_a is the interfacial liquid tension between NAPL$_a$ and water, k_a is the permeability of medium$_a$, n_a is the porosity of medium$_a$, subscript b refers to the NAPL$_b$-water-medium$_b$ system and parameters, and, α is an exponent fitted by scaling the $P_c(s_w)$ data for different NAPL-soil system samples. Kueper and Frind (1991b) used Equation 4-10 to scale

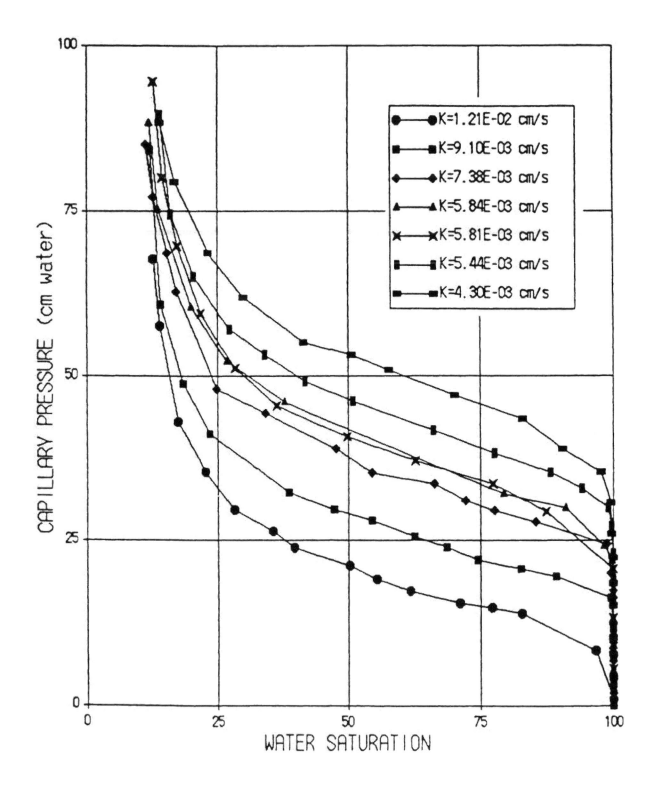

Figure 4-6. Tetrachloroethene-water drainage $P_c(s_w)$ curves determined for seven sands of varying hydraulic conductivity (from Kueper and Frind, 1991b).

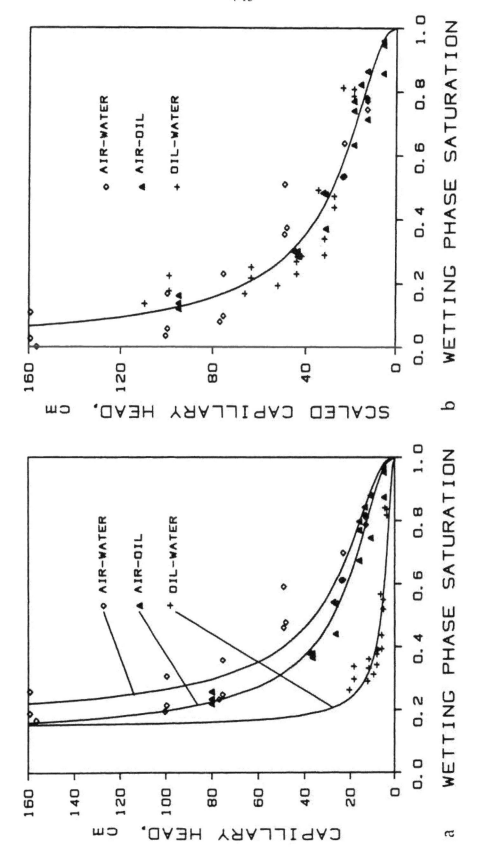

Figure 4-7.   (a) Unscaled and (b) scaled $P_c(s_w)$ relations for air-water, air-benzyl alcohol, and benzyl alcohol-water fluid pairs in the same porous medium (from Parker, 1989).

the data that define the seven $P_c(s_w)$ curves shown in Figure 4-6. The scaled data are plotted with a best-fit Brooks-Corey curve in Figure 4-8. The validity of Equation 4-10 is based on the similarity of pore-size distributions in the scaled media (Brooks and Corey, 1966).

Actual $P_c(s_w)$ relations are more complicated than portrayed in the monotonic curves shown in Figures 4-6 and 4-7 (Lenhard, 1992; Parker, 1989). Changes in capillary pressure with saturation depend on whether the medium is undergoing wetting (imbibition) or drainage of the wetting fluid (Figure 4-9). This capillary hysteresis results from nonwetting fluid entrapment and differences in contact angles during wetting and draining that cause different drying and wetting curves to be followed depending on the prior imbibition-drainage history. During drainage, the larger pores drain the wetting fluid quickly while the smaller pores drain slowly, if at all. As a result of this capillary retention, capillary pressure corresponds to higher saturations on the drainage curve. During wetting, the smaller pores are filled first and the larger pores are least likely to fill with the wetting fluid, thereby leading to a lower capillary pressure curve with saturation. Significant errors can occur by overlooking these hysteretic effects during the simulation of some immiscible flow problems (Lenhard, 1992). Hysteresis is typically ignored, however, during simulation studies because its significance is considered minor compared to uncertainties associated with other parameter estimates (Parker, 1989).

Specific methods to determine entry pressure requirements and capillary pressure-saturation relationships are described in Chapter 10. These measurements provide minimum entry pressure values and average capillary pressure-saturation curves for the samples tested. Utilization of laboratory tests on small samples for interpretation of field-scale phenomena poses a scale problem that requires careful consideration of site conditions, sample size, and sample numbers.

4.5 RESIDUAL SATURATION

During migration, a significant portion of NAPL is retained in porous media, thereby depleting and eventually exhausting the mobile NAPL body. Below the water table, residual saturation (s_r) of NAPL is the saturation (V_{NAPL}/V_{voids}) at which NAPL is immobilized (trapped) by capillary forces as discontinuous ganglia under ambient groundwater flow conditions. In the vadose zone, however, residual NAPL may be more or less continuous depending on the extent to which NAPL films develop between the water and gas phases and thereby interconnect isolated NAPL blobs (Wilson et al., 1990). The physics of oil entrapment and development of methods to minimize s_r by enhanced oil recovery are of great importance to the petroleum industry (Mohanty et al., 1987; Morrow et al., 1988; Chatzis et al., 1988; Wang, 1988; Anderson, 1987c; Chatzis et al., 1983; Melrose and Brandner, 1974; McCaffery and Batycky, 1983). Similarly, s_r has important consequences in the migration and remediation of subsurface DNAPL. Results of a major laboratory investigation of the forces affecting NAPL migration and s_r were recently reported by Wilson et al. (1990).

Residual saturation results from capillary forces and depends on several factors, including: (1) the media pore size distribution (i.e., soil structure, heterogeneity, and grain size distribution), (2) wettability, (3) fluid viscosity ratio and density ratio, (4) interfacial tension, (5) gravity/buoyancy forces, and (6) hydraulic gradients. Residual saturation for the wetting fluid is conceptually different from that for the nonwetting fluid. The nonwetting fluid is discontinuous at s_r, whereas the wetting fluid is not.

In the vadose zone, NAPL is retained as films, wetting pendular rings, wedges surrounding aqueous pendular rings, and as nonwetting blobs in pore throats and bodies in the presence of water (Wilson et al., 1990; Cary et al., 1989b). NAPL will spread as a film between the water and gas phases given a positive spreading coefficient:

$$\Sigma = \sigma_{aw} - (\sigma_{nw} + \sigma_{an}) \qquad (4\text{-}11)$$

where Σ is the spreading coefficient and σ_{aw}, σ_{nw}, and σ_{an} are the interfacial tensions for air-water, NAPL-water, and air-NAPL, respectively (Wilson et al., 1990). Halogenated solvent DNAPLs typically have negative spreading coefficients and will not spread as films in the vadose zone due to their internal cohesion (Wilson et al., 1990).

The capacity of the vadose zone to trap NAPL is sometimes measured and reported as the volumetric retention capacity

$$R = 1000 \, s_r \, n \qquad (4\text{-}12)$$

where R is liters of residual NAPL per cubic meter of media and n is porosity (de Pastrovich et al., 1979;

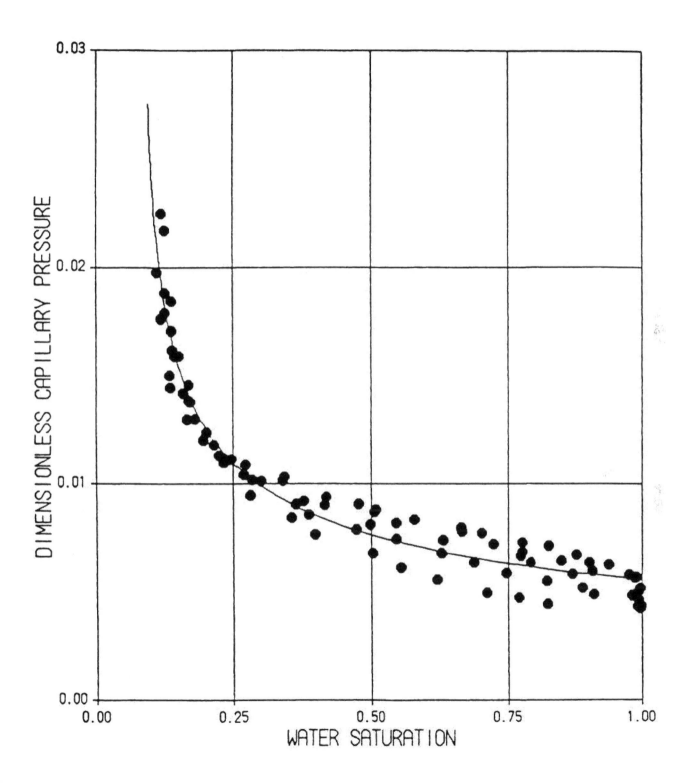

Figure 4-8. Tetrachloroethene-water drainage $P_c(s_w)$ curves (shown in Figure 4-6) for seven
sands of varying hydraulic conductivity scaled using Equation 4-6 (from Kueper
and Frind, 1991b).

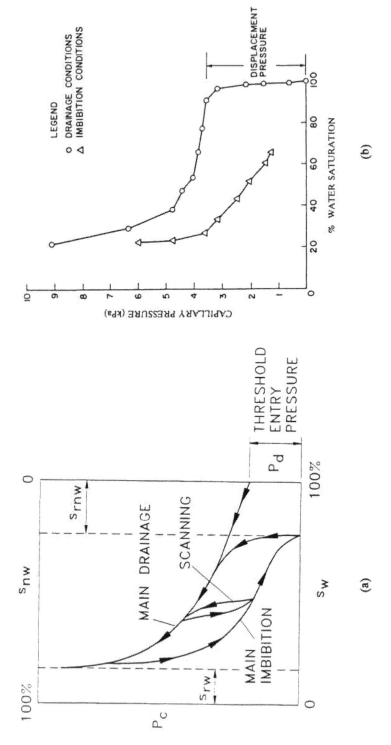

Figure 4-9. (a) Hysteresis in two-fluid $P_c(s_w)$ relations whereby changes in capillary pressure depend on whether the medium is undergoing imbibition (wetting) or drainage of the wetting fluid; in (b), the main drainage and imbibition curves determined for a tetrachloroethene and water in #70 silica sand are shown (from Kueper and McWhorter, 1991). Note that s_{rw} is the irreducible wetting fluid content; s_{rnw} is the nonwetting fluid residual saturation; s_w is the wetting fluid saturation; and s_{nw} is the nonwetting fluid saturation.

Schwille, 1988; Wilson and Conrad, 1984). Residual saturation measurements for a variety of NAPLs and media compiled in Table 4-4 indicate that s_r values typically range from 0.10 to 0.20 in the vadose zone. In general, s_r and retention capacity values in the vadose zone increase with decreasing intrinsic permeability, effective porosity, and moisture content (Hoag and Marley, 1986; Fussel et al., 1981; Schwille, 1988; Anderson, 1988).

Residual saturation values in the saturated zone generally exceed those in the vadose zone because: (1) the fluid density ratio (NAPL:air versus NAPL:water above and below the water table, respectively) favors greater drainage in the vadose zone; (2) as the nonwetting fluid in most saturated media, NAPL is trapped in the larger pores; and, (3) as the wetting fluid (with respect to air) in the vadose zone, NAPL tends to spread into adjacent pores and leave a lower residual content behind, a process that is inhibited in the saturated zone where NAPL is usually the nonwetting fluid (Anderson, 1988). Values of s_r in saturated media generally range from 0.10 to 0.50 (Table 4-5).

Based on laboratory determinations using various NAPLs and saturated soils, Wilson et al. (1990) found that s_r could not be reliably predicted from soil texture because very minor textural differences, such as the inclusion of trace silt or clay in sand, and the presence of heterogeneities may significantly affect s_r values. They also reported, for a given NAPL release volume, that: (1) s_r values in fractures and macropores tend to be less than in homogeneous porous material (see also Schwille, 1988), but can be expected to extend over a larger portion of the aquifer; and (2) s_r values in heterogeneous media containing discontinuous coarse lenses tend to be greater than in homogeneous media, but can be expected to extend over a smaller portion of the aquifer. Similarly, column experiments performed by Powers et al. (1992) entrapped larger volumes of NAPL in graded sand than in uniform sand with the same mean grain size.

On the pore scale, residual NAPL below the water table is immobilized by the snap-off and bypassing mechanisms illustrated in Figure 4-10 (Chatzis et al., 1983). Snap-off occurs in high aspect ratio pores where the pore body is much larger than the pore throat, resulting in single droplets or blobs of residual NAPL (Figure 4-10a). Powers et al. (1992) observed that NAPL entrapped in coarse, uniform sand was distributed primarily as spherical, singlet blobs that were larger than singlets entrapped in finer, uniform sand; and that a much greater

fraction of NAPL is entrapped in large, multi-pore blobs in graded sand. Bypassing is prevalent when wetting fluid flow disconnects the nonwetting fluid, causing NAPL ganglia to be trapped in clusters of large pores surrounded by smaller pores (Figure 4-10b). As a result of these mechanisms, s_r tends to increase with increasing pore aspect ratios and pore size heterogeneity (Chatzis et al., 1983; Powers et al., 1992), and with decreasing porosity, probably due to reduced pore connectivity and a decrease in mobile nonwetting fluid in smaller pore throats (Wilson and Conrad, 1984). Residual saturation is reduced in near-neutral wettability media because the capillary trapping forces are minimized (Anderson, 1987c).

As with saturation in general, determination of field-scale values of s_r presents a considerable sampling problem, in large part, due to the heterogeneity of NAPL distributions and natural media. Higher s_r values will generally be found in the pathways of preferential NAPL transport. A sampling volume on the order of the scale of variability in capillary and permeability properties must be utilized to obtain a true space-averaged measure of residual NAPL saturation (Poulsen and Kueper, 1992). This can be accomplished by analysis of large bulk samples (which may lead to a misinterpretation of NAPL absence if NAPL presence is sparse) or of many small samples (some of which will not contain NAPL).

To determine the potential s_r for a particular porous medium, Wilson et al. (1990) recommend that site-specific s_r measurements be made using an ideal fluid (such as soltrol or decane) having a sufficient density difference with water, low solubility, low volatility, and low toxicity rather than the site-specific NAPL (except if some unusual wetting behavior or NAPL-dependent interaction between phases is expected). Under low capillary number and Bond number conditions (see Glossary for definitions), they found s_r values to be relatively insensitive to fluid properties (Wilson et al., 1990).

A residual source may contaminate groundwater for decades because drinking water standards for many NAPLs are orders of magnitude less than their solubility limits. Water can be contaminated by direct dissolution of residual NAPL and/or by contact with soil gas containing DNAPL volatiles from a residual source in the vadose zone. Combined with practical limitations on residual NAPL recovery (Wilson and Conrad, 1984), the consequences of NAPL dissolution may necessitate perpetual hydraulic containment at some contamination

Table 4-4. Laboratory and field residual saturation data for the vadose zone.

Residual Fluid	Medium	Residual Saturation (s_r) or retention factor (R in L/m³)	Ref.
Water	sand	$s_r = 0.10$	1
Water	loamy sand	$s_r = 0.14$	1
Water	sandy loam	$s_r = 0.16$	1
Water	loam	$s_r = 0.18$	1
Water	silt loam	$s_r = 0.17$	1
Water	silt	$s_r = 0.07$	1
Water	sandy clay loam	$s_r = 0.26$	1
Water	clay loam	$s_r = 0.23$	1
Water	silty clay loam	$s_r = 0.19$	1
Water	sandy clay	$s_r = 0.26$	1
Water	silty clay	$s_r = 0.19$	1
Water	clayey soil	$s_r = 0.18$	1
Gasoline	coarse gravel	R = 2.5	2
Gasoline	coarse sand and gravel	R = 4.0	2
Gasoline	medium to coarse sand	R = 7.5	2
Gasoline	fine to medium sand	R = 12.5	2
Gasoline	silt to fine sand	R = 20	2
Middle distillates	coarse gravel	R = 5.0	2
Middle distillates	coarse sand and gravel	R = 8.0	2
Middle distillates	medium to coarse sand	R = 15	2
Middle distillates	fine to medium sand	R = 25	2
Middle distillates	silt to fine sand	R = 40	2
Fuel oils	coarse gravel	R = 10	2
Fuel oils	coarse sand and gravel	R = 16	2
Fuel oils	medium to coarse sand	R = 30	2
Fuel oils	fine to medium sand	R = 50	2
Fuel oils	silt to fine sand	R = 80	2
Light oil and gasoline	soil	$s_r = 0.18$	3
Diesel and light fuel oil	soil	$s_r = 0.15$	3
Lube and heavy fuel oil	soil	$s_r = 0.20$	3

Table 4-4. Laboratory and field residual saturation data for the vadose zone.

Residual Fluid	Medium	Residual Saturation (s_r) or retention factor (R in L/m^3)	Ref.
Gasoline	coarse sand	$s_r = 0.15$-0.19	4
Gasoline	medium sand	$s_r = 0.12$-0.27	4
Gasoline	fine sand	$s_r = 0.19$-0.60	4
Gasoline	well graded fine-coarse sand	$s_r = 0.46$-0.59	4
Mineral oil	Ottawa sand (dm=0.5mm) [NA]	$s_r = 0.110$	5
Mineral oil	Ottawa sand (dm=0.35 mm) [NA]	$s_r = 0.140$	5
Mineral oil	Ottawa sand (dm=0.25 mm) [NA]	$s_r = 0.172$	5
Mineral oil	Ottawa sand (dm=0.18 mm) [NA]	$s_r = 0.235$	5
Mineral oil	glacial till [NA]	$s_r = 0.15$-0.28	5
Mineral oil	glacial till	$s_r = 0.12$-0.21	5
Mineral oil	alluvium [NA]	$s_r = 0.19$	5
Mineral oil	alluvium	$s_r = 0.19$	5
Mineral oil	loess [NA]	$s_r = 0.49$-0.52	5
Paraffin oil	coarse sand	$s_r = 0.12$	6
Paraffin oil	fine sediments	$s_r = 0.52$	6
Paraffin oil	Ottawa sands	$s_r = 0.11$-0.23	6
DNAPL	sandy soils	$s_r = >0.01$-0.10 $R = >3$-30	7 8
Tetrachloroethene	fracture with 0.2 mm aperture	$R = 0.05$ L/m^2	7
Trichloroethene	medium sand	$s_r = 0.20$	9
Trichloroethene	fine sand	$s_r = 0.19$	9
Trichloroethene	fine sand	$s_r = 0.15$-0.20	9
Trichloroethene	loamy sand	$s_r = 0.08$	10
Soltrol-130	well-sorted, medium-grained, aeolian sand	$s_r = 5.5$-12.2 s_r ave. $= 9.1$	11
Tetrachloroethene	fine to medium beach sand	$s_r = 0.002$-0.20	12

Notes: NA refers to a NAPL-Air unsaturated system (no water); References: 1 = Carsel and Parrish (1988), 2 = Fussell et al. (1981), 3 = API (1980), 4 = Hoag and Marley (1986), 5 = Pfannkuch (1983), 6 = Convery (1979), 7 = Schwille (1988), 8 = Feenstra and Cherry (1988), 9 = Lin et al. (1982), 10 = Cary et al. (1989a), 11 = Wilson et al. (1990), 12 = Poulson and Kueper (1992).

Table 4-5. Laboratory and field residual saturation data for the saturated zone.

Residual Fluid	Medium	Residual Saturation (s_r) or retention factor (R in L/m^3)	Ref.
Mineral oil	sandstone	$s_r = 0.35$-0.43	1
Crude oil	sandstone	$s_r = 0.16$-0.47	1
Crude oil	sandstone	$s_r = 0.26$-0.43	2
Crude oil	petroleum reservoirs	$s_r = 0.25$-0.50	3
Styrene monomer	sandstone	$s_r = 0.11$-0.38	3
Benzene	sand (92% sand, 5% silt, 3% clay)	$s_r = 0.24$	4
Benzyl alcohol	sand (92% sand, 5% silt, 3% clay)	$s_r = 0.26$	4
p-Cymene	sand (92% sand, 5% silt, 3% clay)	$s_r = 0.16$	4
o-Xylene	sand (92% sand, 5% silt, 3% clay)	$s_r = 0.19$	4
1,1,1-Trichloroethane	coarse Ottawa sand	$s_r = 0.15$-0.40	5
Tetrachloroethene	coarse Ottawa sand	$s_r = 0.15$-0.25	5
Kerosene	medium aeolian sand	$s_r = 0.23$-0.29	6
Gasoline	medium aeolian sand	$s_r = 0.27$-0.31	6
n-Decane	medium aeolian sand	$s_r = 0.25$-0.29	6
p-Xylene	medium aeolian sand	$s_r = 0.20$-0.27	6
Tetrachloroethene	medium aeolian sand	$s_r = 0.26$-0.29	6
Soltrol	medium aeolian sand	$s_r = 0.22$-0.37	6
Soltrol	clean coarse fluvial sand	$s_r = 0.16$	6
Soltrol	clean medium beach sand	$s_r = 0.18$	6
Coal tar	siltstone	$s_r = 0.01$-0.03	7
Coal tar	sandstone	$s_r = 0.17$-0.24	7

Notes: References: 1 = Rathmell et al. (1973), 2 = Wang (1988), 3 = Chatzis et al. (1988), 4 = Lenhard and Parker (1987b), 5 = Anderson (1988), 6 = Wilson et al. (1990), 7 = Confidential site.

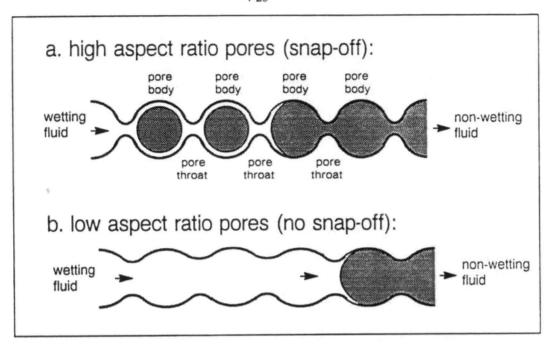

(a)

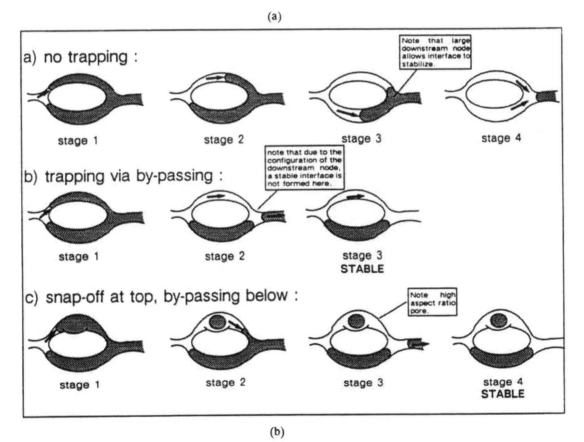

(b)

Figure 4-10.　Sketches illustrating capillary trapping mechanisms: (a) snap-off and (b) by-passing (modified from Chatzis et al., 1983; from Wilson et al., 1990).

sites (Mackay and Cherry, 1989 and Cohen et al., 1987). Unless residual NAPL is replenished by continued contaminant releases or unusual hydraulic conditions, it will tend to be slowly diminished by dissolution, volatilization, and, perhaps, biodegradation. At some sites, remediation may involve partial mobilization of s_r by increasing the prevailing hydraulic gradient or reducing interfacial tension.

4.6 RELATIVE PERMEABILITY

When more than one fluid exists in a porous medium, the fluids compete for pore space. The result is that the mobility is diminished for each fluid. This diminution can be quantified by multiplying the intrinsic permeability by a dimensionless ratio, known as relative permeability. Relative permeability is the ratio of the effective permeability of a fluid at a fixed saturation to the intrinsic permeability; as such, it varies with saturation from zero to one.

Relative permeability relationships are required for numerical simulation of immiscible flow problems. Like capillary pressure, relative permeability can be represented as a function of saturation and also can display hysteresis. Due to difficulties associated with laboratory and field measurement of relative permeability, alternative theoretical approaches are utilized to estimate this function from the more easily measured $P_c(s_w)$ data (Luckner et al., 1989; Parker, 1989; Mualem, 1976). Laboratory data on two-phase relative permeability are typically expressed in terms of phase saturation to a power (n) between 2 and 4 (Faust et al., 1989). For example, Frick (1962) suggested the following relations for unconsolidated sands:

$$k_{rn} = (1 - s_e)^3 \qquad (4\text{-}13)$$

$$k_{rw} = (s_e)^3 \qquad (4\text{-}14)$$

where $s_e = (s_w - s_{wr})/(1 - s_{wr})$, s_w is the water saturation, and s_{wr} is the residual water saturation. An example of relative permeability curves for a water-NAPL system is shown in Figure 4-11a. General features of these curves include the following: (1) the relative permeabilities rarely sum to 1 when both phases are present; (2) k_{rn} is typically greater than k_w at the same saturation for each respective phase; (3) both k_{rn} and k_w go to zero at a finite (residual) saturation; and (4) hysteresis is more prominent for the nonwetting phase than the wetting

phase (Demond and Roberts, 1987). Similar curves are obtained for air-NAPL systems.

Unfortunately, relative permeability data are generally unavailable for DNAPLs found at contamination sites. Lin et al. (1982), however, made laboratory measurements of pressure-saturation relations for water-air and trichloroethene (TCE)-air systems in homogeneous sand columns. These data were converted to two-fluid saturation-relative permeability data by Abriola (1983) using Mualem's theory (1976). Other pressure-saturation data for tetrachloroethene are provided in Kueper and Frind (1991b) (see Figure 4-6). Reviews of two- and three-phase relative permeability data, measurement methods, and governing factors are presented by Honarpour et al. (1986), Saraf and McCaffery (1982), and Demond and Roberts (1987). Several measurement methods are described in Chapter 10.

Three-phase relative permeabilities are required to describe the simultaneous movement of NAPL, water, and air at a point. The functional dependence of relative permeabilities is derived experimentally (Corey et al., 1956; Snell, 1962). However, given the exceptional difficulty and expense of measurement, actual site-specific data and the functional form of three-phase relative permeability are generally not available, particularly for DNAPLs. As a result, theoretical models have been devised to estimate three-phase relative permeability (Stone, 1970; Stone, 1973; Dietrich and Bonder, 1976; Fayers and Matthews, 1984; Parker et al., 1987; Delshad and Pope, 1989). For example, Stone (1973) proposed a model to characterize three-fluid relative permeabilities using data for two-fluid relative permeabilities. With this method, relative permeability data for NAPL are acquired in both water-NAPL and air-NAPL systems. The relative permeability of NAPL in the three-fluid system is then calculated as:

$$k_{rn} = k_{rnw}^* \left[(k_{rnw}/k_{rnw}^* + k_{rw})(k_{rna}/k_{rnw}^* + k_{ra}) - (k_{rw} + k_{ra}) \right] \qquad (4\text{-}15)$$

where k_{rnw}^* is the relative permeability of NAPL at the residual saturation of water in the water-NAPL system; k_{rnw} is the relative permeability of NAPL in the water-NAPL system (a function of water saturation); and k_{rna} is the relative permeability of NAPL in the air-NAPL system (a function of air saturation). This equation can be used to construct a ternary diagram, as given in Figure 4-11b.

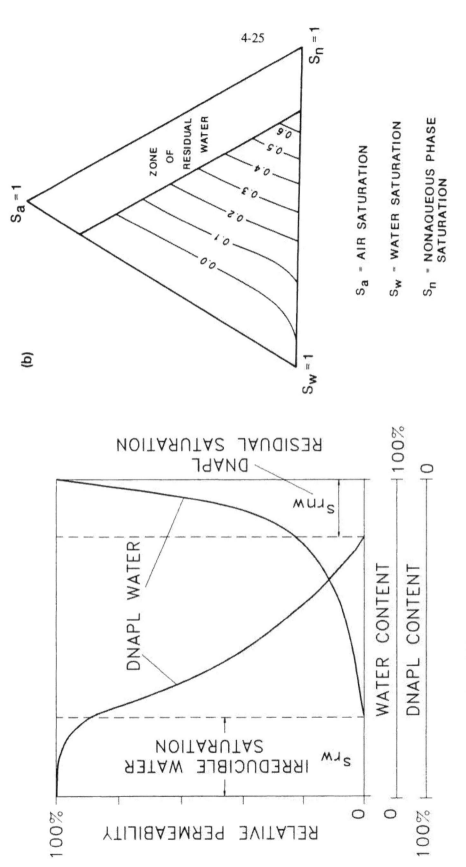

Figure 4-11. (a) Water-NAPL relative permeability (modified from Schwille, 1988); (b) ternary diagram showng the relative permeability of NAPL as a function of phase saturations (from Faust, 1985).

Delshad and Pope (1989) recently assessed seven different three-phase relative permeability models by comparing predicted permeabilities with three sets of experimental data. They found variable and limited agreement between the different models and data sets. Similarly, laboratory relative permeability data often compares poorly with petroleum reservoir production data (Faust et al., 1989). Observations of the control exerted by formation heterogeneity on DNAPL and water flow at contamination sites suggest that relative permeability relations should be based on field-scale data which, in some cases, will tend to be less strongly nonlinear than the laboratory-scale functions (Faust et al., 1989).

4.7 SOLUBILITY

Aqueous solubility refers to the maximum concentration of a chemical that will dissolve in pure water at a particular temperature. Results of laboratory dissolution experiments (Anderson, 1988; Schwille, 1988) show that chemical concentrations approximately equal to aqueous solubility values are obtained in water flowing at 10-100 cm/d through NAPL-contaminated sands. It is widely reported (e.g., Mackay et al., 1985), however, that organic compounds are commonly found in groundwater at concentrations less than ten percent of NAPL solubility limits, even where NAPL is known or suspected to be present.

The discrepancy between field and laboratory measurements is probably caused by heterogeneous field conditions, such as non-uniform groundwater flow, complex NAPL distribution, and mixing of stratified groundwater in a well, and to a lesser extent, NAPL-to-water mass transfer limitations (Feenstra and Cherry, 1988; Mackay et al., 1985; Powers et al., 1991). Various studies suggest that dissolution may be rate limited when NAPL is present as large complex ganglia, groundwater velocities are high, NAPL saturations are low, and/or the mass fraction of soluble species in a NAPL mixture is low (Powers et al., 1992). For halogenated solvents in particular, these chemical and hydrodynamic processes promote the creation of large plumes of groundwater with low chemical concentrations that, however, may greatly exceed drinking water standards (Tables 3-4 and 3-8).

Another factor to consider is that dissolved chemical concentrations will also be less than aqueous solubilities reported for pure chemicals where the NAPL is composed of multiple chemicals. For this case, the effective aqueous solubility of a particular component of the multi-liquid NAPL can be approximated by multiplying the mole fraction of the chemical in the NAPL by its pure form aqueous solubility (Banerjee, 1984; Feenstra et al., 1991; Mackay et al., 1991). Details related to this calculation are presented in Worksheet 7-1. Alternatively, the effective solubility of individual components of a complex NAPL can be determined experimentally (Zalidis et al., 1991; Mackay et al., 1991; Chapter 10).

As identified in Appendix A, DNAPLs vary widely in their aqueous solubility (Appendix A). Solubilities may be obtained from literature, measured experimentally, or estimated using empirical relationships developed between solubility and other chemical properties such as partition coefficients and molecular structure. For example, Lyman et al. (1982) and Kenaga and Goring (1980) present many regression equations that correlate aqueous solubility with K_{ow} (octanol/water) and K_{oc} (organic carbon/water) partition coefficients for various chemical groups. K_{ow} and K_{oc} data for NAPLs are given in Appendix A. Nirmalakhandan and Speece (1988) developed a predictive equation for aqueous solubility based on correlations between molecular structure and solubility of 200 environmentally relevant chemicals. Organic concentrations in water can also be estimated from an equilibrium relationship based on Raoult's Law and Henry's Law (Corapcioglu and Baehr, 1987). For most DNAPL chemicals of interest, multiple aqueous solubilities are reported in the literature (Montgomery and Welkom, 1990; Montgomery, 1991; Lucius et al., 1990; Verschueren, 1983).

Factors affecting solubility include temperature, cosolvents, salinity, and dissolved organic matter. Although the aqueous solubility of most organic chemicals rises with temperature, the direction and magnitude of this relationship is variable (Lyman et al., 1982). Similarly, the effect of cosolvents (multiple organic compounds) on chemical solubility depends on the specific mix of compounds and concentrations. Based on laboratory data and modeling, Rao et al. (1991), however, conclude that solubility enhancement for most organic chemicals will be minor (<20%) unless cosolvent concentrations exceed 2% by volume in pore water. Banerjee (1984) and Groves (1988) describe methods to predict the solubilities of organic chemical mixtures in water based on activity coefficient equations. The aqueous solubility of organic chemicals generally declines with increasing salinity (Rossi and Thomas, 1981 and Eganhouse and Calder, 1973). Dissolved organic matter, such as naturally occurring humic and fulvic acids are known to enhance the solubility of hydrophobic organic compounds in water (Chiou et al., 1986; Lyman et al., 1982).

Subsurface NAPL trapped as ganglia at residual saturation and contained in pools, such as DNAPL trapped in depressions along the top of a capillary barrier, are long-term sources of groundwater contamination. Factors influencing NAPL dissolution and eventual depletion include the effective aqueous solubility of NAPL components, groundwater velocity, NAPL-water contact area, and the molecular diffusivity of the NAPL chemicals in water (Feenstra and Cherry, 1988; Anderson, 1988; Hunt et al., 1988a; Schwille, 1988; Anderson et al., 1992a, b; Pfannkuch, 1984; Miller et al., 1990; Mackay et al., 1991). Pfannkuch (1984) reviewed the literature related to the mass exchange of petroleum hydrocarbons to groundwater. Laboratory studies of LNAPL transfer to water were done by the Working Group (1970), Hoffmann (1969, 1970), Zilliox et al. (1973, 1974), van der Waarden et al. (1971), Fried et al. (1979), Zalidis et al. (1991), and Miller et al. (1990). Schwille (1988), Hunt et al. (1988a), Anderson (1988), and Mackay et al. (1991) conducted experiments to analyze the transfer of DNAPL chemicals into groundwater. Recently, Anderson et al. (1992a, b) evaluated the dissolution of DNAPL from a well-defined residual source and from DNAPL fingers and pools.

Experimental data show that mass exchange coefficients generally: (1) increase with groundwater velocity, except at low velocities where the exchange rate is controlled by molecular diffusion; (2) increase with NAPL saturation; (3) increase with the effective aqueous solubility of the NAPL component; and, (4) decrease with time as NAPL ages (Pfannkuch, 1984; Miller et al., 1990; Mackay et al., 1991; Zalidis et al., 1991; Zilliox et al., 1978). The dissolution process can be rejuvenated, however, by varying hydraulic conditions (i.e., changing groundwater flow directions or rates).

The mass exchange rate (m_x/t), or strength of the dissolved contaminant source, can be expressed as the product of the mass exchange coefficient ($m_x/L^2/t$) and some measure of the contact area (L^2). The contact area of a given mass of residual NAPL ganglia and fingers is more difficult to estimate, but much greater, than that of an equivalent mass of pooled NAPL. Laboratory experiments and theoretical analyses indicate that dissolution of residual NAPL fingers and ganglia will result in groundwater concentrations that are near saturation (Anderson et al., 1992a). These concentrations will then be subject to the dispersion and dilution

processes noted in the second paragraph of this section. Similar experimental and mathematical analyses, however, suggest that dissolution of NAPL pools is mass-transfer limited (Schwille, 1988; Anderson et al., 1992b; Johnson and Pankow, 1992). Consequently, dissolution of residual NAPL fingers and ganglia produces higher chemical concentrations in groundwater and depletes the NAPL source more quickly than dissolution of a NAPL pool of equivalent mass. At many sites, DNAPL pools will provide a source of groundwater contamination long after residual fingers and ganglia have dissolved completely.

Many contamination site DNAPLs are composed of multiple chemicals with varying individual solubilities. At these sites, preferential and sequential loss of the relatively soluble and volatile NAPL components leaves behind a less soluble residue (Senn and Johnson, 1987; Mackay et al., 1991). This weathering causes the ratios of chemicals in the NAPL and dissolved plume to change with time and space. Based on a theoretical analysis of dissolution kinetics and equilibria of sparingly soluble NAPL components and experimental results, Mackay et al. (1991) present equations to estimate: (1) the equilibrium concentrations of dissolved NAPL chemicals in contact with a NAPL of defined composition, (2) changes in the aqueous and NAPL phase chemical concentrations with time due to NAPL depletion by dissolution, and (3) the water flow volume (or time) for a defined depletion of a component within the NAPL mass.

Similarly, relationships have been developed to estimate dissolved chemical concentrations in groundwater and the time required to deplete residual or pooled single-component NAPL sources (Hunt et al., 1988a; Anderson, 1988; Anderson et al., 1992a, b; Azbel, 1981). These models are primarily useful as a conceptual tool to assess the long-term contamination potential associated with subsurface NAPL. For example, the time needed to completely dissolve a NAPL source given an existing or induced interstitial groundwater velocity, v_i, can be estimated as

$$t = m / (v_i n_e C_w A) \qquad (4\text{-}16)$$

where m is the NAPL mass, n_e is the effective porosity, A is the cross-sectional area containing NAPL through which groundwater flow exits with a dissolved NAPL chemical concentration, C_w. The actual dissolution will generally slow with time due to aging and reduction of NAPL-water contact area (Powers et al., 1991). Considering limits to solubility and groundwater velocities, it is obvious that dissolution is an ineffective removal process for significant quantities of many NAPLs.

4.8 VOLATILIZATION

Volatilization refers to mass transfer from liquid and soil to the gaseous phase. Thus, chemicals in the soil gas may be derived from the presence of NAPL, dissolved chemicals, or adsorbed chemicals. Chemical properties affecting volatilization include vapor pressure and aqueous solubility (Appendix A). Other factors influencing volatilization rate are: concentration in soil, soil moisture content, soil air movement, sorptive and diffusive characteristics of the soil, soil temperature, and bulk properties of the soil such as organic-carbon content, porosity, density, and clay content (Lyman et al., 1982).

Volatile organic compounds (VOCs) in soil gas can: (1) migrate and ultimately condense, (2) sorb onto soil particles, (3) dissolve in groundwater, (4) degrade, and/or (5) escape to the atmosphere. Volatilization of flammable organic chemicals in soil can create a fire or explosion hazard if vapors accumulate in combustible concentrations in the presence of an ignition source (Fussell et al., 1981).

The partitioning of volatile chemicals in the vadose zone between the solid, gas, aqueous, and NAPL phases depends on the volatility and solubility of the VOC, the soil moisture content, and the type and amount of soil solids present (Silka and Jordan, 1993). For example, based on experiments with kerosene, Acher et al. (1989) found that adsorption of vapor decreased with increasing soil moisture content. Zytner et al. (1989) observed greater tetrachloroethene (PCE) adsorption to soil with higher organic carbon content resulting in a reduced volatilization rate for both aqueous and pure PCE. Conversely, increasing soil air movement and/or soil temperature elevates the volatilization rate. Volatilization losses from subsurface NAPL are expected where NAPL is close to the ground surface or in dry pervious sandy soils, or where NAPL has a very high vapor pressure (Feenstra and Cherry, 1988).

Estimating volatilization from soil involves (1) estimating the organic partitioning between water and air, and NAPL and air; and (2) estimating the vapor transport from the soil. Henry's Law and Raoult's Law are used to determine the partitioning between water and air, and between NAPL and air, respectively. Vapor transport in the soil is usually described by the diffusion equation and

several models have been developed where the main transport mechanism is macroscopic diffusion (e.g., Lyman et al., 1982; Baehr, 1987). More complex models are also available (e.g., Jury et al., 1990; Falta et al., 1989; Sleep and Sykes, 1989; Brusseau, 1991).

Because a chemical can volatilize from a dissolved state and/or from NAPL, both conditions need be considered to characterize the total amount of chemical that is volatilized. Local equilibrium is typically assumed between the air and other fluids. Henry's Law relates the concentration of a dissolved chemical in water to the partial pressure of the chemical in gas:

$$P = K_H C_W \qquad (4\text{-}17)$$

where P is the partial pressure of the chemical in the gas phase (atm), C_W is the concentration of the chemical in water (mole/m^3), and K_H is Henry's Law constant (atm m^3/mole). Henry's Law is valid for sparingly soluble, non-electrolytes where the gas phase is considered ideal (Noggle, 1985). Henry's Law constants for DNAPL compounds are given in Appendix A. The tendency of a chemical to volatilize increases with an increase in Henry's Law constant.

Raoult's Law can be used to quantify the ideal reference state for the equilibrium between a NAPL solution and air (Corapcioglu and Baehr, 1987). Raoult's Law relates the ideal vapor pressure and relative concentration of a chemical in solution to its vapor pressure over the NAPL solution:

$$P_A = X_A P_A^\circ \qquad (4\text{-}18)$$

where P_A is the vapor pressure of chemical A over the NAPL solution, X_A is the mole fraction of chemical A in the NAPL solution, and P_A° is the vapor pressure of the pure chemical A.

Volatilization represents a source to subsurface vapor transport. Recent studies have examined soil gas advection due to gas pressure and gas density gradients (Sleep and Sykes, 1989; Falta et al., 1989; Mendoza and Frind, 1990a, b; Mendoza and McAlary, 1990). Density-driven gas flow can be an important transport mechanism in the vadose zone that may result in contamination of the underlying groundwater and significant depletion of residual NAPL. Density-driven gas flow is a function of the gas-phase permeability, the gas-phase retardation coefficient, and the total gas density which depends on the NAPL molecular weight and saturated vapor pressure

(Falta et al., 1989). Saturated vapor concentrations and total gas densities calculated for some common NAPLs using the ideal gas law and Dalton's law of partial pressures, respectively, are given in Table 4-6. Density-driven gas flow will likely be significant where the total gas density exceeds the ambient gas density by more than ten percent and the gas phase permeability exceeds 1×10^{-11} m^2 in homogeneous media (i.e., coarse sands and gravel) (Falta et al., 1989; Mendoza and Frind, 1990a, b). Dense gas emanating from NAPL in the vadose zone will typically sink to the water table where it and gas that has volatilized from the saturated zone will spread outward. The pattern of soil gas migration will be strongly influenced by subsurface heterogeneities. Soil gas transport is also discussed in Chapters 5.3 and 8.2.

4.9 DENSITY

Density refers to the mass per unit volume of a substance. It is often presented as specific gravity, the ratio of a substance's density to that of some standard substance, usually water. Density varies as a function of several parameters, most notably temperature. Halogenated hydrocarbons generally are more dense than water, and density increases with the degree of halogenation. According to Mackay et al. (1985), density differences of about 1% influence fluid flow in the subsurface. Density differences as small as 0.1% have been demonstrated to cause contaminated water to sink in physical model aquifers over several weeks (Schmelling, 1992). The densities of most DNAPLs range between 1.01 and 1.65 (1 to 65% greater than water) as shown in Figure 3-1 and Appendix A. Several simple methods for measuring NAPL density are described in Chapter 10.

4.10 VISCOSITY

Viscosity is the internal friction derived from molecular cohesion within a fluid that causes it to resist flow. Following a release, a low viscosity (thin) NAPL will migrate more rapidly in the subsurface than a high viscosity (thick) NAPL assuming all other factors (including interfacial tension effects) are equal. This is because hydraulic conductivity, K, is inversely related to absolute (or dynamic) fluid viscosity, μ, by

$$K = k\rho g/\mu \qquad (4-19)$$

where k is the intrinsic permeability, g is the acceleration due to gravity, and ρ is the fluid density. Absolute

viscosity divided by fluid density is referred to as kinematic viscosity. At sites with multiple DNAPL types, therefore, more distant separate phase migration is usually associated with the less viscous liquids. Subsurface NAPL viscosity can change with time, typically becoming thicker as the more volatile, thinner components evaporate and dissolve from the NAPL mass. Absolute viscosity data for selected NAPLs are provided in Appendix A, and measurement methods are described in Chapter 10.

The NAPL-water viscosity ratio is part of a term used in the petroleum industry known as the mobility ratio. In a water flood, the mobility ratio is defined as the mobility of the displacing fluid (relative permeability/viscosity for water) divided by the mobility of the displaced fluid (relative permeability/viscosity for NAPL). Mobility ratios greater than one favor the flow of water whereas those less than one favor the flow and recovery of NAPL.

During immiscible fluid displacement in a porous medium, the interface between the two fluids may become unstable. Known as viscous fingering, this instability typically arises when a less viscous fluid moves into a more viscous fluid (Chouke et al., 1959; Homsy, 1987). This phenomenon causes fingers of the driving fluid to penetrate the displaced fluid (Figure 4-12). Where viscous fingering begins is also influenced by heterogeneities. These factors are discussed by Kueper and Frind (1988). As a result of viscous fingering, NAPL may not occupy the complete cross-sectional area through which it moves, thus permitting water to flow through and increase dissolution. Additionally, for a given NAPL volume, viscous fingering will promote deeper NAPL penetration than would occur in its absence.

Table 4-6. Vapor concentration and total gas density data for selected DNAPLs at 25°C (from Falta et al., 1989).

Chemical	Molecular Weight, M g/mole	Vapor Pressure kPa (@25°C)	Saturated Vapor Concentration (kg/m³)	Total Gas Density kg/m³
Trichloroethene	131.4	9.9	0.52	1.58
Chloroform	119.4	25.6	1.23	2.11
Tetrachloroethene	165.8	2.5	0.17	1.31
1,1,1-Trichloroethane	133.4	16.5	0.89	1.87
Methylene Chloride	84.9	58.4	2.00	2.50
1,2-Dichloroethene	96.9	43.5	1.70	2.37
1,2-Dichloroethane	99.0	10.9	0.44	1.48
Chlorobenzene	112.6	1.6	0.07	1.23
1,1-Dichloroethane	99.0	30.1	1.20	2.03
Tetrachloromethane	153.8	15.1	0.94	1.93
Air at 1 atm, 25°C	28.6	(101.3)		1.17

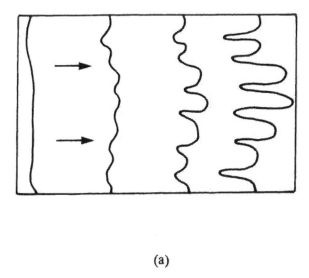

(a)

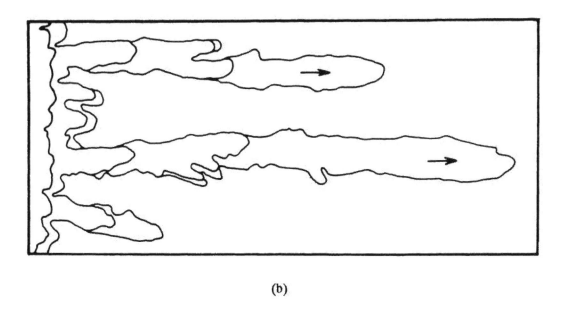

(b)

Figure 4-12. (a) Development of fingering and (b) advanced stages of fingering in Hele-Shaw cell models (from Kueper and Frind, 1988).

5 DNAPL TRANSPORT: PROCESSES, CONCEPTUAL MODELS, AND ASSESSMENT

The properties of fluid and media, and principles of transport, described in Chapter 4 govern the subsurface migration of DNAPL chemicals. At field-scale, migration processes and patterns are controlled by the interaction of these properties and principles with the porous media distribution, hydraulic conditions, and the nature of the DNAPL release. DNAPL migration processes are described briefly in Chapter 5.1. Laboratory and field experiments, modeling studies, and site investigations conducted primarily since 1975 have led to the development of conceptual models of DNAPL chemical transport under varied field conditions. These models are presented and their utility is described in Chapter 5.2. Several quantitative methods for examining site conditions within the context of these conceptual models are discussed in Chapter 5.3. Finally, numerical simulation of immiscible flow problems is briefly addressed in Chapter 5.4.

5.1 OVERVIEW OF DNAPL MIGRATION PROCESSES

DNAPL migration in the subsurface is influenced by (Feenstra and Cherry, 1988): (1) volume of DNAPL released; (2) area of infiltration; (3) time duration of release; (4) properties of the DNAPL; (5) properties of the media; and (6) subsurface flow conditions. Once a DNAPL release occurs, transport mechanisms include: (1) overland flow; (2) immiscible subsurface flow; (3) dissolution and solute transport; and (4) volatilization and vapor transport.

5.1.1 Gravity, Capillary Pressure, and Hydrodynamic Forces

Subsurface DNAPL is acted upon by three distinct forces: (1) pressure due to gravity (sometimes referred to as buoyancy or hydrostatic pressure), (2) capillary pressure, and (3) hydrodynamic pressure (also known as the hydraulic or viscous force). Each force may have a different principal direction of action and the subsurface movement of DNAPL is determined by the interaction of these forces.

The various forces acting on a given mass of DNAPL impart potential energy. The magnitude of this potential energy can be characterized using hydraulic head, pressure, or hydraulic potential. Hydraulic head is a potential function, the potential energy per unit weight of the fluid. Pressure describes the energy on a per unit volume basis, whereas hydraulic potential is the energy per unit mass. All of these are scalar quantities, that is, they are characterized by magnitude only. The rate of change of hydraulic head, pressure, or hydraulic potential with distance is known as the gradient of these quantities, respectively. Gradients of hydraulic head, pressure, or hydraulic potential are used to determine DNAPL movement. In the following discussion, the three forces acting on DNAPL are discussed in terms of head, pressure, and their gradients.

Gravity forces promote the downward migration of DNAPL. The fluid pressure exerted at the base of a DNAPL body due to gravity, P_g, is proportional to the density difference between DNAPL and water ($\rho_n - \rho_w$) in the saturated zone (to account for the buoyancy effect of water), the absolute DNAPL density in the vadose zone, and the DNAPL body height, z_n, such that

$$P_g = z_n\, g(\rho_n - \rho_w) \quad \text{(saturated zone)} \quad (5\text{-}1a)$$

and

$$P_g = z_n\, g\, \rho_n \quad \text{(vadose zone)} \quad (5\text{-}1b)$$

where g is the acceleration due to gravity (9.807 m/s^2). When using British units (e.g., lbs, ft), g must be dropped because weight equals mass multiplied by g. For example, the P_g acting at the base of a 0.5-m thick pool of tetrachloroethene below the water table is

$$P_g = 0.5 \text{ m} * 9.807 \text{ m/s}^2 * (1620 \text{ kg/m}^3 - 1000 \text{ kg/m}^3)$$

$$P_g = 3040 \text{ kg/m}*\text{s}^2 = 3040 \text{ Pa.}$$

This pressure (P_g) can be converted to an equivalent pressure head of water,

$$h_g = z_n g(\rho_n - \rho_w)/(g\rho_w) \quad (5\text{-}2)$$

$$h_g = 0.5 \text{ m} * [(620 \text{ kg/m}^3) / 1000 \text{ kg/m}^3] = 0.31 \text{ m.}$$

Additionally, the hydraulic gradient due to gravity, i_g, can be calculated as

$$i_g = (\rho_n - \rho_w)/\rho_w \quad (5\text{-}3)$$

which equals 0.620 for tetrachloroethene in water. The gravity force that drives DNAPL flow is greater in the vadose zone where the density difference equals the DNAPL density than in the saturated zone and increases with depth within a DNAPL body.

As described in Chapter 4.4, capillary pressure resists the migration of nonwetting DNAPL from larger to smaller pore openings in water-saturated porous media. Capillary pressure effects can be illustrated by considering a hydrostatic system where DNAPL movement is affected only by gravitational and capillary forces. The following discussion of capillary pressure effects in a hydrostatic system is adapted from Arthur D. Little, Inc. (1982).

The radius of a spherical DNAPL globule at rest in a pore body approximates that of the pore (Figure 5-1a). For this case, capillary pressure will be exerted uniformly on the DNAPL globule such that

$$P_c = 2 \sigma \cos \phi / r \qquad (5-4)$$

where σ is the interfacial tension between DNAPL and water, ϕ is the contact angle, and r is the pore radius.

If the DNAPL globule is halfway through the underlying pore throat (Figure 5-1b), then the capillary pressure acting on the globule bottom will exceed that acting on its upper surface. The resulting upward capillary pressure gradient is given by

$$i_{cp} = (P_{ct} - P_{cp}) / z_n \qquad (5-5)$$

$$i_{cp} = [(2\sigma\cos\phi/r_t) - (2\sigma\cos\phi/r_p)] / z_n \qquad (5-6)$$

$$i_{cp} = (2\sigma\cos\phi/z_n) (1/r_t - 1/r_p) \qquad (5-7)$$

where i_{cp} is the capillary pressure gradient, P_{ct} and P_{cp} are the capillary pressures exerted on the nonwetting fluid in the pore throat and pore body respectively, r_t is the pore throat radius, r_p is the pore body radius, and z_n is the height of the DNAPL globule.

If the DNAPL globule is centered within the pore throat (5-1c), then pore radii (and capillary pressures) at the upper and lower ends are equal, and the globule can sink in response to the gravity gradient.

Finally, if the DNAPL globule is significantly through the pore throat (Figure 5-1d), then a downward capillary pressure gradient will exist, and both gravity and capillary pressure will push the globule down into the pore body.

DNAPL migration can occur in a hydrostatic system, therefore, where the downward gravitational pressure or gradient exceeds the resisting capillary pressure or gradient:

$$[z_n g(\rho_n - \rho_w)] > [(2\sigma\cos\phi/r_t) - (2\sigma\cos\phi/r_p)] \qquad (5-8)$$

or

$$[(\rho_n - \rho_w)/\rho_w] > [(2\sigma\cos\phi/z_n) (1/r_t - 1/r_p)] \qquad (5-9)$$

The gravity and capillary pressures and gradients are equal where DNAPL is at rest in a hydrostatic system.

In the field, the capillary pressure exerted downward on top of a continuous DNAPL body will equal: (1) the threshold entry pressure of the host medium if there is no DNAPL above the DNAPL body, or (2) zero if the top of the DNAPL body was last under imbibition conditions and is overlain by residual saturation (see Figure 4-9).

The critical height of DNAPL, z_n, in a host medium required to overcome capillary resistance and penetrate an underlying finer water-saturated medium is given by

$$z_n = [(2\sigma\cos\phi) (1/r_{finer} - 1/r_{host})] / [g(\rho_n - \rho_w)] \qquad (5-10)$$

where there is no DNAPL above the DNAPL body (Kueper and McWhorter, 1991). If the top of the DNAPL body has last been under imbibition conditions and is overlain by residual DNAPL that was trapped at the trailing edge of a sinking DNAPL body, then

$$z_n = (2\sigma\cos\phi/r_{finer}) / [g(\rho_n - \rho_w)] \qquad (5-11)$$

and r_{finer} refers to pore radii at the top of the underlying finer medium (Kueper and McWhorter, 1991). For example, given this condition, the upward P_c resisting entry of a PCE column from coarse sand to an underlying silt layer where $\sigma = 0.044$ N/m (44 dynes/cm), $\phi = 35°$, and the silt layer $r = 0.008$ mm, can be estimated using Equation 5-4 as

$$P_c = [2*0.044 \text{ N/m} *(\cos 35)] / 0.000008 \text{ m} = 9011 \text{ Pa}.$$

Disregarding hydraulic gradients, this capillary pressure is sufficient, below the water table, to halt the downward movement of a PCE body with a thickness as great as 1.48 m (based on Equation 5-11).

The significance of pore size variation and DNAPL height is depicted in Figure 5-2. DNAPL globule A is retained in the pore space because the gravitational force, although sufficient to distort the globule bottom, is offset by the upward capillary pressure gradient. Due to its greater height and gravity force, DNAPL globule B will migrate downward to the underlying finer layer, unless it

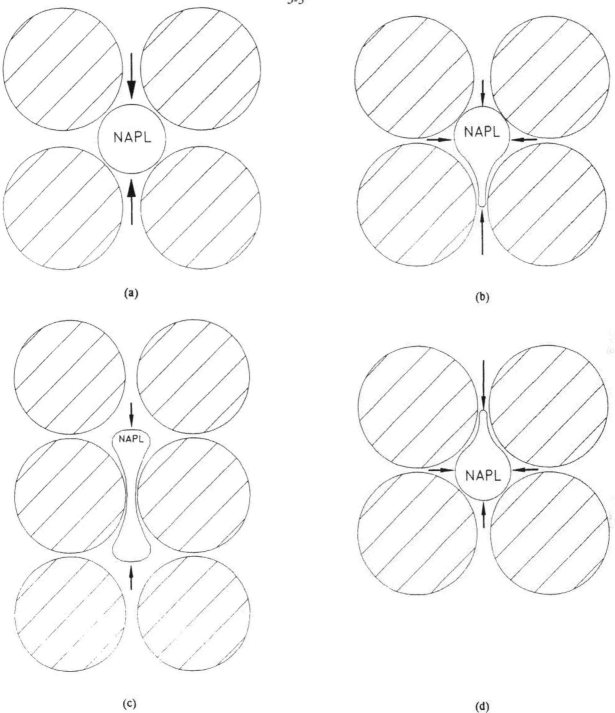

Figure 5-1. (a) Spherical DNAPL globule at rest in a pore space within water saturated media; (b) a nonspherical DNAPL globule at rest halfway through an underlying pore throat because the downward gravity force is balanced by the upward capillary force; (c) centered within the pore throat with equal capillary force from above and below, the DNAPL globule will sink through the pore throat due to the gravity force; and, (d) if the DNAPL globule is primarily through the pore throat, then the capillary and gravity forces will both push the globule downward (modified from Arthur D. Little, Inc., 1982). Note that arrows represent the magnitude of capillary forces.

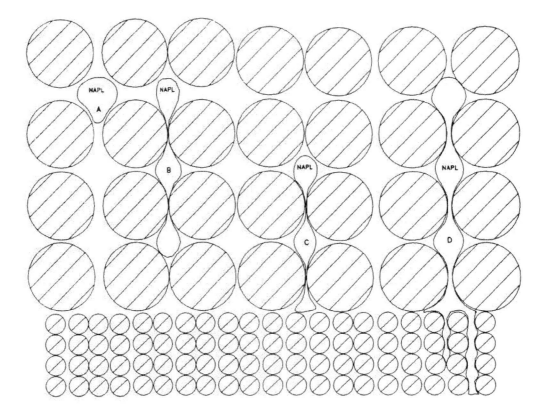

Figure 5-2. The effect of pore size and associated capillary pressure on DNAPL body height (modified from Arthur D. Little, Inc., 1982).

becomes exhausted by residual saturation. DNAPL globule C also has the same height and gravity force as globule B, but it has been immobilized by the increased capillary pressure gradient caused by the underlying finer layer. Lastly, DNAPL globule D, with sufficient height to exceed the resisting upward capillary pressure gradient, is migrating downward into the underlying finer media.

The hydrodynamic force due to hydraulic gradient can promote or resist DNAPL migration and is usually minor compared to gravity and capillary pressures. The control on DNAPL movement exerted by the hydrodynamic force rises with: (1) decreased gravitational pressure due to reduced DNAPL density and thickness, (2) decreased capillary pressure due to the presence of coarse media, low interfacial tension, and a relatively high contact angle; and (3) increasing hydraulic gradient.

Neglecting capillary pressure effects, the upward hydraulic gradient, i_b, and head difference, Δh, necessary to prevent downward DNAPL migration due to gravity are given by:

$$i_b = (\rho_n - \rho_w) / \rho_w \qquad (5\text{-}12)$$

and

$$\Delta h = z_n(\rho_n - \rho_w) / \rho_w \qquad (5\text{-}13)$$

Hydraulic gradients and head differences required to overcome capillary and/or gravity forces are further discussed in Chapter 5.3 and can be calculated using equations given in Table 4-2.

DNAPL migration will occur if and where the sum of the driving forces (gravity and possibly hydrodynamic effects) exceed the restricting forces (capillary pressure and possibly hydrodynamic effects). The pattern of DNAPL migration will be greatly influenced by the capillary properties and distribution of heterogeneous subsurface media.

5.1.2 DNAPL Migration Patterns

Descriptions of DNAPL transport processes are provided by Schwille (1988), Wilson et al. (1990), Mercer and Cohen (1990), Huling and Weaver (1991), USEPA (1992), Feenstra and Cherry (1988), and others. Key aspects of subsurface DNAPL flow phenomena are highlighted below and illustrated in Chapter 5.2.

5.1.2.1 DNAPL in the Vadose Zone

When released to the subsurface, gravity causes DNAPL to migrate downward through the vadose zone as a distinct liquid. This vertical migration is typically accompanied by lateral spreading due to the effects of capillary forces (Schwille, 1988) and medium spatial variability (e.g., layering). Even small differences in soil water content and grain size, such as those associated with bedding plane textural variations, can provide sufficient capillary resistance contrast to cause lateral DNAPL spreading in the vadose zone. Alternatively, downward movement will be enhanced, and lateral spreading limited, by dry conditions and the presence of transmissive vertical pathways for DNAPL transport (i.e, root holes, fractures, uniform coarse-grained materials, and bedding planes with high-angle dip).

As it sinks through the vadose zone, a significant portion of DNAPL is trapped in the porous media at residual saturation due to interfacial tension effects as described in Chapter 4. This entrapment depletes and, given a sufficiently small release or thick vadose zone, may exhaust the mobile DNAPL body above the water table. Residual saturation values measured for NAPLs in variably-saturated soils typically range from 0.05 to 0.20 (Table 4-4), but may be much less if averaged spatially over a zone where DNAPL has moved through only a fraction of the sampled volume (Poulsen and Kueper, 1992).

For a given DNAPL release volume, the depth of infiltration will be influenced by the area over which the release occurs and the release rate. Two experimental releases of 1.6 gallons of tetrachloroethene (PCE) into the vadose zone of a slightly stratified sand aquifer are described by Poulsen and Kueper (1992). The extent of PCE penetration resulting from each release was carefully mapped during excavation of the infiltration area as shown in Figure 5-3. PCE spilled instantaneously onto bare ground through a 1.1 ft^2 steel cylinder migrated to a depth of 7 ft (Figure 5-3a). A second release of 1.6 gallons was made by slowly dripping PCE on 0.16 in^2 of ground surface over 100 minutes. PCE from this drip release infiltrated to the water table at a depth of 10.8 ft (Figure 5-3b), 3.8 ft deeper than PCE penetration from the instantaneous spill. This difference was attributed to: (1) the smaller infiltration area utilized for the drip release; and, (2) a higher residual PCE content at shallow depth when ponding from the spill release increased the gravity force and induced PCE movement into a higher

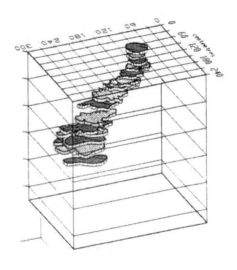
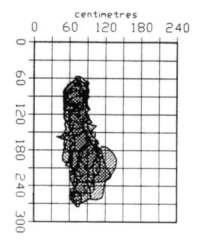

(a)

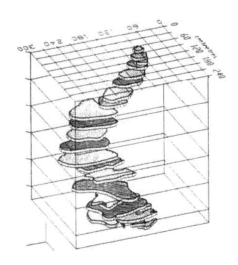
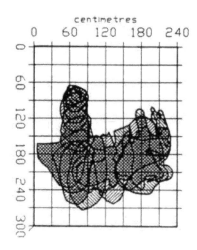

(b)

Figure 5-3. The overall mapped outline and plan views of PCE migration from (a) an instantaneous release, and (b) a drip release of 1.6 gallons of PCE which penetrated 2.0 and 3.2 m into the Borden sand, respectively (reprinted with permission from ACS, 1992).

proportion of shallow sand laminations (Poulsen and Kueper, 1992).

PCE from each release was distributed very heterogeneously as stringers at the millimeter scale due to the influence of sand bedding stratification. DNAPL flowed selectively along coarser-grained layers and did not enter finer-grained layers. The influence of very small-scale stratification and associated capillary effects on DNAPL infiltration was greatest for the drip release where the gravity force was less due to the absence of PCE ponding. Residual PCE saturations of the laminations containing DNAPL were in the range of 0.02 to 0.20. The results of Poulsen and Kueper (1992) suggest that small DNAPL releases on the order of only a few gallons have the potential to penetrate to depths of many feet below ground surface within hours or days.

DNAPL is retained at residual saturation as films, wetting pendular rings, wedges surrounding aqueous pendular rings, and as nonwetting blobs and ganglia in the presence of water within the vadose zone (Chapter 4.5). This residual DNAPL will dissolve slowly into infiltrating precipitation and will be a long-term source for groundwater contamination (Chapter 4.7). That is, each recharge event (infiltration of water that reaches the water table) will transport contaminants to the water table. Thus, DNAPL immobilized within the vadose zone can result in multiple repeated incidents of dissolved contaminant releases to groundwater.

In addition, some of DNAPL will volatilize and form a vapor extending beyond the separate phase liquid (Chapters 4.8 and 5.3). These vapors can condense on soil water and the water table, also causing additional groundwater contamination. Most dense organic solvents have high vapor pressures and, where DNAPL exists in the vadose zone, a plume of solvent vapor develops in the soil air surrounding the DNAPL source. Modeling studies indicate that contaminated vapors can diffuse tens of yards or more from a DNAPL source in the vadose zone within a period of weeks to months (Mendoza and McAlary, 1990; Mendoza and Frind, 1990a, 1990b). Where high vapor pressure compounds with relative vapor densities significantly greater than air are present in high permeability media (i.e., coarse sand and gravel), dense vapors can sink by advection through the vadose zone to the water table and then dissolve in groundwater (Falta et al., 1989). Field experiments involving trichloroethene (TCE) vapor transport in a sand formation confirm the modeling study findings noted above (Hughes et al., 1990). These experiments show

that significant groundwater plumes may form over a period of weeks from solvent vapor sources in a thin (less than 10 ft thick) vadose zone. Vapor transport can cause shallow groundwater contamination in directions opposite to groundwater and/or DNAPL flow. The resulting groundwater contamination plumes can have high dissolved chemical concentrations, but tend to be very thin in vertical extent, and occur close to the water table.

In summary, only a small volume of DNAPL is required to penetrate most vadose zones. Although stratigraphic layering will cause some lateral migration, penetration through the vadose zone can be fairly rapid (on the order of days). Residual DNAPL above the water table will provide a continuing source of groundwater contamination via vapor transport and dissolution processes.

5.1.2.2 DNAPL in the Saturated Zone

Upon encountering the capillary fringe, DNAPL will tend to spread laterally and accumulate until the gravitational pressure developed at the base of the accrued DNAPL exceeds the threshold entry pressure of the underlying water-saturated medium. When this occurs, DNAPL will displace water and continue its migration under pressure and gravity forces. Preferential spreading will occur where DNAPL encounters relatively permeable layers, fractures, or other pathways that present less capillary resistance to entry than underlying less permeable strata.

Given sufficient volume, DNAPL will typically migrate downward until it reaches a barrier layer upon which it may continue to flow laterally under pressure and gravity forces. Transport of DNAPL upon a capillary barrier, therefore, will be governed in large part by the barrier layer slope. DNAPL may be immobilized as a reservoir of continuous immiscible fluid if the capillary barrier forms a bowl-shaped stratigraphic trap. Multiple DNAPL reservoirs of varying dimensions may develop in stratigraphic traps at sites with abundant DNAPL and complex stratigraphy. Given sufficient accumulation, DNAPL will overflow discontinuous traps.

In the absence of a stratigraphic trap, mobile DNAPL will continue to migrate over the surface of the barrier layer. If the barrier layer slopes in a direction that varies from that of the hydraulic gradient, DNAPL will move in a different direction than groundwater flow and solute transport (unless the hydraulic force is sufficient to control the DNAPL flow direction, which is unusual).

Determining the slope and location of low permeability layers, therefore, can be critical to evaluating DNAPL migration potential.

Many fine-grained layers are inadequate capillary barriers to DNAPL migration due to the presence of preferential pathways which allow spreading DNAPL to sink into lower formations. For example, as DNAPL spreads above a fine-grained layer, it may encounter and enter fractures, root holes, stratigraphic windows, burrow holes, inadequately sealed wells or borings, etc. DNAPL migration may occur through hairline fractures that are as small as 10 microns in diameter. As noted in Chapter 4.4, the potential for DNAPL penetration of progressively finer pore openings increases proportionally to the overlying DNAPL column thickness and the DNAPL-water density difference. Fracture networks are commonly associated with relatively shallow stiff clayey soils and nearly all bedrock formations.

As a result of these processes, DNAPL will be present in the saturated zone as pools and disconnected globules and ganglia within relatively coarse pathways that are bounded by fine-grained capillary barriers. A finite DNAPL source will eventually be immobilized by residual saturation and/or in stratigraphic traps. Mobile and immobile DNAPL in the saturated zone will dissolve in flowing groundwater as described in Chapter 4.7 and thereby act as a long-term source of groundwater contamination.

5.2 CONCEPTUAL MODELS

The development and utilization of conceptual models to explain geologic processes and environments has long been the province of geoscientists. Contamination site investigators routinely formulate conceptual models of chemical migration to guide characterization and clean-up efforts. Although site conditions, DNAPL properties, and release characteristics are variable, these parameters generally conform to certain types of hydrogeologic environments and releases.

Conceptual model development involves integrating knowledge of site conditions, physical principles that govern fluid flow and chemical transport, and past experiences with similar problems. Field activities are nearly always guided by some degree of site conceptualization. Typically, conceptual models are refined to conform with new information as it becomes available.

Most of the fundamental physical processes affecting the subsurface migration of DNAPL chemicals were examined by Freiderich Schwille in laboratory experiments conducted in Germany between 1977 and 1984 (Schwille, 1988). Based on his experiments using saturated, variably-saturated, fractured, and porous media, Schwille developed several conceptual models to illustrate DNAPL flow, vapor transport of volatilized DNAPL chemicals, and groundwater transport of dissolved DNAPL chemicals. Others, most notably researchers at the Waterloo Centre for Groundwater Research (WCGR, 1991), have further developed and refined these DNAPL conceptual models. Several DNAPL conceptual model illustrations are provided in Table 5-1.

DNAPL conceptual models are utilized to assess:

- site characterization priorities,

- the utility of alternative subsurface characterization methods,

- site data,

- the potential for separate phase DNAPL migration,

- the potential for vapor transport of DNAPL chemicals,

- the potential for dissolution of DNAPL chemicals and dissolved chemical transport,

- chemical distributions associated with these transport mechanisms,

- cross-contamination risks associated with characterization and remedial activities, and,

- the potential effectiveness of alternative remedial actions.

Several quantitative methods are presented to examine various site characterization issues within the context of these conceptual models in following section. In Chapter 6, field investigation objectives and activities are discussed within the framework provided by these models.

5.3 HYPOTHESIS TESTING USING QUANTITATIVE METHODS

Based on the properties of fluid and media and transport processes described in Chapters 4 and 5.1, and the

Table 5-1. Conceptual models of DNAPL transport processes (modified with permission from ACS, 1992).

CASE	ILLUSTRATION
Case 1: DNAPL Release to Vadose Zone Only After release on or near the surface, DNAPL moves vertically downward under the force of gravity and soil capillarity. Because only a small amount of DNAPL was released, all of the mobile DNAPL is eventually trapped in pores and fractures in the vadose zone. Infiltration through the DNAPL zone leaches soluble organic constituents from the residual DNAPL and transports them to the water table, thereby producing a dissolved organic contaminant plume in the aquifer. Migration of gaseous vapors by diffusion and density flow also act as a source of dissolved organics to groundwater.· Contaminated vapors are leached by infiltration which recharges the water table and sink to contact the saturated zone.	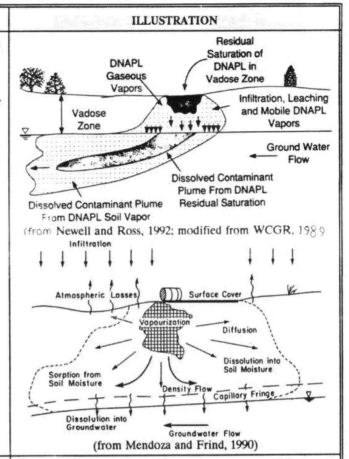 (from Newell and Ross, 1992; modified from WCGR, 1989 (from Mendoza and Frind, 1990)
Case 2: DNAPL Release to the Vadose and Saturated Zones If enough DNAPL is released at or near the surface, it can migrate through the vadose zone, overcome the capillary resistance provided by water-saturated pores at the capillary fringe, and sink into the saturated zone because it is denser than water. DNAPL migration will continue until the mobile DNAPL is trapped at residual saturation by capillary mechanisms and/or in pools above stratigraphic traps. Groundwater flowing past the trapped DNAPL leaches soluble components from the DNAPL, thereby creating a dissolved contaminant plume downgradient from the DNAPL zone. As with Case 1, water infiltrating from the source zone also carries dissolved chemicals to the aquifer and contributes further to the dissolved plume.	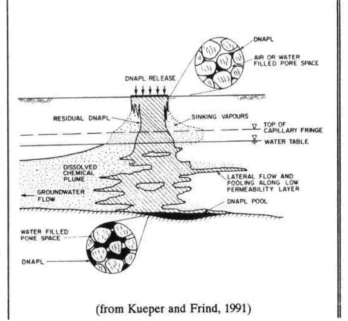 (from Kueper and Frind, 1991)

Table 5-1. Conceptual models of DNAPL transport processes (modified with permission from ACS, 1992).

CASE	ILLUSTRATION
Focus: **DNAPL spreading on the capillary fringe** As demonstrated during experiments in porous and fractured media (Schwille, 1988), DNAPL penetration is resisted by the capillary fringe which results in lateral spreading.	 (modified from Schwille, 1988)
Focus: **Effect of Layering on DNAPL penetration, residual saturation, and dissolved chemical migration** Within the saturated zone, lateral spreading of DNAPL is promoted just above finer layers and generally increases with decreasing permeability and grain size. DNAPL saturation typically increases at the base of coarser layers overlying finer layers. The rate of dissolved chemical migration with groundwater increases with layer permeability.	 (modified from Schwille, 1988)
Case 3: **DNAPL Pools and Effect of Low-Permeability Capillary Barriers** Mobile DNAPL will continue to sink downward until it is trapped at residual saturation (Cases 1 and 2) or until low-permeability stratigraphic units are encountered which create capillary barriers upon which DNAPL pools. In this figure, a perched DNAPL pool fills up and then spills over the lip of the low-permeability lens. The spill-over point (or points) can be some distance away from the original source, greatly complicating the process of tracking the DNAPL migration. Also see Figure 4-4.	 (from Newell and Ross, 1992)

Table 5-1. Conceptual models of DNAPL transport processes (modified with permission from ACS, 1992).

CASE	ILLUSTRATION
Case 4: Composite Site In this case, mobile DNAPL migrates downward through the vadose zone, producing a dissolved chemical plume in the upper aquifer. Although a DNAPL pool is formed on the fractured clay unit, the fractures are large enough to permit vertical migration downward to the deeper aquifer (also see Case 5). DNAPL pools in a surface depression in the underlying capillary barrier and a second dissolved chemical plume is formed.	 (from Newell and Ross, 1992)
Case 5: Fractured Rock or Fractured Clay System DNAPL introduced into a fractured rock or fractured clay system follows a complex pathway based on the distribution of fractures in the original matrix. The number, density, size, and direction of the fractures often cannot be determined due to the heterogeneity of the fractured system and the lack of economical formation characterization technologies. Relatively small volumes of DNAPL can penetrate deeply into fractured systems due to the low retention capacity of the fractures and the ability of some DNAPLs to migrate through very small (<20 microns) fractures. Many clayey units act as fractured media with preferential pathways for vertical and horizontal DNAPL migration. **Focus: DNAPL dissolution, dissolved chemical migration, and matrix diffusion in fractured media** DNAPL contained in fractures will dissolve and be transported through the fracture network with groundwater, and will also diffuse into and sorb onto the porous inter-fracture matrix. Residual saturation and adsorbed chemicals both provide long-term sources for groundwater contamination.	 (from Kueper and McWhorter, 1992) (from Mackay and Cherry, 1989)

conceptual models provided in Table 5-1, it is apparent that several fate and transport issues are common to many DNAPL sites. Posed as questions, these issues include the following:

- How much DNAPL is required to sink through the vadose zone (Chapter 5.3.1)?

- How long will it take DNAPL released at or near the ground surface to sink to the water table (Chapter 5.3.2)?

- What thickness of DNAPL must accumulate on the capillary fringe to cause DNAPL to enter the saturated zone (Chapter 5.3.3)?

- Will a finer-grained layer beneath the contamination zone act as a capillary barrier to continued downward migration of DNAPL? What minimum DNAPL column or body height is required to enter a particular capillary barrier beneath the water table (Chapter 5.3.4)?

- If DNAPL is perched above a finer-grained capillary barrier layer, what size fracture or macropore will permit continued downward migration into (or through) the capillary barrier (Chapter 5.3.5)?

- What DNAPL saturation at the base of the host medium must be attained for DNAPL to enter the underlying finer-grained capillary barrier (Chapter 5.3.6)?

- What upward hydraulic gradient will be required to prevent continued downward migration of DNAPL (Chapter 5.3.7)?

- What upslope hydraulic gradient will be required to prevent continued downslope movement of DNAPL along the base of a dipping fracture or the base of a coarser layer underlain by a dipping finer layer (Chapter 5.3.8)?

- What will be the stable DNAPL pool length that can exist above a sloping capillary barrier or sloping fracture below the water table (Chapter 5.3.9)?

- What will be the stable DNAPL height and area after spreading above an impenetrable flat-lying capillary barrier (Chapter 5.3.10)?

- What is the volume of DNAPL contained below the water table within porous or fractured media (Chapter 5.3.11)?

- How will fluid viscosity and density affect the velocity and distance of DNAPL migration (Chapter 5.3.12)?

- What hydraulic gradient will be required to initiate the lateral movement of a DNAPL pool or globule (Chapter 5.3.13)?

- How long does DNAPL in the saturated zone take to dissolve completely (Chapter 5.3.14)?

- Given a DNAPL source of dissolved groundwater contamination, how do you determine the movement of dissolved chemical plumes (Chapter 5.3.15)?

- Given a DNAPL source of vapor contamination in the vadose zone, how do you determine the relative movement of the vapor plume (Chapter 5.3.16)?

- How can the chemical composition of a dissolved plume associated with a DNAPL source be estimated (Chapter 5.3.17)?

- What is the equivalent mass/volume of DNAPL contained within a dissolved groundwater plume (Chapter 5.3.18)?

- What is the relationship between concentrations in soil gas and groundwater (Chapter 5.3.19)?

- Given a DNAPL source in the vadose zone, how do you determine the movement of a vapor plume? What are the conditions that favor vapor transport away from a DNAPL source in the vadose zone that would allow soil-gas monitoring (Chapter 5.3.20)?

Relatively simple quantitative methods that can be used to test hypotheses regarding these issues, in particular, by making bounding type calculations, are described by example below.

5.3.1 How much DNAPL is required to sink through the vadose zone?

The capacity of the vadose zone to trap DNAPL can be calculated as

$$V_{nR} = R \ V_{mm} = 1000 \ s_r \ n \ V_{mm} \qquad (5\text{-}14)$$

or

$$V_n = s_r n V_m \qquad (5-15)$$

where: V_{nR} is the liters of DNAPL retained in a volume of media measured in cubic meters, V_{mm}; V_n is the volume of DNAPL retained in a volume of media, V_m (any consistent units); R is the volumetric retention capacity in liters of residual DNAPL per cubic meter of media; s_r is residual saturation; and n is porosity. Measured values of R and s_r for variably-saturated media are given in Table 4-4. Equation 5-15 is solved for a range of s_r, n, and V_m values in Figure 5-4.

For example, how much DNAPL would have to be released over a 1 m² area to reach a water table 15 m below ground surface? If we assume a residual saturation of 0.1, a porosity of 0.3, and no lateral spreading as the DNAPL sinks, then only 0.45 m³ of DNAPL would be required to penetrate 15 m to the water table. For a given DNAPL release volume, the depth of penetration will increase with decreases in n, s_r, lateral spreading, and mass loss due to processes such as volatilization and dissolution. The presence of macropores and fractures may facilitate deep penetration of small DNAPL volumes. Although lateral spreading caused by stratified media will generally slow DNAPL penetration in the vadose zone, this calculation demonstrates that leaks on the order of tens of gallons can reach the water table.

5.3.2 How long will it take DNAPL released at or near the ground surface to sink to the water table?

The rate of DNAPL infiltration in the subsurface may be extremely rapid. For example, in laboratory experiments, Schwille (1988) observed tetrachloroethene to sink through 2 ft of variably saturated coarse sand in 10 minutes and through 3 ft of saturated coarse sand in 60 minutes. At this rate of penetration (with no lateral diversion), for example, it would take a DNAPL such as PCE approximately 5 hours to penetrate 60 ft of coarse sand in the vadose zone. The actual penetration time will most likely be greater because of soil heterogeneities and will vary with soil conditions and DNAPL properties (i.e., density and viscosity). However, this calculation shows that for a sufficient volume, DNAPL can reach a relatively deep water table in days to weeks, as opposed to years.

5.3.3 What thickness of DNAPL must accumulate on the capillary fringe to cause DNAPL to enter the saturated zone?

As illustrated in Table 5-1 (Case 2), upon reaching the capillary fringe above the water table, sinking DNAPL will tend to be obstructed and spread laterally until a sufficient DNAPL thickness has accumulated to exceed the threshold entry pressure at the capillary fringe. DNAPL entry will typically occur through the largest pore connections beneath the area of accumulation. At some sites, DNAPL entry will be facilitated by heterogeneous wetting characteristics of the medium (e.g., a portion of sediment containing a high organic matter content conducive to DNAPL entry; see Chapter 4.3) and/or enhanced DNAPL wetting of solid surfaces due to the presence of surfactant contaminants. The critical height, z_n, of DNAPL required for downward entry of DNAPL through the capillary fringe can be estimated by

$$z_n = (2\sigma\cos\phi) / (rg\rho_n) \qquad (5-16)$$

where σ is the interfacial tension between the DNAPL and water, ϕ is the wetting contact angle, r is the pore radius, g is gravitational acceleration, and ρ_n is the density of DNAPL.

Given $\sigma = 0.040$ N/m, $\phi = 35°$, and, $\rho_n = 1300$ kg/m³, Equation 5-16 is solved for pore radii from 0.0001 to 1 mm in Figure 5-5. The DNAPL thickness, z_n, required to enter the saturated zone varies significantly with pore size, but is relatively insensitive to interfacial tension (which typically varies within a factor of about 3, between 0.015 and 0.50), DNAPL density (which typically varies within a factor of < 2, from 1.01 to 1.70), and contact angle (unless ϕ is > 60°). As shown, substantial DNAPL thicknesses must accumulate above the capillary fringe to penetrate water-saturated clay and silt pores (i.e., approximately >20 m and 0.8-20 m of DNAPL, respectively). Subsurface samples from the top of the saturated zone should be examined carefully for DNAPL presence at suspected DNAPL sites. In the absence of macropores or solid surfaces that are not strongly water-wet, silt and clay layers can prevent DNAPL from penetrating the water table. Typically, however, shallow fine-grained materials contain fractures and other macropores.

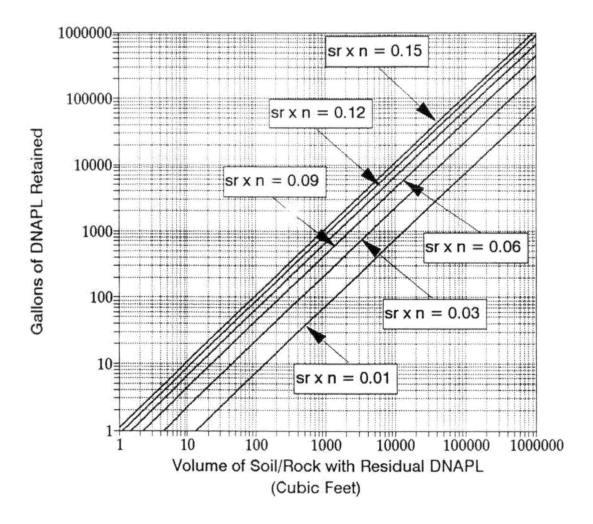

Figure 5-4. DNAPL volume retained in the vadose zone as a function of residual saturation, s_r, effective porosity, n, and contamination zone volume.

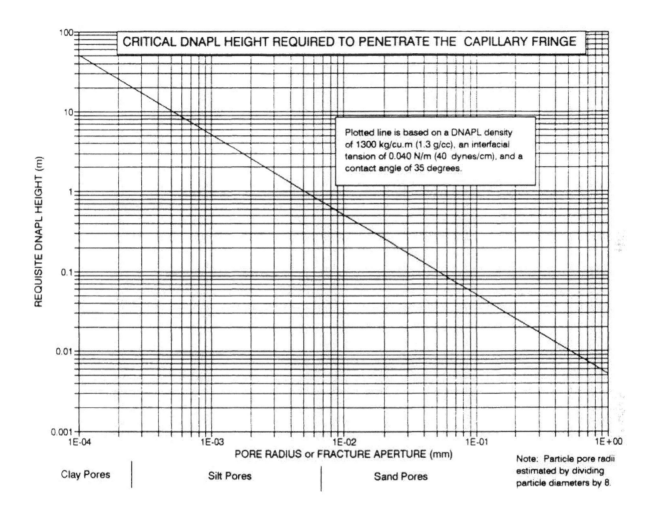

Figure 5-5. Critical DNAPL height required to penetrate the capillary fringe as a function of pore radius given a DNAPL density of 1300 kg/m^3, an interfacial tension of 0.040 N/m, and a contact angle of 35 degrees.

5.3.4 Will a finer-grained layer beneath the contamination zone act as a capillary barrier to continued downward migration of DNAPL? What minimum DNAPL column or body height is required to enter a particular capillary barrier beneath the water table?

The thickness of continuous DNAPL required to penetrate a finer-grained capillary barrier below the water table exceeds that needed to enter the saturated zone within a similar unit due to the reduced density difference between DNAPL and water compared to DNAPL and air. As explained in Chapter 5.1, the critical height of DNAPL, z_n, required to penetrate an underlying finer layer below the water table will depend on the saturation and direction (drainage or imbibition) of the $P_c(s_w)$ curve being followed at the top of the DNAPL column.

Below the water table, for cases where residual DNAPL is above the continuous DNAPL body, and the hydraulic gradient is much less than the DNAPL-water gravity gradient, z_n can be estimated by

$$z_n = (2\sigma\cos\phi) / [r_{finer}g(\rho_n - \rho_w)] \qquad (5\text{-}17)$$

where r_{finer} refers to pore radii in the underlying capillary barrier. Where there is no DNAPL above the DNAPL body, then

$$z_n = [(2\sigma\cos\phi)(1/r_{finer} - 1/r_{host})] / [g(\rho_n - \rho_w)] \qquad (5\text{-}18)$$

where r_{host} is the pore radii in the host medium.

Equation 5-17 is solved for $\sigma = 0.040$ N/m, $\phi = 35°$, $\rho_n = 1300$ kg/m^3, and r from 0.001 to 10 mm in Figure 5-6. As shown, substantial DNAPL thicknesses may be required to penetrate clay, silt, and fine sand pores. Below the water table, z_n is sensitive to DNAPL density, due to the reduced density-difference between immiscible fluids within the saturated zone, in addition to pore radius (and contact angle if $\phi > 60$). This is illustrated in Figure 5-7. Relatively dense DNAPLs, such as highly chlorinated solvents, therefore, have a greater vertical migration potential than less dense DNAPLs such as creosote and coal tar (Figure 3-2). The influence of vertical hydraulic gradients on the migration of DNAPL into capillary barriers is discussed in Chapter 5.3.7.

5.3.5 If DNAPL is perched above a finer-grained capillary barrier layer, what size fracture or macropore will permit continued downward migration into (or through) the capillary barrier?

Neglecting hydraulic gradients, the critical DNAPL thickness, z_n, required for entry into a fracture with an aperture, b, can be estimated as

$$z_n = (2\sigma\cos\phi) / (bg\rho_n) \qquad (5\text{-}19)$$

at the top of the zone of saturation, and as

$$z_n = (2\sigma\cos\phi) / [bg(\rho_n - \rho_w)] \qquad (5\text{-}20)$$

where residual DNAPL overlies the continuous DNAPL body within the saturated zone. To calculate z_n above a fracture where there is no residual DNAPL over the DNAPL column, substitute b for r in Equation 5-18. Values of z_n calculated using Equations 5-19 and 5-20 for a range of apertures sizes given $\rho_n = 1300$ kg/m^3, $\sigma = 0.040$ N/m, and $\phi = 35°$ at the top of and within the saturated zone are graphed in Figures 5-5 and 5-6, respectively. Due to the irregular geometry of fractures, the actual DNAPL thickness required to enter a particular fracture will likely be intermediate between that calculated using the fracture radius, r, or aperture, b.

If the thickness of a DNAPL body that is perched atop a capillary barrier is known from boring data, then the maximum aperture which will resist DNAPL entry can be calculated by rearranging Equations 5-19 and 5-20.

Once DNAPL enters a vertical fracture, it will readily enter finer and finer fractures with depth due to the increase in DNAPL column height with depth. At hydrostatic equilibrium, the increased gravity force with depth is countered by increased capillary resistance provided by the corresponding fracture apertures which prevent DNAPL entry (Figure 5-8).

5.3.6 What saturation must be attained at the base of a host medium for DNAPL to enter an underlying finer-grained capillary barrier?

Several methods are available to determine or estimate whether or not DNAPL has or will penetrate a capillary barrier. Direct sampling of water and solids within and below the capillary barrier can be conducted, but may pose a significant risk of causing chemical migration. If the thickness of a DNAPL body, z_n, at the base of the

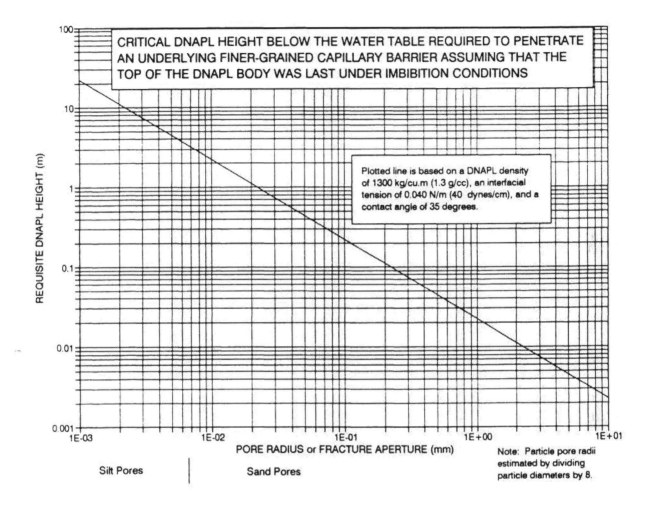

Figure 5-6. Critical DNAPL height required to penetrate a capillary barrier beneath the water table as a function of pore radius given a DNAPL density of 1300 kg/m³, an interfacial tension of 0.040 N/m, and a contact angle of 35 degrees.

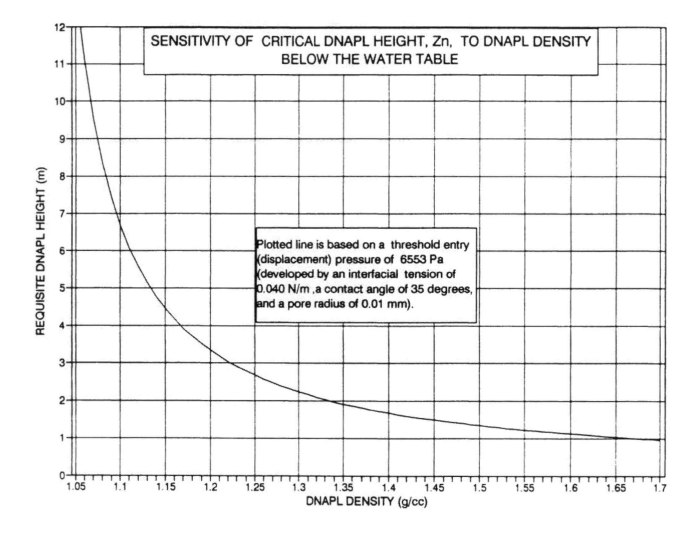

Figure 5-7. Sensitivity of critical DNAPL height to DNAPL density below the water table.

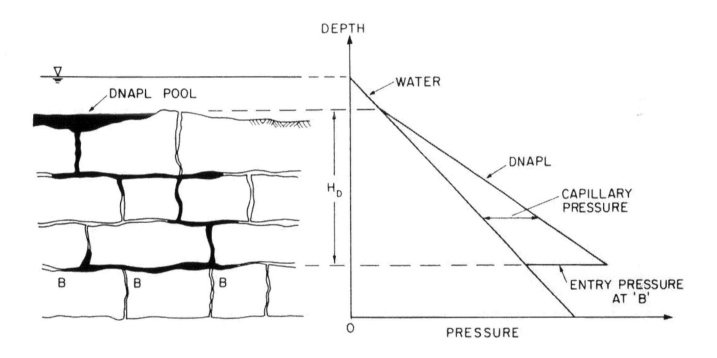

Figure 5-8. Pressure profiles in a fracture network for DNAPL at hydrostatic equilibrium (from Kueper and McWhorter, 1991).

host medium can be determined, then the capillary pressure at this point can be estimated (by assuming hydrostatic equilibrium) as

$$P_c = g\, z_n\, (\rho_n - \rho_w) \qquad (5\text{-}21)$$

and inferences can be made regarding the likelihood that DNAPL has or will enter the barrier layer based on estimates of the threshold entry pressure of the capillary barrier.

A third approach is to: (1) determine the DNAPL saturation in samples of the host medium taken from just above the capillary barrier; (2) estimate the associated capillary pressure, P_c, above the barrier layer based on a measured or estimated $P_c(s_w)$ curve for the host medium; and, then (3) compare the inferred P_c to a threshold entry pressure, P_d, measured or estimated for pores or fractures in the underlying capillary barrier. If the P_c at the base of the host medium exceeds the barrier layer P_d, then DNAPL probably has or will penetrate the finer-grained underlying layer. Such a comparison utilizing $P_c(s_w)$ curves for a sand layer overlying a silt layer is given in Figure 5-9.

5.3.7 What upward hydraulic gradient will be required to prevent continued downward migration of DNAPL?

As described in Chapter 5.1, the three driving forces that act concurrently on subsurface DNAPL are the gravity gradient, the capillary pressure gradient, and the hydraulic gradient. Groundwater flow driven by upward vertical hydraulic gradients will prevent or slow the downward movement of DNAPL. Shallow recovery wells and drains, and/or deeper injection wells, can be used to create or increase upward vertical hydraulic gradients, particularly across an aquitard that separate two aquifers. This hydraulic barrier concept has been likened to using an upward-blowing fan to suspend a ping-pong ball in air. It was apparently first considered to contain sinking DNAPL at the S-Area Landfill in Niagara Falls, New York (Guswa, 1985; Cohen et al., 1987), and, more recently, has been evaluated for containment of DNAPL beneath disposal basins (Hedgecoxe and Stevens, 1991) and within fractured media (Kueper and McWhorter, 1991).

Neglecting the capillary pressure gradient, the upward hydraulic gradient, i_h, and associated hydraulic head

difference, Δh, needed to prevent DNAPL from sinking vertically downward due to gravity are given by:

$$i_h = (\rho_n - \rho_w) / \rho_w \qquad (5\text{-}22)$$

and

$$\Delta h = z_n(\rho_n - \rho_w) / \rho_w \qquad (5\text{-}23)$$

where z_n is the thickness of the DNAPL body.

The hydraulic gradient required to halt DNAPL sinking due to gravity increases linearly with DNAPL density as shown in Figure 5-10. As a result, vertical containment of low density DNAPLs such as coal tar and creosote is generally much more feasible than for denser DNAPLs such as highly chlorinated solvents.

In addition to the density gradient, capillary pressure will influence whether or not DNAPL will migrate vertically into and through a finer-grained layer. Considering both the density and capillary pressure gradients in a simple one-dimensional system, the steady-state upward vertical hydraulic gradient required to prevent DNAPL sinking through a capillary barrier is given by

$$i_h = \Delta h/L = [(\rho n - \rho w)/\rho w] + [(P_c - P_d)/(\rho_w gL)] \qquad (5\text{-}24)$$

where P_c is the capillary pressure at the base of the DNAPL pool overlying the capillary barrier, P_d is the threshold entry pressure of the capillary barrier, and L is the thickness of the capillary barrier. P_c at the base of the DNAPL pool can be estimated using Equation 5-21 or based on measurement of DNAPL saturation at this interface as described in Chapter 5.3.6 if the $P_c(s_w)$ curve is known.

Equation 5-24 is illustrated and solved in Figure 5-11. A negative i_h value is the downward hydraulic gradient that must be exceeded to overcome capillary pressure and cause downward DNAPL migration through the capillary barrier. As shown, the capillary pressure gradient (the second term on the right side of Equation 5-24) is inversely proportional to aquitard thickness, L. Vertical hydraulic gradients required to prevent DNAPL sinking through a capillary barrier, therefore, decrease as the thickness of the barrier layer increases.

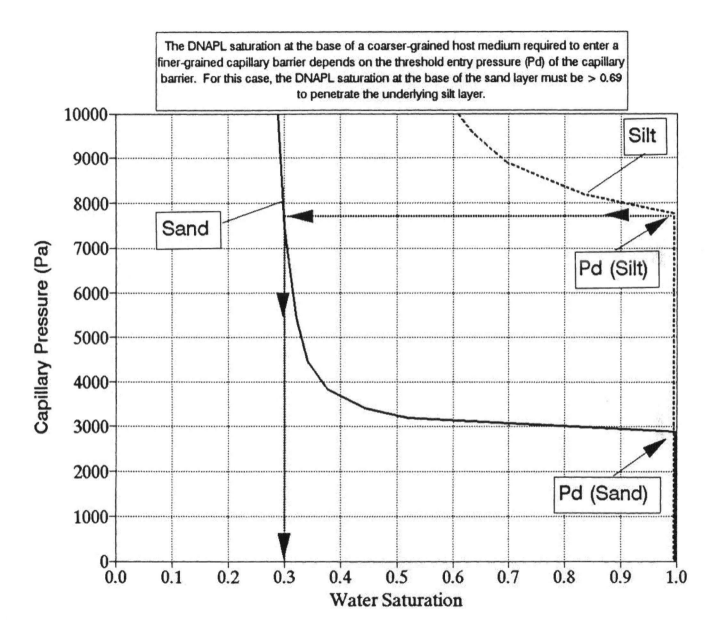

Figure 5-9. Comparison of $P_c(s_w)$ curves to determine the DNAPL saturation required in an overlying coarser layer to enter an underlying finer liner.

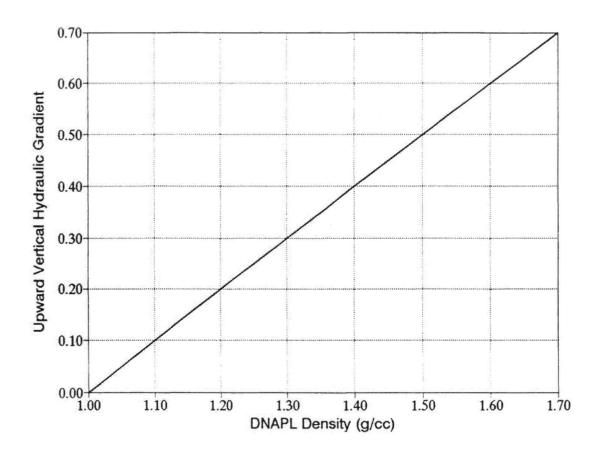

Figure 5-10. Neglecting the capillary pressure gradient, the upward vertical hydraulic gradient required to prevent DNAPL sinking in the saturated zone is a function of DNAPL density.

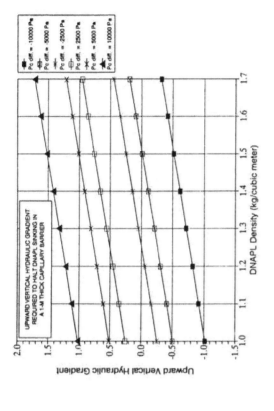

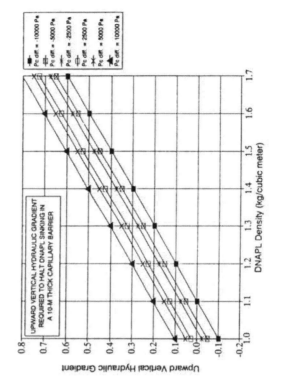

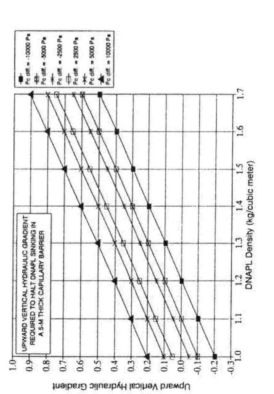

Figure 5-11. Considering both the density and capillary pressure gradients, the upward vertical hydraulic gradient required to prevent DNAPL sinking through a capillary barrier is a function of DNAPL density and the capillary pressure difference between the base of the overlying coarser layer and the threshold entry pressure of the underlying finer layer.

5.3.8 What upslope hydraulic gradient will be required to prevent continued downslope movement of DNAPL along the base of a dipping fracture or the base of a coarser layer underlain by a dipping finer layer?

The hydraulic gradient needed to halt DNAPL movement may be greatly reduced if DNAPL is sinking along an inclined plane (i.e., a bedding plane, joint, or sloping fine-grained layer). For this case, neglecting the capillary pressure gradient, the hydraulic gradient measured parallel to the inclined plane needed to arrest DNAPL movement and the associated hydraulic head difference are given by

$$i_h = [(\rho_n - \rho_w) \sin \theta] / \rho_w \qquad (5\text{-}25)$$

and

$$\Delta h = [(\rho_n - \rho_w) z_n \sin \theta] / \rho_w \qquad (5\text{-}26)$$

where θ is the inclined plane dip angle in degrees and z_n is the length of the DNAPL body measured parallel to the inclined plane. For example, the requisite upslope hydraulic gradient measured along an inclined plane surface with a dip of 15° to arrest the downward flow of DNAPL with a density of 1.18 is 0.047. The diminution of upslope hydraulic gradient required to prevent downslope DNAPL movement is shown as a function of density gradient and slope as shown in Figure 5-12. Measured vertically, however, the required hydraulic gradient is still equal to the density gradient.

5.3.9 What will be the stable DNAPL pool length that can exist above a sloping capillary barrier or sloping fracture below the water table?

At hydrostatic equilibrium, the stable pool length, L, measured parallel to a capillary barrier or fracture with a dip angle of θ in degrees can be estimated as

$$L = P_d / [(\rho_n - \rho_w) g \sin \theta] \qquad (5\text{-}27)$$

where P_d is the threshold entry pressure of the host medium or fracture (WCGR, 1991). This scenario is illustrated and Equation 5-27 is solved for a range of DNAPL density and capillary barrier dip angles in Figure 5-13. As shown, the stable DNAPL pool length increases with DNAPL density and dip angle.

5.3.10 What will be the stable DNAPL height and area after spreading above an impenetrable flat-lying capillary barrier?

DNAPL will mound and spread along an impenetrable capillary barrier below the water table until the threshold entry pressure of the host medium resists further spreading. If the capillary barrier is flat-lying, the stable height of the DNAPL pool can be estimated by

$$z_n = (2\sigma \cos\phi) / [r_{host} g (\rho_n - \rho_w)] \qquad (5\text{-}28)$$

For a given volume, V_n, of DNAPL released, the maximum area, A_m, of DNAPL spreading above the impenetrable capillary barrier can be estimated by: (1) subtracting an estimate of the DNAPL volume, V_r, retained at residual saturation (and, if applicable, in stratigraphic traps) above the stable DNAPL pool from the volume of DNAPL released; and then, (2) dividing the remaining DNAPL volume by the stable DNAPL pool height and by an estimate of DNAPL residual saturation, s_r, in the saturated zone:

$$A_m < (V_n - V_r) / (z_n s_r) \qquad (5\text{-}29)$$

The actual area of spreading will likely be less than the calculated value of A_m because the DNAPL saturation in the pool will exceed the s_r value. Given a homogeneous host medium and neglecting hydraulic forces, DNAPL can be expected to spread out radially from a source mound. In heterogeneous media, however, the pattern of the DNAPL spreading is typically very irregular and difficult to define.

5.3.11 What is the volume of DNAPL contained below the water table within porous or fractured media?

This question is similar to that posed in Chapter 5.3.1 for the vadose zone, and Equation 5-15 applies, except that the DNAPL saturation may exceed residual saturation. Residual saturation data for the saturated zone are provided in Table 4-5. In general, more DNAPL is immobilized in the saturated zone than in the vadose zone (Chapter 4.5). Using the same example as that in Chapter 5.3.1, but with a DNAPL residual saturation of 0.3, then 93 gallons of DNAPL would be trapped in a volume of 4 ft X 4 ft X 35 ft. This is three times that trapped in an equivalent volume of vadose zone (given the assumed values of residual saturation).

In fractured media, DNAPL presence may be largely confined to fractures. To estimate the volume of DNAPL

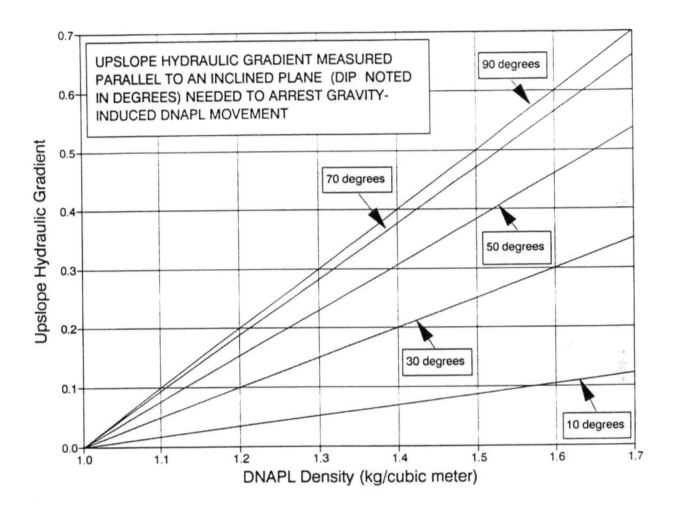

Figure 5-12. Neglecting the capillary pressure gradient, the upslope hydraulic gradient required to arrest DNAPL movement downslope along an inclined capillary barrier is a function of DNAPL density and capillary barrier dip.

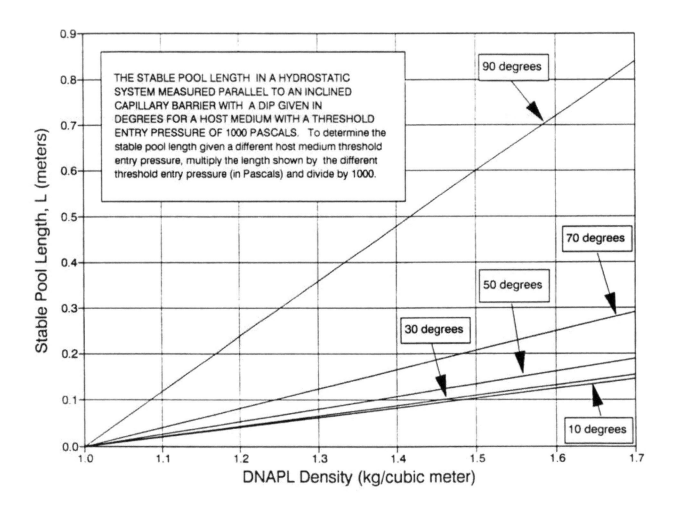

THE STABLE POOL LENGTH IN A HYDROSTATIC
SYSTEM MEASURED PARALLEL TO AN INCLINED
CAPILLARY BARRIER WITH A DIP GIVEN IN
DEGREES FOR A HOST MEDIUM WITH A THRESHOLD
ENTRY PRESSURE OF 1000 PASCALS. To determine the
stable pool length given a different host medium threshold
entry pressure, multiply the length shown by the different
threshold entry pressure (in Pascals) and divide by 1000.

Figure 5-13. The stable DNAPL pool length above an inclined capillary barrier is a function of
DNAPL density and capillary barrier dip.

contained within a particular volume of fractured media, it is necessary to evaluate fracture porosity. Semi-regular fracture patterns, such as those depicted in Figure 5-14, are discernable in many fractured clay and rock units. Using the fracture porosity equations provided in Figure 5-14, it is possible to calculate fracture porosities and void volumes given estimates of fracture spacing and aperture. For example, fracture porosity is shown as a function of these parameters for the matches model (Figure 5-14b) in Figure 5-15. Estimates of the volume of DNAPL present in fractures can be obtained by multiplying the void volume by a DNAPL saturation estimate. Where present, the distribution of DNAPL in fractures is typically very complex and not amenable to accurate volume calculation. Volume calculation is further complicated because some DNAPL typically will migrate from the fractures into larger pore openings (i.e., root holes, dissolution cavities, sand laminations, etc.) that intersect the fracture walls.

5.3.12 How do fluid viscosity and density affect the velocity and distance of DNAPL migration?

The velocity and distance of DNAPL migration will be controlled, in part, by DNAPL density and viscosity. The rate of DNAPL sinking generally increases with increasing DNAPL density (and gravity gradient) and decreasing DNAPL viscosity. As a result, chlorinated solvents sink much more rapidly through the subsurface than coal tar/creosote (Figure 3-1).

The lateral migration of DNAPL is also affected by DNAPL fluid density and viscosity. As noted in Chapter 4.10, hydraulic conductivity is directly related to fluid density and inversely related to fluid viscosity. Given the wide range of DNAPL viscosities (Figure 3-1), the rate and distance of DNAPL movement due to gravity and/or hydraulic gradients may be significantly greater for low viscosity (thin) DNAPLs than high viscosity (thick) DNAPLs. Where multiple DNAPLs are present at a contamination site, consideration should be given to the implications of variable DNAPL density and viscosity on transport potential. Measurement on DNAPL samples collected from the subsurface are recommended because chemical aging and mixing (with other DNAPLs or water) can modify DNAPL properties.

5.3.13 What hydraulic gradient will be required to initiate the lateral movement of a DNAPL pool or globule?

The hydraulic gradient, i_h, across a DNAPL pool or globule required to initiate lateral DNAPL movement can be estimated as

$$i_h > P_d / (\rho_w \, g \, L) \qquad (5\text{-}30)$$

where P_d is the threshold entry pressure of the host medium or fracture and L is the length of the DNAPL pool or globule perpendicular to the hydraulic gradient (WCGR, 1991). This scenario is illustrated and Equation 5-30 is solved for a range of P_d and L values in Figure 5-16. As shown, the requisite hydraulic gradient increases with (1) increasing P_d (i.e., increasing interfacial tension, decreasing contact angle, and decreasing pore radius) and (2) decreasing pool/globule length. The hydraulic gradient required to sustain DNAPL movement typically increases with time and distance of DNAPL movement. This is because the length of the pool/globule is shortened as residual DNAPL is retained at its trailing edge.

Residual DNAPL also can be mobilized by increasing hydraulic gradients. The capillary number, N_c, the ratio of capillary to viscous forces, provides a measure of the propensity for DNAPL trapping and mobilization. It is defined as the product of intrinsic permeability, water density, gravitational acceleration constant, and hydraulic gradient divided by the interfacial tension. The critical value, N_c^*, of the capillary number is defined as the value at which motion of some of the DNAPL blobs is initiated. Based on experimental data, Wilson and Conrad (1984) noted a strong correlation between displacement of residual DNAPL and the capillary number when the hydraulic gradient was greater than that producing the critical value of the capillary number. The hydraulic gradient necessary to initiate blob mobilization for various permeabilities and interfacial tensions is shown in Figure 5-17. As may be seen, in very permeable media (e.g., gravel or coarse sand), it is theoretically possible to obtain sufficient hydraulic gradients to remove all DNAPL blobs. In soils of medium permeability (e.g., fine to medium sand), some of the residual can be hydraulically removed. In less permeable media, removal is not possible, unless surfactants are used to drastically reduce interfacial tension.

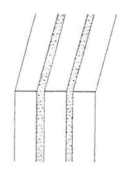

(a) Slides Model

$$n_f = \frac{b}{a}$$

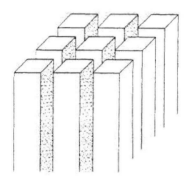

(b) Matches Model

$$n_f = \frac{2b}{a}$$

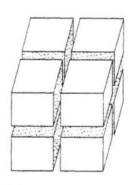

n_f = fracture porosity

a = fracture spacing

b = fracture aperture

(c) Cubes Model

$$n_f = \frac{3b}{a}$$

Figure 5-14. Fracture porosity equations for the slides, matches, and cubes fracture models where a is the fracture spacing and b is the fracture aperture.

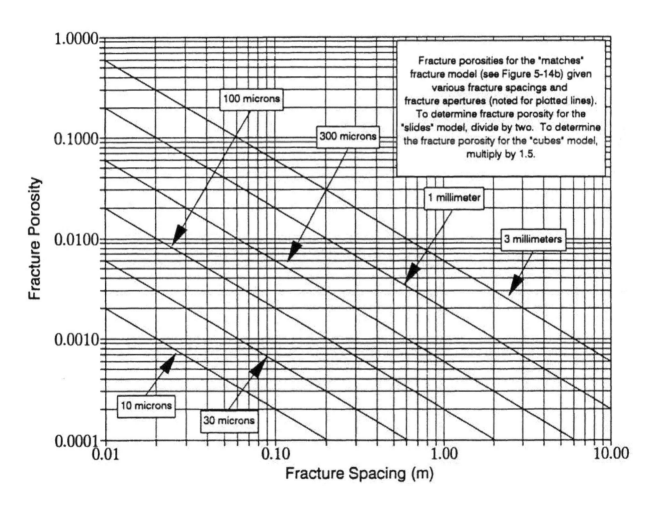

Figure 5-15. Fracture porosity is a function of fracture spacing and aperture.

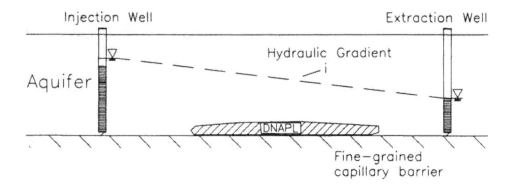

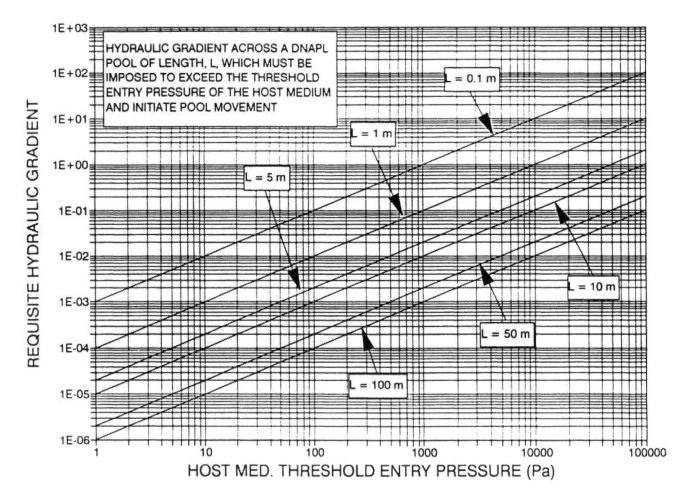

Figure 5-16. The hydraulic gradient required to initiate lateral movement of a DNAPL pool or globule is directly proportional to the threshold entry pressure of the host medium and inversely proportional to pool length.

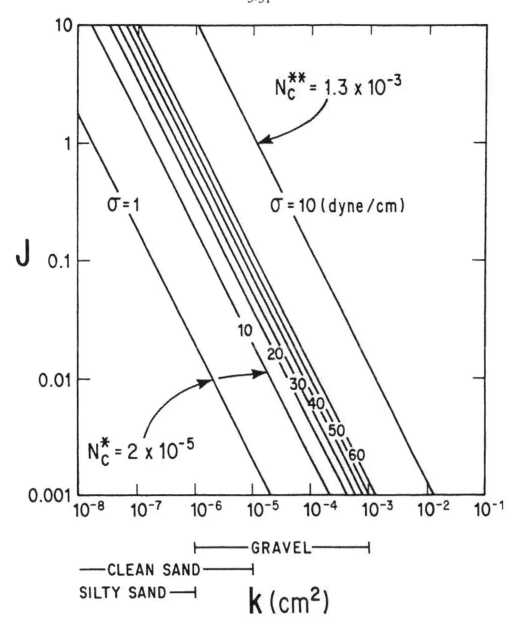

Figure 5-17. Hydraulic gradient, J, necessary to initate DNAPL blob mobilization (at N_c^*) in soils of various permeabilities, for DNAPLs of various interfacial tensions, σ. The upper curve represents the gradient necessary for complete removal of all hydrocarbons (N_c^*), with σ = 10 dynes/cm (from Mercer and Cohen, 1990; after Wilson and Conrad, 1984).

5.3.14 How long does DNAPL in the saturated zone take to dissolve completely?

As discussed in Chapter 4.7, complete dissolution of DNAPL in the saturated zone can take decades or centuries due to limits on chemical solubility, groundwater velocity, and vertical dispersion. DNAPL pools will be particularly long-lasting, compared to residual DNAPL ganglia and fingers, because of their low water-DNAPL contact area.

Most simply, the time, t, needed to completely dissolve a DNAPL source can be estimated as

$$t = m / (v_i n C_w A) \qquad (5-31)$$

where m is the DNAPL mass, v_i is the average interstitial groundwater velocity, n is the effective porosity, and, A is the cross-sectional area containing DNAPL through which groundwater flow exits with a dissolved DNAPL concentration, C_w.

For example, consider a 1 m^3 volume of sandy soil with a residual DNAPL content of 30 L/m^3. If it is assumed that the hydraulic conductivity is 10^{-3} cm/s, the hydraulic gradient is 0.01 and the porosity is 0.30, then groundwater flows through this hypothetical sandy soil at a rate of 0.03 m/d. Furthermore, assuming that DNAPLs dissolve into groundwater to 10% of their solubility, then for PCE (density of 1.63 g/cm^3 and solubility of 200 mg/L), approximately 744 years would be required to dissolve the DNAPL PCE.

DNAPL dissolution from residual and pool sources was recently examined by Anderson et al. (1992a) and Johnson and Pankow (1992), respectively. For rectangular DNAPL pools, Johnson and Pankow (1992) defined a surface-area-averaged mass-transfer rate, M_a ($M/L^2/T$), as

$$M_a = [(4D_v v_i)/(\pi L_p)]^{1/2} C_{SAT} n \qquad (5-32)$$

where D_v is the coefficient of vertical dispersion (L^2/T), v_i is the average interstitial groundwater velocity (L/T), L_p is the length of a DNAPL pool in the direction of groundwater flow (L), n is the effective porosity, and C_{SAT} is the saturation concentration (M/L^3). The coefficient of vertical dispersion is calculated as

$$D_v = D_e + (v_i \, \alpha_v) \qquad (5-33)$$

where D_e is the effective aqueous diffusion coefficient (L^2/T), and α_v is the vertical transverse dispersivity (L).

Assuming that the areal dimensions of the pool do not vary during dissolution, the time, t_d (T), required for complete DNAPL pool dissolution can be estimated as (Johnson and Pankow, 1992)

$$t_d = P_h L_p n \rho_n s_n / M_a \qquad (5-34)$$

where P_h is the pool height, ρ_n is the DNAPL density (M/L^3), and s_n is the DNAPL saturation. For example, assuming for a particular TCE-sand system that C_{SAT} is 1100 g/m^3, n is 0.35, s_n is 1.0, D_e is 2.7 X 10^{-10} m^2/s, α_v is 0.00023 m, and P_h is 0.01 L_p, Johnson and Pankow (1992) calculated TCE pool dissolution times for four pool lengths as shown in Figure 5-18.

5.3.15 Given a DNAPL source of dissolved groundwater contamination, how do you determine the movement of a dissolved plume?

A dissolved plume generally moves with the flowing groundwater. Thus, a site investigation needs to determine groundwater flow directions in three dimensions using water-level data from monitor wells. Dissolved chemicals from DNAPLs also sorb to the soil matrix to varying degrees. The rate of movement is retarded due to this process, which may be characterized by the retardation factor, R_f. The following example shows how the retardation factor is used to determine the relative movement of three different chemicals that may form dissolved plumes from DNAPLs.

At a hypothetical site, 1,1,1-trichloroethane (TCA), trichloroethene (TCE), and methylene chloride (MC) have been released into the subsurface from a tank storage area causing development of dissolved TCA, TCE, and MC plumes. Which of these chemicals is expected to move the fastest in the groundwater?

An indicator of a chemical's tendency to partition between groundwater and soil is the organic carbon partition coefficient, K_{oc}. Values of K_{oc} for DNAPL chemicals are given in Appendix A. This coefficient is related to the distribution coefficient, K_d, by the following:

$$K_d = K_{oc} f_{oc} \qquad (5-35)$$

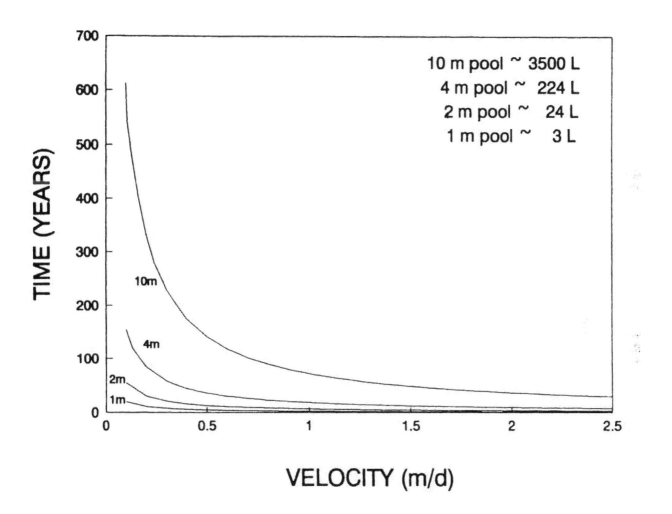

Figure 5-18. Dissolution time versus average interstitial groundwater velocity for four different TCE pool lengths (reprinted with permission from ACS, 1992).

where f_{oc} is the fraction of total organic carbon content in terms of grams of organic carbon per gram of soil. In natural soils, f_{oc} values range from <0.001 to >0.05. Assuming a f_{oc} of 0.017, the distribution coefficients for TCA, TCE, and MC are 2.58, 2.14, and 0.15 ml/g, respectively, as shown in Table 5-2.

The distribution coefficient is related to the retardation factor, R_f, by

$$R_f = [1 + (\rho_b/n)\,K_d] \qquad (5-36)$$

where n is porosity and ρ_b is the bulk density of the porous media. Bulk mass density is related to particle mass density, ρ_s, by the following:

$$\rho_b = \rho_s\,(1-n) \qquad (5-37)$$

where $\rho_s = 2.65$ g/cm^3 for most mineral soils. If porosity is 0.3, then by Equation 5-37, $\rho_b = 1.85$ g/cm^3, and by Equation 5-36, the retardation factors for TCA, TCE, and MC are 16.9, 14.2, and 1.93, respectively (Table 5-2).

Therefore, the relative order of transport velocity in groundwater is MC, followed by TCE, and then TCA. That is, if all three chemicals were released to groundwater at the same time, MC should create the largest dissolved plume and move the fastest, whereas dissolved TCA should move the slowest and have the smallest plume.

5.3.16 Given a DNAPL source of vapor contamination in the vadose zone, how do you determine the movement of the vapor plume?

There are models available for evaluating this question, as exemplified by the analysis provided in Chapter 5.3.20. To examine the relative movement of various chemicals from a complex DNAPL source, consider the same chemicals that were used in the dissolved plume example: 1,1,1-trichloroethane (TCA), trichloroethene (TCE), and methylene chloride (MC). Which of these chemicals would be expected to migrate farthest from the source area in the vapor plume?

Although diffusion is the primary transport mechanism in the vadose zone, the effects of sorption and dissolution reactions can retard the vapor front. Potential localized density effects are noted in Chapters 4.8 and 8.4. The retardation factor for the vapor phase is defined as (Mendoza and McAlary, 1990)

$$R_a = 1 + n_w/(n_a K'_H) + \rho_b\,K_d/(n_a K'_H) \qquad (5-38)$$

where n_w is the water-filled porosity (dimensionless), n_a is the air-filled porosity, K'_H is Henry's Law constant (dimensionless, see Equation 5-44), ρ_b is the soil bulk density (M/V), and, K_d is the solid-liquid distribution coefficient (mg/g solid per mg/ml liquid).

This retardation factor is constant if the water content of the soil does not change and is analogous to the retardation factor, R, for the movement of chemicals in the saturated zone. The second term in Equation 5-38 represents the partitioning of the chemicals from the gas phase to the water phase. The third term represents the partitioning from the gas phase, through the water phase, to the solid phase. Henry's Law constant values are provided in Appendix A.

As shown in Table 5-2, vapor phase retardation factors and the relative order of movement for TCA, TCE, and MC are calculated based on K'_H values of 0.599, 0.379, and 0.084, respectively, and by assuming that $\rho_b = 1.85$ g/cm^3, and $n_a = n_w = 0.50$. The relative order of movement in soil gas is MC, followed by TCA, and then TCE. This order differs from that in groundwater.

5.3.17 What will be the chemical composition of a dissolved plume associated with a DNAPL source?

Several approaches are available to characterize the composition and concentrations of chemicals dissolved from a DNAPL source into groundwater. The most direct and reliable approach is to have groundwater sampled downgradient and close to the DNAPL source submitted for comprehensive chemical analysis. A second approach is to conduct an equilibrium dissolution study by placing a sample of the DNAPL in a sample of the local groundwater (or tapwater). A few options for performing dissolution studies are provided in Chapter 12.

If samples cannot be obtained, or if it is desirable to predict the dissolved chemical concentrations over time, then theoretical calculations can be made based on knowledge of DNAPL source chemistry. For DNAPLs comprised of a mixture of chemicals, the effective solubility of each component in groundwater can be estimated using Worksheet 7-1 in Chapter 7. Based on the effective solubility concept, Mackay et al. (1991) present equations to estimate: (1) the equilibrium concentrations of dissolved chemicals in groundwater in

Table 5-2. Parameters and values used to calculate the relative order of transport velocity for TCA, TCA, and MC and Chapter 5.3.15 (reprinted from ACS, 1992).

Chemical	K_{oc} (ml/g)	K_d (ml/g)	R	K'_H	R_a	Relative Order of Movement in Groundwater	Relative Order of Movement in Vapor
1,1,1-Trichloroethane (TCA)	152	2.58	16.9	0.599	55.8	3	2
Trichloroethene (TCE)	126	2.14	14.2	0.379	73.3	2	3
Methylene Chloride (MC)	8.8	0.15	1.93	0.084	34.9	1	1

contact with a NAPL of defined composition; (2) changes in aqueous and NAPL phase chemical concentrations with time due to NAPL depletion by dissolution; and (3) the water flow volume (or time) required for a defined depletion of the component within the NAPL mass. For example, the dissolution characteristics of a mix of chlorobenzene, 1,2,4-trichlorobenzene, and 1,2,3,5-tetrachlorobenzene as a function of water leaching with time are shown in Figure 5-19.

5.3.18 What is the equivalent mass/volume of DNAPL contained within a dissolved groundwater plume?

Mackay and Cherry (1989) show several examples of organic plumes in sand-and-gravel aquifers that extend 0.5 to 1.0 km from the source area (Table 5-3). Using these plumes, they compute an equivalent DNAPL volume in terms of liters or drums of DNAPL. As depicted in Table 5-3, only a few liters of NAPL are required to create large dissolved plumes. This is further illustrated by the following example.

Groundwater at a hypothetical site has been contaminated with trichloroethene (TCE) concentrations which vary between 100 to 10,000 µg/L. The affected water-table aquifer is a sand overlying a silty clay. The aquifer has a saturated thickness of approximately 10 ft, and the facility occupies an area of 680 ft by 130 ft (approximately two football fields). If TCE mass in soil above the water table and separate-phase TCE mass are ignored, what is the dissolved and sorbed mass of TCE in the aquifer underlying the facility? What is this equal to in terms of DNAPL TCE volume?

The total mass (M_T) is equal to the dissolved mass plus the sorbed mass, or

$$M_T = (C\ n\ V_T) + (S\ \rho_b\ V_T) \qquad (5-39)$$

where C is the dissolved concentration (ML^{-3}), n is porosity, V_T is the total volume (L^3), S is the mass of the sorbed chemical per unit mass of solid, and ρ_b is the bulk mass density of the porous medium (ML^{-3}).

For linear sorption, the sorbed mass, S, is related to the dissolved mass through the distribution coefficient (K_d):

$$S = K_d\ C \qquad (5-40)$$

Substitution of (5-40) into (5-39) and rearranging terms yields

$$M_T = n\ C\ V_T\ R \qquad (5-41)$$

where R is the retardation factor defined by

$$R = [1 + (\rho_b/n)K_d] \qquad (5-42)$$

Thus, Equation 5-41 shows that the total mass is equal to the dissolved mass times the retardation factor.

For the site involving TCE, V_T = 680 ft X 130 ft X 10 ft = 884,000 ft^3. Further, it is estimated that the sand has a porosity of 0.3; thus, there are 265,200 ft^3 of water in the saturated media. Using the definition of concentration, the amount of mass required to cause a concentration of 1 µg/L in the water throughout the site (265,200 ft^3 = 7,509,668 L) is 7.51 grams of TCE. TCE has a density of 1.46 g/cm³. Using the density, the volume of separate-phase (DNAPL) TCE required to cause a concentration of 1 µg/L throughout the site is only 5.14 cm³ (or 0.00514 L or 0.00136 gallons). This does not include sorption.

The sorbed phase can be estimated using the organic carbon partition coefficient (K_{oc}) for TCE, which is 126 ml/g. The fraction of organic carbon (f_{oc}) is assumed to be approximately 0.001. The distribution coefficient is calculated as

$$K_d = f_{oc}\ K_{oc} = 0.126 \qquad (5-43)$$

which is used in Equation 5-42 to calculate a retardation factor of 1.78.

Using this information, the volume required to produce a range of TCE concentrations throughout the site is presented in Table 5-4. The volume is expressed in gallons of DNAPL TCE. As may be seen, only a small amount of DNAPL TCE is required to cause a rather large plume; less than two gallons, if not sorbed, will contaminate the entire site to above 1,000 µg/L, where the maximum contaminant level (MCL) for TCE is 5 µg/L. If the TCE is sorbed, the volume of DNAPL TCE required to cause the same levels of concentration in the dissolved phase is approximately doubled.

This example illustrates two important points. First, small amounts of DNAPL can potentially contaminant large volumes of water above MCLs. Second, at many sites, there is an attempt to estimate mass in place, and use this to estimate clean-up times. This calculation, while interesting, has a large uncertainty associated with

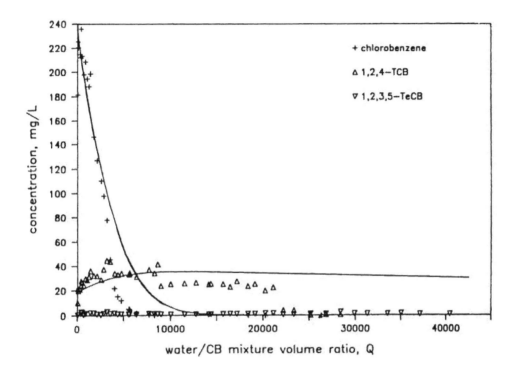

Figure 5-19. Measured and predicted dissolution characteristics for a mixture of chlorobenzenes: 37% chlorobenzene, 49% 1,2,4-trichlorobenzene, 6.8% 1,2,3,5-tetrachlorobenzene, 3.4% pentachlorobenzene, and 3.4% hexachlorobenzene (from Mackay et al., 1991). The water/CB (chlorobenzene) volume ratio, Q, is the volume of water to which the chlorobenzene mixture was exposed divided by the initial volume of chlorobenzene mixture.

Table 5-3. Equivalent DNAPL mass associated with some relatively well-documented organic contaminant plumes in sand-gravel aquifers (modified from Mackay and Cherry, 1989).

SITE LOCATION AND PLUME MAP	PRESUMED SOURCES	PREDOMINANT DNAPL CONTAMINANTS	PLUME VOLUME (LITERS)	ESTIMATED CHEMICAL MASS DISSOLVED IN PLUME (AS EQUIVALENT DNAPL VOLUME IN LITERS OR 55-GAL DRUMS)
Ocean City, NJ	chemical plant	Trichloroethene 1,1,1-Trichloroethane Tetrachloroethene	5,700,000,000	15,000 (72 drums)
Mountain View, California	electronics plant	Trichloroethene 1,1,1-Trichloroethane	6,000,000,000	9800 (47 drums)
Cape Cod, Ma.	sewage infiltration beds	Trichloroethene Tetrachloroethene	40,000,000,000	1500 (7 drums)
Gloucester, Ont.	special waste landfill	1,4-Dioxane Freon 113	102,000,000	190 (0.9 drum)
San Jose, Cal.	electronics plant	1,1,1-Trichloroethane Freon 113 1,1-Dichloroethene	5,000,000,000	130 (0.6 drum)
Denver, Colorado	trainyard, airport	1,1,1-Trichloroethane Trichloroethene Dibromochloropropane	4,500,000,000	80 (0.4 drum)

0   5 km

Flow ⟶

Table 5-4. Volumes of TCE required to produce a range of TCE concentrations in 884,000 ft^3 of aquifer assuming porosity = 0.3 and K_d = 0.126.

Dissolved TCE (μg/L)	DNAPL TCE (gal)	
	Dissolved Phase	Dissolved and Sorbed
100	0.1359	0.2419
1,000	1.359	2.419
10,000	13.59	24.19

Note: a barrel usually contains 55 gallons.

it. For this example, any mass or volume of DNAPL that might be present at the site (e.g., as residual saturation) could greatly exceed the estimated mass in place. This, in turn, could have significant consequences on remediation and extend clean-up times.

5.3.19 What is the relationship between concentrations in soil gas and groundwater?

The interaction of soil gas and groundwater can be complex. If equilibrium conditions exist, then Henry's Law can be used to describe this relationship and used to address a number of issues. For example, assuming that the aqueous and gaseous phases are in equilibrium, and that a vapor plume of TCE has migrated away from a residual DNAPL source in the vadose zone, what gas concentrations are required to produce aqueous concentrations of 5 $\mu g/L$?

To answer this question, Henry's Law can be used, but in a form slightly different from that provided in Equation 4-17. A second method of defining Henry's Law constant is:

$$K'_H = C_a/C_w \qquad (5-44)$$

where C_a is the molar concentration in air (mole/m^3), C_w is the molar concentration in water (mole/m^3), and K'_H is the alternate form of Henry's Law constant (dimensionless).

Equations 5-44 and 4-17 are related using the ideal gas law as follows:

$$K'_H = K_H/(R_g T) = 41.6 \, K_H \text{ at } 20°C \qquad (5-45)$$

where T is the temperature of water (°K), and R_g is the ideal gas constant (8.20575×10^{-5} atm-m^3/mol-K).

Using Appendix A, the Henry's Law constant for TCE is 9.10×10^{-3} atm-m^3/mol. Using Equation 5-45, this converts to a dimensionless Henry's Law constant of 0.379. Substituting this into Equation 5-44 indicates that a gas concentration of only about 2 $\mu g/L$ would be in equilibrium with groundwater containing 5 $\mu g/L$ TCE. The water containing TCE might be a small layer on top of the water table. Thus, monitor well results that sample over a larger vertical distance would yield lower concentrations.

5.3.20 Given a DNAPL source in the vadose zone, how can you evaluate the movement of a vapor plume? What are the conditions that favor vapor transport away from a DNAPL source in the vadose zone that would allow soil-gas monitoring?

These questions have been addressed by several modeling studies (e.g., Silka, 1986; Mendoza and McAlary, 1990) and are answered using an analytical solution below.

Assuming that DNAPL fully penetrates the vadose zone, then vapors from this source are transported away by diffusion in the radial direction. This transport is affected by vapor-phase retardation. For this simplified situation, the vapors are assumed not to interact with the water table (lower boundary) or the land surface (upper boundary). Furthermore, the vadose zone is assumed to be infinite in extent.

The differential equation governing unsteady, diffusive, radial flow is given by

$$\partial^2 C_a/\partial r^2 + [1/r(\partial C_a/\partial r)] = (R_a/D^*)(\partial C_a/\partial t) \qquad (5-46)$$

where the air-filled porosity, n_a, is assumed to be constant, and where

$$D^* = D\tau_a \qquad (5-47)$$

$$\tau_a = n_a^{2.333}/n_t^2 \text{ (Millington and Quirk, 1961)} \qquad (5-48)$$

and, (Mendoza and McAlary, 1990)

$$R_a = 1 + n_w/(n_a K'_H) + \rho_b K_d/(n_a K'_H) \qquad (5-49)$$

Parameter definitions and values used in this assessment are provided in Table 5-5.

To solve this equation, it is assumed that the vadose zone initially contains no chemical vapors. Therefore, the initial conditions are $C_a = 0$ for all r at t = 0. The boundary away from the DNAPL source is assumed to be far enough away that it remains uncontaminated. Therefore, $C_a = 0$ at $r = \infty$ for $t \geq 0$.

The boundary at the DNAPL is assumed to be maintained at a constant concentration as the DNAPL generates chemical vapors according to Raoult's Law (Equation 4-18). That is, the volume concentration in the gas phase is:

$$C_g = mol/vol = n/V \qquad (5-50)$$

Table 5-5. Parameter values for assessment of TCE diffusion and nomenclature used in Chapter 5.3.20.

PARAMETER	VALUE
total porosity, n_t	0.3
bulk density, ρ_b	1.65 g/ml
bulk water content, n_w	0.06
air-filled porosity, n_a	0.24
tortuosity factor, τ_a	0.40
air diffusion coefficient, D	8×10^{-6} m^2/s
effective diffusion coefficient, D*	3.2×10^{-6} m^2/s
dimensionless Henry's Law constant, K'$_H$	0.39
distribution coefficient, K_d	0.01 ml/g
retardation factor, R_a	1.82
source concentration, C_s	3.28 mol/m^3
source radius, r_s	1 m
ideal-gas constant, R	8.2057×10^{-5} m^3 atm/(mol °K)
vapor pressure of the pure solvent, P°$_A$	0.079 atm
temperature (°K), T	293.15
radial distance from source, r	variable
time, t	variable
mole fraction, X_a	1
concentration in air, C_a	computed, variable

From the ideal gas equation,

$$PV = nR_gT \qquad (5\text{-}51)$$

or

$$C_g = n/V = P/R_gT \qquad (5\text{-}52)$$

But P is total pressure or the sum of the partial pressures of each component. Thus, from Raoult's Law,

$$C_{gA} = X_A P^o_A /R_gT \qquad (5\text{-}53)$$

Thus, the initial and boundary conditions are as follows:

$$C_a (r,0) = 0 \qquad (5\text{-}54)$$

$$C_a (r_s, t) = C_s \text{ (from Equation 5-53)} \qquad (5\text{-}55)$$

$$C_a (\infty, t) = 0 \qquad (5\text{-}56)$$

The general solution for this governing equation subject to the above initial conditions as solved using LaPlace Transform techniques can be found in Carslaw and Jaeger (1959). The solution, as presented in Carslaw and Jaeger (1959), is given in integral form for the full time domain and in asymptotic series form for the early time domain. To generate answers covering the full time domain, the integral form of the solution comprised of the zero-order Bessel functions,

$$C_a = C_s + 2C_s /\pi \int_0^\infty \exp (-D^*u^2t/R_a)^*$$

$$\frac{J_o (ur) Y_o (ur_s) - Y_o (ur) J_o (ur_s)}{J_o^2 (ur_s) + Y_o^2 (ur_s)} \quad \frac{du}{u} \qquad (5\text{-}57)$$

needs to be numerically integrated.

To circumvent potential problems associated with trying to numerically integrate a solution comprised of oscillatory functions, the answers can also be generated by numerical inversion of the solution in the LaPlace domain. The governing equation in the LaPlace domain (Carslaw and Jaeger, 1959) is as follows:

$$\partial^2\overline{C}_a/\partial r^2 + [1/r(\partial\overline{C}_a/\partial r)] = pR_a/D^*; \ \overline{C}_a=0, \ r>r_s \quad (5\text{-}58)$$

where

$$\overline{C}_a (\infty,p) = 0 \qquad (5\text{-}59)$$

$$\overline{C}_a (r_s, p) = C_s/p \qquad (5\text{-}60)$$

and its solution is of the form

$$C_a = [C_s K_o (qr)] / [p K_o (qr_s)] \qquad (5\text{-}61)$$

where

$$q = (p R_a /D^*)^{\frac{1}{2}} \qquad (5\text{-}62)$$

Herein the solution is generated by taking the above solution in the LaPlace domain and inverting it numerically using the DeHoog et al. (1982) algorithm.

To perform some example calculations, TCE is considered and data are used, in part, from Mendoza and McAlary (1990). Data are presented in Table 5-5. The source concentration in Table 5-5 is in mol/m³; this can be converted to mg/L using the molecular weight of TCE (131.5 g/mol) to give a source concentration 433 mg/L gas. The source radius is assumed to be 1 m; that is, r$_s$ = 1. The problem defined by these data is referred to as the base case.

The results of the base case are shown in Figure 5-20(b), which contains plots of vapor concentration versus radial distance for three times after emplacement of the DNAPL source. As may be seen, immediately next to the DNAPL source, the vapor concentrations approach the source concentration. A few meters away from the DNAPL source, the concentrations decrease rapidly. As expected, the concentrations increase away from the source with time.

These results are helpful in estimating how far vapors will extend from a DNAPL source, and how fast they will migrate. This analytical solution is based on a number of assumptions: (1) diffusion only, as described by Fick's second law, (2) partitioning coefficients are linear and the system is at local equilibrium, (3) use of Millington-Quirk tortuosity (Equation 5-48), (4) soil properties are uniform, (5) soil system is isothermal and chemical properties are constant, (6) the chemical is conservative, and (7) there is no interaction with upper and lower boundaries. In general, these assumptions allow the vapors to migrate farther and faster than if recharge events and condensation on the water table were considered.

It is of interest to see how sensitive the solution is to variations in parameters. For example, vapor diffusion

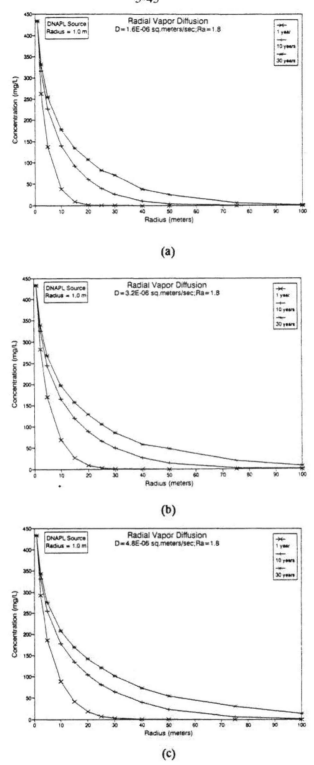

(a)

(b)

(c)

Figure 5-20. Radial vapor diffusion after 1, 10 , and 30 years from a 1.0 m radius DNAPL source for a vapor retardation factor of 1.8 and diffusion coefficient of (a) 1.6 X 10^{-6} m^2/sec, (b) 3.2 X 10^{-6} m^2/sec, and (c) 4.8 X 10^{-6} m^2/sec.

coefficients are provided in Appendix A. As shown, for most DNAPL chemicals, the diffusion coefficients are similar. If the diffusion coefficient is increased and decreased by 50%, the results in Figure 5-20(a) and (c) are obtained, respectively. The concentrations are similar to those in Figure 5-20(b), with the higher diffusion yielding higher concentrations.

Another parameter that can vary is the source concentration. For this problem, the solution is linear. Therefore, as the source concentration changes, the solution can be scaled either up or down depending on whether the source concentration increases or decreases.

The final parameter that affects the solution is the vapor retardation factor. As indicated in Equation 5-49, retardation increases with increasing water content and increasing distribution coefficient, and decreases with increasing air-filled porosity and increasing Henry's Law constant. Less mobile vapors have a high R_a and more mobile vapors have a low R_a. The most mobile situation occurs when $K_d = n_w = 0$ and $R_a = 1$. The effect of varying the vapor retardation factor is demonstrated in Figure 5-21 for $R_a = 1$ (a), $R_a = 10$ (b), and $R_a = 100$ (c). The unretarded case is shown in Figure 5-21(a), in which the vapor concentrations are transported the farthest. As expected, with increasing R_a, transport is more limited.

Thus, for soil-gas monitoring, the optimal conditions are those that promote vapor transport. These include (1) high vapor pressure of the pure solvent to allow a high source concentration, (2) low water content (dry conditions), and (3) high Henry's Law constant.

5.4 NUMERICAL SIMULATION OF IMMISCIBLE FLUID FLOW

Although petroleum reservoir simulators have been used to model immiscible fluid flow for more than 20 years (Peaceman, 1977; Crichlow, 1977), with few exceptions, it is only within the past decade that multiphase flow codes have been utilized to examine NAPL contamination problems (e.g., Arthur D. Little, Inc., 1983; Abriola and Pinder, 1985a, b; Guswa, 1985; Faust, 1985b; Faust et al., 1989; Osborne and Sykes, 1986; Kuppusamy et al., 1987; Parker and Lenhard, 1987a; Lenhard and Parker, 1987b; and Kueper and Frind, 1991a, b). Similarly, within the past few years, a variety of simulators have been developed to examine vapor transport from subsurface NAPL sources (e.g., Baehr, 1987; Silka, 1986; Mendoza

and Frind, 1990a, b). Reviews of multiphase flow codes are provided by Abriola (1988) and Camp Dresser and McKee (1987). The following discussion is adapted from Mercer and Cohen (1990) and Abriola (1988).

Early recognition of NAPL movement in groundwater as a two-fluid flow phenomenon is attributed to van Dam (1967). Later, several models were developed to describe mathematically the flow of NAPL in the subsurface (Mull, 1971, 1978; Dracos, 1978; Holzer, 1976; Hochmuth and Sunada, 1985). Common to each of these is the assumption of negligible capillarity (piston-like flow).

Brutsaert (1973) presents an early code used to examine multifluid well flow that accounts for capillarity. The model is radial and based on a finite-difference approximation. Later, Guswa (1985) developed a one-dimensional (vertical) finite-difference, two-fluid flow simulator. Faust (1985b) extended this work to accommodate two dimensions as well as a static air phase, a necessary step to simulate NAPL flow in the vadose zone. A model similar to Faust's model (1985), which did not consider an air phase, was applied to the Hyde Park Landfill, Niagara Falls, New York by Osborne and Sykes (1986). Abriola and Pinder (1985a,b) developed a two-dimensional model that also considers volatilization and dissolution. A similar model is presented in Corapcioglu and Baehr (1987) and Baehr and Corapcioglu (1987). Subsequently, Parker and Lenhard (1987) and Lenhard and Parker (1987a) incorporated hysteretic constitutive relations. More recently, a three-dimensional model that extends Faust's (1985b) model is described in Faust et al. (1989); Kueper and Frind (1991a) presented a two-dimensional, vertical slice model formulated in terms of wetting phase saturation and pressure; and Guarnaccia et al. (1992) developed a two-dimensional, two-phase code to simulate NAPL emplacement and subsequent removal through dissolution in near-surface saturated environments.

Because of previous reviews on immiscible flow models, a detailed review is not provided herein. The basic governing equations are presented for completeness, along with constitutive relationships that concern many of the properties discussed in the Chapter 4. This discussion follows closely that presented by Abriola (1988).

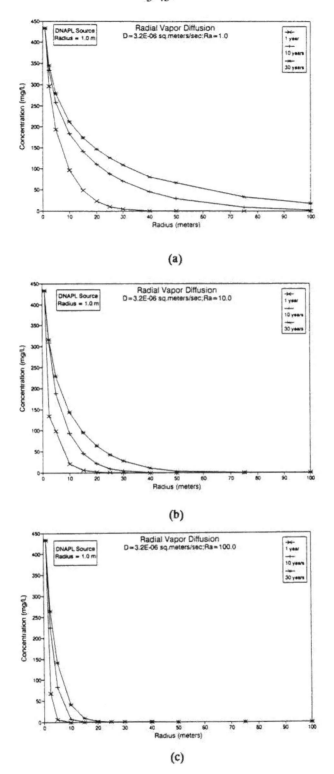

(a)

(b)

(c)

Figure 5-21. Radial vapor diffusion after 1, 10 , and 30 years from a 1.0 m radius DNAPL source for a diffusion coefficient of 3.2 X 10⁻⁶ m²/sec and vapor retardation factors of (a) 1, (b) 10, and (c) 100.

5.4.1 Mass Balance Equations

The equation development begins with the mass balance equation for species i in phase α, where α stands for soil, air, water, and NAPL or a subset of these. A species is defined as a specific chemical that is present in one or more phases. The mass balance equation is written as (Abriola, 1988):

$$\frac{\partial}{\partial t}(e_\alpha\rho^\alpha\omega_i^\alpha) + \nabla \cdot (e_\alpha\rho^\alpha\omega_i^\alpha v^\alpha) - \nabla \cdot J_i^\alpha = S_i^\alpha + R_i^\alpha \quad (5\text{-}63)$$

where: v^α is the mass average velocity of the α phase; ω_i^α is the mass fraction of species i in the α phase; e_α is the fraction of volume occupied by the α phase; ρ^α is the intrinsic mass density of the α phase; J_i^α is the non-advective flux of a species i in the α phase; S_i^α represents the exchange of mass of species i due to interphase diffusion and/or phase change; R_i^α represents an external supply of species i to the α phase; and ∇ is the differential operator.

The first and second terms in Equation 5-63 represent mass accumulation of species i in phase α and mass movement due to advection of the phase, respectively. Mass transport due to non-advective effects (i.e., dispersion and diffusion) is incorporated by the third term. The first term on the right side of Equation 5-63, S_i^α, is a source/sink term to account for phase changes; and the second term, R_i^α accounts for the destruction or creation of the species due to biological or chemical transformations.

Equation 5-63 is constrained by:

$$\sum_i \omega_i^\alpha = 1 \quad (5\text{-}64)$$

and

$$\sum_\alpha e_\alpha = 1 \quad (5\text{-}65)$$

based on the definitions of mass and volume fraction. Also, when mass is lost from one phase due to interphase exchange, an equal amount of mass is gained by another phase, or:

$$\sum_\alpha S_i^\alpha = 0 \quad (5\text{-}66)$$

A mass balance relationship for a specific phase may be developed by summing over all species present in the phase or alternatively summing over all phases based on Equation 5-63.

5.4.2 Immiscible Flow Equations

Equation 5-63 can be simplified by assuming that there is no mass exchange between phases and no chemical or biological transformations and by summing over all species to yield (Abriola, 1988):

$$\frac{\partial}{\partial t}(\rho^\alpha e_\alpha) + \nabla \cdot (\rho^\alpha e_\alpha v^\alpha) = 0 \quad (5\text{-}67)$$

where use has been made of constraint Equation 5-64. The non-advective flux terms, which deal with relative motion of the species within a phase, also sum to zero. Equation 5-67 has been used to model the flow of NAPLs with chemical or physical properties that can be considered spatially invariant.

In general, one equation is written for each of the four phases: soil(s), air(a), water(w), and NAPL(N). If the porous medium is incompressible (porosity is constant in time), then the soil equation is not needed. Similarly, assuming the gas phase remains at atmospheric pressure, the gas equation also can be deleted to yield:

$$n\frac{\partial}{\partial t}(s_\alpha\rho^\alpha) + \nabla \cdot (\rho^\alpha s_\alpha n v^\alpha) = 0 \quad \alpha = w,N \quad (5\text{-}68)$$

where n is porosity and s_α is the saturation of the α phase ($e_\alpha = ns_\alpha$). Finally, if the fluids are treated as incompressible, then:

$$n\frac{\partial}{\partial t}(s_\alpha) + \nabla \cdot (s_\alpha n v^\alpha) = 0 \quad \alpha = w,N \quad (5\text{-}69)$$

5.4.3 Compositional Equations

For the interphase transfer of mass (i.e., the formation of a dissolved plume or the transport of organic vapors), balance equations for each species are written. The species balance equations are obtained by summing Equation 5-63 over all phases to yield (Abriola, 1988):

$$\sum_\alpha \frac{\partial}{\partial t}\{(\rho^\alpha e_\alpha\omega_i^\alpha) + \nabla \cdot (\rho^\alpha e_\alpha v^\alpha\omega_i^\alpha) - \nabla \cdot J_i^\alpha\} = \sum_\alpha R_i^\alpha \quad (5\text{-}70)$$

where α = a,s,w,N and constraint 5-66 has been incorporated. The right side of Equation 5-70 is zero for nonreactive species. The number of equations that are required depend upon the number of species. If the soil matrix is rigid, the soil species equation may be deleted. Solving Equation 5-70 yields fluid distributions and compositions in time and space.

5.4.4 Constitutive Relations

Assuming it is valid, Darcy's law may be substituted into a system of mass balance equations (such as Equation 5-68 or 5-70) to derive the equations governing multiphase fluid flow in a porous medium. For example, consider NAPL flow in a rigid matrix (Faust et al., 1989):

$$n\frac{\partial}{\partial t}(s_\alpha\rho^\alpha) - \nabla \cdot [\rho^\alpha \frac{kk_{r\alpha}}{\mu_\alpha} \cdot (\nabla P^\alpha - \rho^\alpha g)] = 0 \qquad (5-71)$$

Equation 5-71 is formulated in the unknowns P^α, which are continuous in space. Porosity n and intrinsic permeability k are assumed known properties of the matrix. Generally, viscosity (μ_α), a weak function of pressure, is assumed constant. For the incompressible fluid case, ρ^α is a constant. In general, however, density will depend on the fluid pressure, P^α. Thus, fluid density may be expanded in terms of fluid pressure by incorporating β^α, the compressibility of the α phase. For slightly compressible fluids, β^α is essentially constant.

Capillary pressure and relative permeability, which can exhibit hysteretic behavior, are generally considered functions of saturation as described in Chapter 4. For example, based on the van Genuchten $k_{r\alpha}$ - s_α - P_c model described by Luckner et al. (1989) and Guarnaccia et al. (1992):

$$s_{we} = (s_w - s_{wr}) / (1.0 - s_{wr}), \quad (0 \le s_{we} \le 1) \qquad (5-72)$$

$$s_{ne} = [(1 - s_{nr}) - s_w)] / (1.0 - s_{nr}), \quad (0 \le s_{we} \le 1) \quad (5-73)$$

$$s_{ne} = 1 / [1 + (a\ h_c)^n]^m \qquad (5-74)$$

$$k_{rw}(s_w) = (s_{we})^{\frac{1}{2}} [1-(1-s_{we}^{1/m})^m]^2 \qquad (5-75)$$

$$k_{rn}(s_w) = (s_{ne})^{\frac{1}{2}} [1-(1-s_{ne})^{1/m}]^{2m} \qquad (5-76)$$

where s_{we} is the effective wetting phase saturation, s_{ne} is the effective nonwetting phase saturation, s_{wr} is the

wetting phase residual saturation, s_{nr} is the nonwetting phase residual saturation, h_c = $P_c/\rho_w g$ is the capillary pressure head (L), η (dimensionless) and a (L^{-1}) are porous medium dependent parameters to be fit to data, and m = (1 - 1/η).

The compositional model Equation 5-70 has additional considerations due to the dependence of properties on composition. For example, viscosity and density may be functions of composition. There are also two additional equation terms, those accounting for non-advective flux and chemical reaction, which must be evaluated. The non-advective flux term is commonly assumed to have a Fickian form. This is the form used in most existing solute transport models and accounts for both molecular diffusion and hydrodynamic dispersion effects (Bear, 1979). There is no general functional form that may be specified for the reaction terms that appear in Equation 5-70. Simple decay rates can be directly substituted in Equation 5-70 or a system of chemical reaction equilibrium or rate equations can be solved to determine these terms.

Finally, for a compositional model, expressions are needed to relate mass fractions of a given species within all the phases. Generally, the assumption of local equilibrium is used to develop these relations (e.g., for sorption, see Valocchi, 1985). Local equilibrium implies that within some relatively short time scale, contiguous phases reach a thermodynamic equilibrium. Thus, the mass fraction of a species in one phase can be related to the mass fractions of the same species in other phases via partition equations such as (Abriola, 1988):

$$\omega_i^\alpha = K_i^{\alpha\beta}\omega_i^\beta \qquad (5-77)$$

where $K_i^{\alpha\beta}$ is the partition coefficient of species i between the α and β phases. Partition coefficients are defined as functions of phase compositions and pressures and may be determined from solubility relations and Henry's Law constants (Appendix A). See, for example, Corapcioglu and Baehr (1987), Baehr and Corapcioglu (1987), and Baehr (1987).

5.4.5 Model Utility

Although NAPL models can, in theory, simulate a variety of problems, the data required for applications are generally lacking. Examples of the application, and limitations, of using an immiscible flow code to simulate

DNAPL migration at two hazardous waste sites in Niagara Falls, New York are given by Faust et al. (1989). Heterogeneity influences NAPL flow and solute transport; however, spatial variability of pore size (affecting displacement pressures) and intrinsic permeability are rarely sufficiently defined to permit accurate prediction. Therefore, use of multiphase flow codes is limited by the difficulties associated with measuring field-scale $P_c(s_w)$ and $K_r(s_w)$ relations and characterizing media heterogeneity. As noted in Chapter 4, the theoretical description of mass transfer in porous media has not been adequately developed, and also adds uncertainty to any modeling approach. Currently, immiscible flow models are used most frequently for hypothesis testing in a conceptualization mode, particularly at NAPL contamination sites where extensive research has already been conducted.

6 DNAPL SITE CHARACTERIZATION OBJECTIVES/STRATEGIES

Site characterization, a process following the scientific method, is performed in phases (see Figure 6-1). During the initial phase, a hypothesis or conceptual model of chemical presence, transport, and fate is formulated based on available site information and an understanding of the processes that control chemical distribution. The potential presence of DNAPL at a site should be considered as part of this early hypothesis. A variety of DNAPL conceptual models are described in Chapter 5.

Based on the initial hypothesis, a data collection program is designed in the second phase to test and improve the site conceptual model and thereby facilitate risk and remedy assessment. As such, site characterization efforts should focus on obtaining data needed to implement potentially feasible remedies. After analyzing the newly acquired data within the context of the initial conceptual model, an iterative step of refining the hypothesis is performed using the results of the analysis, and additional data may be collected. As knowledge of the site increases and becomes more complex, the working hypothesis may take the form of either a numerical or analytical model. Data collection continues until the hypothesis is proven sufficiently.

During implementation of a remedy, the subsurface system often is stressed. This provides an opportunity to monitor and not only learn about the effectiveness of the remediation, but to learn more about the subsurface. Therefore, remediation (especially pilot studies) should be considered part of site characterization, yielding data that may allow improvements to be made in the conduct of the remediation effort. Specific objectives, strategies, and concerns related to DNAPL site investigation are discussed in this chapter.

6.1 DIFFICULTIES AND CONCERNS

The difficulty in evaluating chemical presence, transport, risk, and remediation at DNAPL sites is compounded by the following factors (USEPA, 1992).

- [1] "The relative importance of the forces that control the rate, flow direction, and ultimate fate of DNAPL is different from the relative importance of those that control the distribution of dissolved phase plumes. DNAPL behavior is only loosely coupled to that of groundwater. Movement of DNAPLs is remarkably sensitive to the capillary properties of the subsurface, and the distribution of those properties controls the distribution of the DNAPL. Thus, knowledge of geologic conditions is relatively more important than knowledge of hydrologic conditions for adequate characterization of DNAPL sites."

- [2] "Obtaining a detailed delineation of subsurface DNAPL distribution is difficult and may be impractical using conventional site characterization techniques. DNAPL migrates preferentially through relatively permeable pathways (soil and rock fractures, root holes, sand layers, etc.) and is influenced by small-scale heterogeneities (such as bedding dip and slight textural changes) due to density, capillary forces, and viscous forces. As a result, the movement and distribution of DNAPL is difficult to determine even at sites with relatively homogeneous soil and a known, uniform DNAPL source. This difficulty is compounded by fractured bedrock, heterogeneous strata, complex DNAPL mixtures, etc. The relative importance of small-scale heterogeneities may depend on the volume of the release [i.e., diminish with increasing release volume]."

- [3] "DNAPL in fractured media poses exceptionally difficult problems for site investigation and remediation because fracture networks are complex, DNAPL retention capacity (mass of DNAPL per volume of rock) is generally small, and the depth to which DNAPL may penetrate can be very large."

- [4] "Failure to directly observe DNAPL at a site does not mean it does not exist. Often, only very low aqueous concentrations of DNAPL constituents are detected in monitor wells at known DNAPL sites." These concentrations, however, may greatly exceed drinking water standards.

- [5] "DNAPLs can be broadly classified on the basis of physical properties such as density, viscosity, and solubility. Of the various types of DNAPLs found in the subsurface, chlorinated solvents and creosote/coal tar are apparently the most common. These two types of DNAPLs, however, present groundwater and remediation problems of a very different nature due to the differences in their physical properties. Some of the conclusions applicable to one are not generally applicable to the other. Physical characteristics can guide the choice of characterization and remediation options."

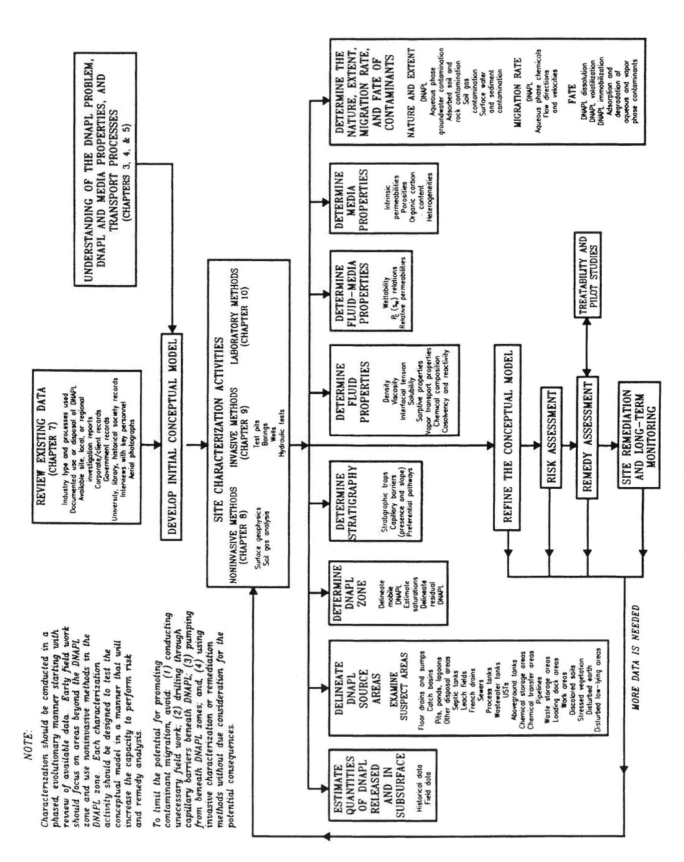

Figure 6-1. DNAPL site characterization flow chart.

•₆ "The risk of causing DNAPL remobilization must be assessed during site investigation. Conventional drilling technologies have a high potential for promoting vertical DNAPL movement. The appropriate investigation strategy is dependent on site-specific conditions, including the geology and the type of DNAPL."

•₇ "Nonintrusive, low-risk investigation methods such as surface geophysical techniques and reviews of site history and existing data should first be used to develop and improve the conceptual model of DNAPL presence and lessen the risks associated with subsequent drilling. Currently, surface geophysical methods capable of delineating DNAPL and the availability of geophysicists training in investigating DNAPL problems are extremely limited. However, even routine, noninvasive geophysical techniques can be used to evaluate site geology, which has a great influence on DNAPL migration."

•₈ "Site characterization should be a continuous, iterative process, whereby each phase of investigation and remediation is used to refine the conceptual model of the site. The time required to define site characteristics cannot easily be reduced because of the heterogeneous, site-specific nature of subsurface environments and the evolutionary [investigatory] process required. However, the information required to implement early containment of dissolved-phase contamination is less extensive that required to design final remedial alternatives."

•₉ "The complexity of the DNAPL problem dictates that site investigators have a sophisticated knowledge of DNAPL contaminant hydrology. Most decisions regarding site investigations and cleanup depend on site-specific conditions."

In essence, the impetus to characterize the complexities associated with DNAPL presence at contamination sites derives from three fundamental truths.

• The behavior of subsurface DNAPL cannot be adequately defined by investigating miscible contaminant transport due to differences in properties (fluid and media) and principles that govern DNAPL and solute transport (see Chapters 4 and 5).

• DNAPL can persist for decades or centuries as a significant source of groundwater and soil vapor contamination.

• Without adequate precautions or understanding of DNAPL presence and behavior, site characterization activities, such as drilling and aquifer testing, may result in expansion of the DNAPL contamination zone and increased remedial costs.

Thus, characterization of DNAPL presence, transport, and fate is required to perform an adequate assessment of site risks and remedies, and to minimize the potential for inducing unwanted DNAPL migration during remedial activities.

6.2 OBJECTIVES AND STRATEGIES

The ultimate objective of characterizing a contaminated site is to be able to assess risk and select appropriate remedial measures. Specific objectives of a DNAPL site investigation are incorporated within the DNAPL site characterization flow chart given in Figure 6-1. Selected aspects of site characterization objectives and strategies are discussed below.

6.2.1 Regulatory Framework

Regulations drive cleanups, hence, site characterization, at hazardous waste sites. Two laws that provide the framework for site investigation and remediation at hazardous waste sites are the Resource Conservation and Recovery Act (RCRA) and the Comprehensive Environmental Response, Compensation, and Liability Act (CERCLA), or Superfund. RCRA deals with active sites, whereas CERCLA addresses the cleanup of inactive and abandoned hazardous waste sites.

Under RCRA, site characterization occurs during the RCRA Facility Investigation (RFI). The RFI is used to evaluate the nature and extent of the release of hazardous waste and hazardous constituents and to gather necessary data to support a Corrective Measure Study (CMS). The CMS is used to develop and evaluate corrective measures and recommend the final corrective measure. Under CERCLA, the analogous process is known as a Remedial Investigation/Feasibility Study (RI/FS). The purpose of the RI/FS is to assess site conditions and evaluate alternatives to the extent necessary to select a remedy.

Guidance for conducting site characterization studies pursuant to CERCLA and RCRA regulations is provided by USEPA (1988) and USEPA (1989a), respectively. Emphasis is placed on rapid characterization of site risk and selection of a remedy or corrective measure that meets certain criteria. For groundwater cleanups, the criteria are risk based and include Maximum Contaminant Levels (MCLs), Alternate Concentration Limits (ACLs), detection limits, and natural (background) water quality. These criteria are well below the aqueous solubility of chemicals forming DNAPLs. Therefore, characterizing DNAPL prior to site remedy selection is critical to setting remedial objectives and ensuring that clean-up goals are met.

6.2.2 Source Characterization

Objectives of source characterization include determination of: mobile and residual DNAPL distributions, DNAPL volumes, DNAPL composition and fluid properties, stratigraphic controls on DNAPL movement, etc. (Figure 6-1). As indicated in Figure 6-1, site conditions, especially stratigraphy, should be examined initially away from suspected DNAPL source areas. Stratigraphic exploration can be used to identify capillary barriers that may effectively limit the downward movement of DNAPL. If possible, suspected DNAPL source areas should first be investigated using noninvasive techniques (Chapter 8) such as soil gas analysis and surface geophysical methods to reduce the risk of mobilizing contaminants.

Depending on site conditions, soil gas analysis may be used to reveal volatile DNAPL source areas. Surface geophysical methods including ground penetrating radar, complex resistivity, and electromagnetic induction methods have successfully detected aqueous and nonaqueous-phase organic contaminants at a very limited number of sites. More routinely, surface geophysical methods are employed to provide information on the stratigraphic layering and boundaries, depths to groundwater and bedrock, disturbed earth limits, and buried waste presence. Such information can be used to focus invasive characterization of source areas and to reduce the risk of causing undesirable DNAPL movement.

Invasive subsurface exploration methods, such as drilling and test pit excavation, will continue to be relied upon for characterizing DNAPL source areas. Subsurface samples should be examined carefully, and analyzed chemically as needed, as drilling progresses to identify the presence of DNAPL pools and residual zones, and barrier layers. As a general rule, borings should be discontinued

upon encountering a barrier layer beneath a DNAPL zone (Chapter 9).

6.2.3 Mobile DNAPL Delineation

Above residual saturation, DNAPL will flow unless it is immobilized in a stratigraphic trap or by hydrodynamic forces. Mobile DNAPL is a moving contaminant source from which chemicals dissolve into groundwater and volatilize into soil gas. Dissolution may continue for decades or centuries due to the undiluted nature of the DNAPL and limits on solubility and groundwater flow rates. DNAPL migration can greatly expand the vertical and horizontal extent of subsurface contamination. This is because gravity may force DNAPL to sink to depths and in directions that are not downgradient hydraulically from the DNAPL entry location (the initial source area). Therefore, a primary focus of many DNAPL site characterization studies will be to delineate the presence of mobile DNAPL to facilitate assessment of containment and recovery options.

Unfortunately, the subsurface DNAPL distribution may defy definition, particularly at sites with heterogeneous strata, fractured media, and multiple DNAPL release locations. Experience at DNAPL sites indicates that the distribution of highly mobile chlorinated solvents is typically much more difficult to define than that of less mobile creosote/coal tar. This appears related to: (1) the larger volumes of creosote/coal tar typically released at wood-treating and manufactured gas facilities versus the smaller volume of chlorinated solvents released at many industrial and waste disposal sites; (2) the difficulty observing chlorinated solvents (which are often colorless) versus the relative ease of observing brown or black creosote/coal tar in the field; and/or, (3) the much higher ratio of density to viscosity of many chlorinated solvents compared to creosote/coal tar which promotes more rapid and distant movement of the former from the DNAPL entry locations. Conversely, creosote/coal tar are more likely to remain near the release areas. Benefits and risks of characterizing the subsurface distribution of mobile DNAPL must be made on a site-specific basis.

6.2.4 Nature and Extent of Contamination

As indicated in USEPA (1988), the ultimate objective of site characterization is to determine the nature and extent of contamination so that informed decisions can be made regarding remediation based on the determined level of risk. As a source of multimedia contamination, the nature and extent of separate phase DNAPL and related chemical contamination needs to be determined in the affected media. DNAPL chemicals may contaminate soil, bedrock, groundwater, surface water, surface water sediments, and air.

Focusing on groundwater, this assessment includes analysis of groundwater flow and chemical transport. Emphasis is often placed on the latter component through intensive groundwater quality monitoring. This information is used to help: (1) identify contaminants, (2) determine the distribution and concentration of contaminants, (3) determine sources of contaminants, and (4) determine the contaminant phase - dissolved, sorbed or nonaqueous.

In determining the nature and extent of groundwater contamination, it is important to estimate the hydrogeologic characteristics that influence the contaminant distribution. These characteristics include: (1) stratigraphy, (2) hydraulic properties of aquifers and confining beds, (3) hydraulic gradients, (4) recharge/discharge rates, and (5) sorption potential. If fractures are present, an evaluation should be made of their orientation, spacing, and vertical and lateral extent. The spatial distribution of hydraulic properties, gradients, and contaminants can be very complex, requiring vertical as well as horizontal assessment. Groundwater contamination is often conceptualized as forming plumes. At some sites, however, heterogeneities in hydraulic properties (especially in fractured media) and a complex distribution of contaminant sources may result in very erratic contaminant distributions.

6.2.5 Risk Assessment

The Risk Assessment Guidance for Superfund (RAGS) (USEPA, 1989b) presents the general approach to conducting a risk assessment. Although this approach can be applied to DNAPL sites, the document does not discuss DNAPLs. The baseline risk assessment discussed in RAGS consists of four main steps. Relevant information identified through data collection and evaluation (step 1) is used to develop exposure (step 2) and toxicity (step 3) assessments. Risk characterization (step 4) summarizes and integrates both toxicity and exposure steps into quantitative and qualitative expressions of risk.

The risk assessment determines if the site warrants remedial action and, if so, helps determine risk-based, site-specific remediation goals. For groundwater, preliminary remediation goals at CERCLA sites are often established based on readily available chemical-specific applicable or relevant and appropriate requirements (ARARs) such as maximum contaminant levels (MCLs) for drinking water. For chlorinated solvents that have MCLs, their aqueous solubilities exceed their respective MCLs (Table 3-4). Thus, at DNAPL sites, a risk assessment will most likely require some form of remediation.

6.2.6 Remedy Assessment

Site characterization data is ultimately used to evaluate the design and operation of remedial measures at DNAPL sites. Specific data requirements vary between alternative remedial technologies. Therefore, acquisition of site characterization data should be phased and become more focused with respect to remedy selection as the site conceptual model becomes more refined.

Groundwater restoration in the DNAPL zone is impractical, in large part, because there are no proven remedial technologies for completely removing subsurface DNAPL in reasonable time frames (except for excavation where feasible). Depending on site-specific conditions, the objectives of DNAPL site remediation may include:

- removal of mobile DNAPL;

- excavation of the DNAPL contamination zone;

- containment of the DNAPL contamination zone;

- containment of contaminated groundwater beyond the DNAPL contamination zone; and,

- restoration of contaminated groundwater beyond the DNAPL zone to permit beneficial uses.

USEPA (Clay, 1992) recently recommended that remedial strategies at DNAPL sites include:

- expeditious containment of aqueous phase plumes and extraction of mobile DNAPL where possible;

- phased implementation of carefully monitored and documented remedial measures; and,

- modification of remedial measures and goals as warranted based on the monitoring data.

Various remedial technologies that are potentially applicable at DNAPL contamination sites are discussed in Table 6-1. Technologies for DNAPL recovery are generally unproven under field-scale conditions; and there are no methods available, or series of methods, which have been demonstrated to completely remove DNAPL from the subsurface.

DNAPL recovery operations are ongoing at relatively few sites. Off-the-shelf remedial measures used at DNAPL sites have typically been limited to excavation, cut-off walls, and relatively simple pump-and-treat systems. There are, however, many DNAPL recovery and treatment technologies being developed which have undergone limited laboratory and pilot-scale field testing (Table 6-1). A major goal of ongoing research is to develop DNAPL recovery and treatment systems that can be used in series to optimize site cleanup. USEPA is currently funding a long-term research effort by the Robert S. Kerr Environmental Research Laboratory to evaluate innovative remedial technologies and develop guidance documents related to DNAPL site remediation and performance assessment (Clay, 1992).

Removal of mobile DNAPL generally involves some form of pumping, which requires knowledge of permeability and DNAPL distribution and properties. Options for in-situ treatment of groundwater and residual DNAPL include biodegradation and enhanced oil recovery (EOR) technologies (Table 6-1). For both of these options, the spatial distribution of permeability must be estimated to optimize delivery of nutrients, oxygen, or EOR fluids to the matrix and to allow movement of microbes or contaminants. The presence and characteristics (e.g., density, viscosity, and interfacial tension) of DNAPL need to be determined to assess the feasibility of EOR. If EOR is used to mobilize DNAPL at residual saturation, DNAPL flow must be controlled carefully. Otherwise, previously clean portions of the subsurface may become contaminated during remediation.

To design an efficient hydraulic containment system, the components of the groundwater flow system (i.e., permeability distribution, hydraulic gradients, boundary conditions, etc.) must be defined. Aquifer sorption properties, including the distribution of organic carbon content, must be estimated to assess dissolved chemical removal efficiency and clean-up times. For each chemical or chemical class, data required for remediation decisions include characteristics related to: (1) leaching potential (e.g., water solubility, organic carbon partition coefficient); (2) volatilization potential (e.g., vapor pressure, Henry's Law constant); (3) degradation potential (e.g., half life, degradation products); and (4) chemical reactivity (e.g., hydrolysis half life, chemical kinetics). Additionally, the potential for chemical precipitation and plugging should be estimated using major cation/anion chemistry of the groundwater when considering the feasibility of injection well and/or surface treatment options.

Characterization data and considerations relevant to the assessment of remedial options that are potentially applicable to DNAPL contamination sites are further described in Table 6-1.

Table 6-1. Remedial options potentially applicable to DNAPL Contamination sites (modified from USEPA, 1992).

METHOD	APPLICATION	PROCESS	EFFECTIVENESS / ADVANTAGES / LIMITATIONS	EXPERIENCE	REFERENCES
Hydraulic Containment	Hydraulic containment is used to prevent the undesired migration of chemicals through the saturated zone.	Hydraulic containment of DNAPL and dissolved chemicals can be achieved by pumping groundwater from wells and/or drains. Fluid flow control can be augmented by injecting water through wells and/or drains, and by the installation of physical barriers (cut-off walls and landfill covers). Monitor wells are utilized to determine whether or not the specified hydraulic gradients have been obtained and chemical migration has been arrested.	Long-term hydraulic containment will be needed at many DNAPL sites because residual and trapped pools of DNAPL are long-term sources of groundwater contamination. The effectiveness of a hydraulic containment system depends largely on the adequacy of the design and operation of the system. Containment is eased where there is a continuous barrier layer that prevents downward DNAPL migration. Downward migration can be arrested at some sites by creating an upward hydraulic gradient into the DNAPL zone that exceeds the density difference between DNAPL and water. Vertical hydraulic containment of DNAPL, however, has yet to be demonstrated in the field. The main drawbacks to hydraulic containment systems are cost and the need for long-term operation.	The components of a hydraulic containment system (wells, drains, cut-off walls, and covers) have been widely used for contamination site remediation and other applications. Although hydraulic containment is generally a proven migration control technology, its success depends on adequate design and implementation.	Cherry et al. (1990), Cohen et al. (1987), Mackay and Cherry (1989), Mercer et al. (1990)
Containment using Physical Barriers	Capillary and low permeability barriers (fine-grained walls) can be constructed to limit NAPL migration.	Low permeability, fine-grained barrier walls (i.e., slurry walls, concrete walls, sheet piling with grouted joints, etc.) can be constructed to impede the lateral migration of non-wetting DNAPL below the water table. Where possible, barrier walls should be keyed into a low permeability, capillary barrier layer beneath the DNAPL contamination zone.	Barrier walls can provide cost-effective control over NAPL migration in favorable settings. Barrier walls have not been tested, however, to determine their capacity for long-term impedance of NAPL migration. Small fractures or openings will facilitate DNAPL breakthrough. The long-term integrity of engineered subsurface barriers is not well-known. Consideration must be given to the compatibility of barrier wall materials with subsurface chemicals, the potential for inducing migration during wall construction, and changes to the hydrogeologic system effected by wall emplacement.	Cutoff walls have been installed as part of containment systems at many sites.	Cherry et al. (1990); Sale et al. (1988)

Table 6-1. Remedial options potentially applicable to DNAPL Contamination sites (modified from USEPA, 1992).

METHOD	APPLICATION	PROCESS	EFFECTIVENESS / ADVANTAGES / LIMITATIONS	EXPERIENCE	REFERENCES
Product Recovery by Pumping	Mobile NAPL can be pumped from wells or drains.	Mobile NAPL can be pumped from wells and drains utilizing single pumps to extract total fluids or NAPL only, or using individual pumps to withdraw water and NAPL separately.	Wells should be placed in stratigraphic traps to optimize recovery where NAPL pools are present. Long-term recovery is increased by maintaining a maximum thickness and saturation of NAPL adjacent to the well. This can be achieved by pumping NAPL and water separately. Pumping water above a DNAPL pool causes DNAPL upwelling which works to increase the formation transmissivity to DNAPL flow. A dual pumping system can be operated using wells or drains. Overpumping the NAPL may result in truncation of the NAPL layer at the well edge and significantly reduce the formation transmissivity to NAPL flow. Pumping may cause NAPL to enter previously uncontaminated sections, thereby enlarging the contaminated zone. In shallow, unconfined formations, it will generally not be possible to significantly diminish the NAPL residual saturation by increasing hydraulic gradients alone. Pumping can be used to remove mobile NAPL and reduce the potential for continued NAPL migration, however. Vacuum-enhanced pumping may be a way to increase gradients for improved mobilization and hydraulic control.	A great deal of experience has been acquired by the oil industry pumping LNAPL from crude oil reservoirs and by the environmental industry pumping petroleum products from contaminated shallow formations. Little documentation is available, however, regarding DNAPL product recovery at contamination sites.	Blake et al. (1990), Ferry et al. (1986), McWhorter (1991), Sale et al. (1988 and 1989), Sale and Piontek (1988), Schmidtke et al. (1987), Villaume (1991), Wisniewski et al. (1985)
Soil Flushing by Flooding with Water, Steam, Surfactants, Alkaline Agents, Polymers, and Cosolvents [see specific flood EOR technologies below]	In situ soil flushing can be used to enhance recovery of NAPLs, adsorbed chemicals, and dissolved chemicals from the saturated or unsaturated zone.	Injection wells or drains, or surface application delivery systems are used to flood the contaminated zone with flushing solutions and sweep the contaminants to recovery wells or drains (i.e., surfactants, cosolvents, alkaline agents, polymers, steam, etc.). Drains are typically used to effect line-drive or five-spot sweeps; wells are typically used to effect line-drive or five-spot sweeps. The flood enhances recovery of NAPL by reducing interfacial tension, reducing NAPL viscosity, lowering the mobility ratio, increasing solubility, and/or increasing hydraulic gradients (i.e. raising the capillary number). Recovery of adsorbed and/or dissolved chemicals is also enhanced by these processes. Displaced NAPL and chemicals are recovered by pumping wells and/or drains. At the conclusion of the flood, the flushing solution can be displaced to the recovery system by injecting water via the delivery system. Soil flushing may be used as an intermediate process in a train of remedial measures.	Soil flushing is most effective in permeable, uniform media. It can be used to speed the permanent reduction or removal of contaminants from the subsurface. Heterogeneous and low permeability soils will generally result in reduced sweep efficiency, longer project duration, and less successful recovery. It will generally not be possible to remove all NAPL from sites with substantial NAPL presence. Soil flushing can, however, be used to reduce the NAPL residual saturation to levels below the immobile saturation at ambient site conditions. The movement of contaminants mobilized by the flood must be carefully controlled to prevent detrimental migration. Consideration must also be given to: the toxicity, nature and fate of the flushing solution and potential adverse reactions (permeability reduction, impairment of biodegradation rates, etc.) caused by the solution. Site-specific bench and pilot-scale field tests are generally recommended prior to implementation of a field-scale remedial project. The site-specific application of EOR methods may prove to be effective within a train of treatment measures to remediate NAPL-contaminated sites.	Although the oil industry has made extensive use of flooding technology to enhance oil recovery, few in-situ soil flushing operations have been conducted at contamination sites. In general, the application of soil flushing technologies to remediate contamination sites is at the pilot-test stage and the effectiveness of these technologies in environmental applications is unknown.	Sims (1990), USEPA (1990a and 1990b) [see specific flooding technologies below]

Table 6-1. Remedial options potentially applicable to DNAPL Contamination sites (modified from USEPA, 1992).

METHOD	APPLICATION	PROCESS	EFFECTIVENESS / ADVANTAGES / LIMITATIONS	EXPERIENCE	REFERENCES
EOR Using Water Flooding	Waterflooding can be used to increase the recovery of NAPL from the saturated zone.	Referred to as secondary recovery by the oil industry, waterflooding involves the injection of water in wells or drains to hydraulically sweep NAPL toward production wells. Recovery can be enhanced because injection/extraction systems (i.e., line-drive and five spot systems) allow for the development and sustenance of increased hydraulic gradients and flow rates, elimination of dead zones, and overall improved flow control management.	Refer to the soil flushing comments.	Although routinely utilized for secondary recovery by the oil industry, waterflooding has been utilized to recover NAPL at only a few environmental contamination sites.	Anderson (1987c), Donaldson et al. (1989), Rathmell et al. (1973), Sale and Piontek (1988), Sale et al. (1989), Sale et al. (1988)
EOR Using Thermal Methods (Steam or Hot Water Flooding)	Thermal methods (steam or hot water flooding) can be used to increase NAPL recovery.	High-temperature steam is injected via wells into the contamination zone. The steam yields heat to the formation and condenses into a zone that acts as a hot water flood. Coupled with the continuous injection of steam behind it, this hot water drives NAPL to the recovery wells. NAPL recovery is enhanced because: (1) the NAPL becomes less viscous and more mobile upon heating; (2) NAPL solubility may be increased by the higher temperatures; (3) volatile NAPL vaporizes, moves ahead of the hot water and then condenses to form a NAPL bank; and, (4) the increased NAPL saturation in the NAPL condensate bank provides increased NAPL transmissivity and, under favorable conditions, results in a snowball effect.	The application of thermal methods to enhance DNAPL recovery at contamination sites may become more popular based on the successful use of steam injection by the oil industry, the encouraging results of limited lab and pilot-scale testing of steam injection for DNAPL recovery, and the fact that additional chemicals need not be injected to recover the contaminants. Heating may convert DNAPL to LNAPL, thereby promoting mobility due to buoyancy forces. Costs may be high due to heat loss and the need to heat large volumes of subsurface materials. Refer to the soil flushing comments.	There were approximately 200 active thermal EOR projects in the United States in 1986. Steam and hot water displacement of NAPLs at contamination sites has been evaluated in a few laboratory and pilot field studies.	Blevins et al. (1984), Boberg (1988), Donaldson et al. (1989), Doscher and Ghassemi (1981), Goyal and Kumar (1989), Hunt et al. (1988a and 1988b), Leuchner and Johnson (1990), Mandl and Volek (1969), Menegus and Udell (1985), Miller (1975), Offeringa et al. (1981), Prats (1989), USEPA (1990b), Volek and Pryor (1972), Willman et al. (1961), Yortsos and Gavalas (1981)

Table 6-1. Remedial options potentially applicable to DNAPL Contamination sites (modified from USEPA, 1992).

METHOD	APPLICATION	PROCESS	EFFECTIVENESS / ADVANTAGES / LIMITATIONS	EXPERIENCE	REFERENCES
EOR Using Surfactant - Water Flooding	Surfactant flooding can be used to increase NAPL recovery during a flood operation.	Surfactant solution is injected as a slug in a flooding sequence to decrease the interfacial tension between NAPL and water by several orders of magnitude (i.e., from 20-50 dynes/cm to less than 0.01 dynes/cm). The development of ultra-low interfacial tension effects a commensurate 3-5 orders-of-magnitude increase in the capillary number (N_{ca}) which is the ratio of viscous to capillary forces, sometimes expressed as $N_{ca} = \mu v / \sigma \phi$ where μ and v are the viscosity and Darcy velocity of the displacing fluid, σ is the interfacial tension, and ϕ is the pore volume. Ultra-low interfacial tension and higher capillary numbers improve the NAPL displacement efficiency of a flood, promote the coalescence of NAPL ganglia and development of a NAPL bank in front of the surfactant slug, and result in increased NAPL recovery and reduced NAPL residual saturation. Surfactant flooding can also enhance NAPL recovery by causing increased NAPL wetting, solubilization, and emulsification. Surfactants used in EOR operations by the oil industry include petroleum sulfonates, synthetic sulfonates, ethoxylated sulfonates, and ethoxylated alcohols.	The high cost of surfactant chemicals has limited the commercial application of surfactant flooding by the oil industry. Using surfactant solutions to enhance NAPL recovery at contamination sites, however, may be more attractive given the higher costs associated with waste site remediation. At many sites, reducing interfacial tension will be the only practical way to mobilize residual NAPL. Refer to the soil flushing comments.	There were approximately 30 active EOR field-scale projects using surfactant injection in the U.S. in 1960. Surfactant flooding by the oil industry for EOR is limited by its high cost relative to other EOR methods. The use of surfactant flooding to enhance NAPL recovery at contamination sites is in its infancy. Surfactants were used in a soil flushing solution of alkaline, polymer, and surfactant agents to boost DNAPL recovery at the Laramie Tie site.	Akstinat (1981), Belkirch (1991), Donaldson et al. (1989), Ellis et al. (1985), Flumerfelt et al. (1981),Fountain (1991), Gogarty (1983), Heselink and Faber (1981), Manji and Stasiuk (1988), Nash (1987), Nelson et al. (1984), Neustadter (1984), Novosad (1981), Pitts et al. (1989), Salager et al. (1979), Shah (1981), Sharma and Shah (1989) Tuck et al. (1988)
EOR Using Alkaline Water Flooding	Alkaline flooding can be used to increase NAPL recovery during a flood operation.	Alkaline waterflooding is an EOR process where inexpensive caustics such as sodium carbonate, sodium silicate, sodium hydroxide, and potassium hydroxide are mixed with the injection water. The alkaline agents raise the pH of the flood and react with organic acids that are present in oil. This reaction generates surfactants at the oil-water interface and leads to improved oil recovery due to (1) greatly reduced interfacial tension, (2) emulsification effects, and (3) wettability reversals, which can mobilize entrapped oil ganglia. Alkaline waterflooding may be used in conjunction with other EOR methods; for example, it is reported to be an effective preflush for surfactant-polymer floods.	NAPLs must have acidic components to react with the alkali agents to form surfactants. Alkaline flooding is relatively cheap compared to some other EOR methods, but alkali consumption due to reaction with porous media may be a limiting factor. Refer to the soil flushing comments.	Numerous alkaline waterflood EOR projects have been conducted by the oil industry (approximately 40 were reported begun in the U.S. between 1979 and 1981). Use of alkaline flooding to boost NAPL recovery at contamination sites is in its infancy. Alkaline agents were used in a soil flushing solution of alkaline, polymer, and surfactant agents (A-P-S) to enhance DNAPL recovery at the Laramie Tie Plant.	Breit et al. (1981), Campbell (1981), Castor et al. (1981), Donaldson et al. (1989), Janssen-VanRoesmalen and Heselink (1981), Kumar et al. (1989), Mayer et al. (1983), Nelson et al. (1984), Pitts et al. (1989), Sale et al. (1989), Surkalo (1990)

Table 6-1. Remedial options potentially applicable to DNAPL Contamination sites (modified from USEPA, 1992).

METHOD	APPLICATION	PROCESS	EFFECTIVENESS / ADVANTAGES / LIMITATIONS	EXPERIENCE	REFERENCES
EOR Using Polymer Water Flooding	Polymer flooding can be used to increase NAPL recovery during a flood operation.	Polymers are large molecules (molecular weight greater than 200 with at least 8 repeating units) that can be dispersed in a waterflood to increase the viscosity of the flood, thereby reducing the mobility ratio and improving the volumetric sweep efficiency (NAPL recovery). The mobility ratio is defined as the mobility of the displacing fluid (relative permeability/viscosity for waterflood) divided by the mobility of the displaced fluid (relative permeability/viscosity for NAPL). Lower mobility ratios favor NAPL displacement and recovery. An effective polymer will impart a high viscosity at low concentration. Only two types of polymers are commonly used by the oil industry (polyacrylamides and polysaccharides). In EOR operations, polymer flooding is often used as part of a phased injection sequence consisting of: (1) a preflush to adjust the pH and salinity of the reservoir; (2) surfactants and/or alkaline agents to reduce interfacial tension; (3) polymer solution to increase viscosity and improve displacement efficiency, and (4) waterflood to displace the mobilized oil and EOR solutions.	The advantage of polymer flooding is that it improves the volumetric sweep efficiency of a water flood process. By itself, polymer flooding will not, however, mobilize trapped residual NAPL. Potential limitations include: the risk of reduced injectivity caused by wellbore plugging, increased project durations and slowed recovery due to the lower absolute flood mobility, polymer degradation, excessive cost. Refer to the soil flushing comments.	Polymer flooding had been initiated at about 180 field-scale EOR projects in the U.S. by the mid 1980s. There is debate in the oil industry, however, regarding whether or not polymer flooding by itself provides more than a small incremental recovery. Use of polymers to boost NAPL recovery at contamination sites is in the early development stage. Polymers were used in a soil flushing solution of alkaline, polymer, and surfactant agents (A-P-S) to enhance DNAPL recovery at the Laramie Tie Plant.	Caenn et al. (1989), Chauvcteau and Zaitoun (1981), Donaldson et al. (1989), Heaselink and Faber (1981), Labaste and Vio (1981), Lin et al. (1987), Littman (1988), Pitts et al. (1989), Sale et al. (1989), Shah (1981), Surkalo et al. (1986), Yen et al. (1989)
EOR Using Chemically-Enhanced Dissolution	Cosolvents can be used to increase the dissolution and recovery of NAPL and adsorbed chemicals from the subsurface.	Cosolvents injected into a contamination zone via wells or drains increase the dissolution of NAPLs and adsorbed chemicals. Continued flooding of the contamination zone with cosolvents or another flood (water, polymers, etc.) drives the elutriate to production wells or drains. The elutriate may be treated and recycled through the system.	Refer to the soil flushing comments.	Miscible flooding with carbon dioxide and/or hydrocarbon solvents has been tested at numerous sites for EOR by the oil industry. Several bench- and pilot-scale studies have been conducted on in-situ cosolvent flushing technology at contamination sites with variable success.	Blackwell (1981), Fountain (1991), Groves (1988), Mehdizadeh et al. (1989), Nash (1987), Nash and Traver (1986), Rao et al. (1991), Sayegh and McCaffery (1981), Taber (1981), USEPA (1990a and 1990b)

Table 6-1. Remedial options potentially applicable to DNAPL Contamination sites (modified from USEPA, 1992).

METHOD	APPLICATION	PROCESS	EFFECTIVENESS / ADVANTAGES / LIMITATIONS	EXPERIENCE	REFERENCES
Pumping Chemicals Dissolved in Groundwater (Pump-and-treat)	Dissolved chemicals can be removed from the saturated zone by pumping groundwater.	Contaminated groundwater is pumped from wells or drains using conventional technology. Recovery rates can be optimized by fine-tuning pumping rates, well locations, etc.	Pump-and-treat is typically utilized as part of a hydraulic containment or aquifer restoration program. DNAPL, where present, will be a long-term source of groundwater contamination that will prevent restoration. Due to the relatively low solubility of most DNAPLs, pumping is generally not an effective method for removing DNAPL from the subsurface.	Recovery of dissolved chemicals by pumping is a widely-used, proven technology, but not for removal of NAPL.	Anderson et al. (1987, 1992a and 1992b), Feenstra and Cherry (1988), Hunt et al. (1988a), Johnson (1991), Johnson and Pankow (1992), Keely (1989), Mackay and Cherry (1989), Mercer et al. (1990), Mercer and Cohen (1990), Schwille (1988)
In-Situ Aeration in the Saturated Zone (Air Sparging and UVB Wells)	Air stripping can be applied below the water table to remove volatile contaminants from the saturated zone.	Volatile contaminants below the water table can be stripped by injecting air through wells. Vaporized volatiles move with the air to the unsaturated zone and are recovered using a vacuum extraction system. Another in-situ groundwater stripping process is known as the Underpressure-Vaporizer-Well (UVB) method in which contamianted groundwater is stripped by air at negative pressures in a special filtered well. The contaminated gas is collected and treated at the well head.	In-situ air injection will be most effective removing low molecular weight, volatile compounds. Little documentation is available regarding the effectiveness, advantages or limitations of air sparging and UVB wells.	The application of in-situ air stripping of groundwater is very limited. The UVB Well method is currently in use at several sites in Germany.	Blake et al. (1990), Herrling et al. (1990), Herrling and Buermann (1990)

Table 6-1. Remedial options potentially applicable to DNAPL Contamination sites (modified from USEPA, 1992).

METHOD	APPLICATION	PROCESS	EFFECTIVENESS / ADVANTAGES / LIMITATIONS	EXPERIENCE	REFERENCES
Vacuum Extraction (VE)	VE can be used to remove volatile chemicals from the unsaturated zone and to prevent uncontrolled migration of volatile chemicals in soil gas. If the water table is lowered, VE can be used to remove residual NAPL from below the original water table elevation.	Using unsaturated zone wells equipped with blowers or vacuum pumps, air is forced through soils contaminated with volatile chemicals. The air flow generates advective vapor fluxes that change the vapor-liquid equilibrium, inducing volatilization of contaminants. The resulting vapors are collected and treated. Positive differential pressure systems induce vapor flow away from the control points and negative differential pressure systems induce vapor flow toward control points. Experience had demonstrated that generation of negative differential gas pressures typically provides the most favorable field results.	Vacuum extraction is most effective removing low molecular weight, volatile chemicals (dimensionless Henry's Law Constant > 0.01) from homogeneous, permeable media. Intermittent vacuum extraction operation is generally more efficient than constant operation. Vacuum extraction systems can be installed using off-the-shelf components and conventional drilling methods. Vacuum extraction is less effective at removing volatile chemicals from heterogeneous and low permeability soils and is ineffective removing volatile chemicals from the saturated zone. Because it induces water table upwelling, vacuum extraction can result in groundwater contamination where chemicals are located just above the water table. Groundwater recovery wells may be necessary. Alternatively, lowering the water table to allow volatile chemical recovery by vacuum extraction may promote DNAPL remobilization and sinking.	In-situ vacuum extraction processes have been employed at more than 100 contamination sites in the United States.	Agrelot et al. (1985), Ardito and Billings (1990), Baehr et al. (1989), Blake et al. (1990), Blake and Gates (1986), Crow et al. (1985 and 1987), DiGiulio and Cho (1990), Dunlap (1984), Giertie et al. (1990), Hutzler et al. (1989), Johnson et al. (1988, 1990a and 1990b), Jury et al. (1990), Mackay et al. (1990), Marley and Hoag (1984), Massmann (1989), McClellen and Gillham (1990), O'Connor et al. (1984), Pedersen and Curtis (1991), Rathfelder (1989), Rathfelder et al. (1991), Regalbuto et al. (1988), Sims (1990), Stephanatos (1988), Texas Res.Inst. (1984), Thornton and Wootan (1982), Wilson et al. (1987), USEPA (1989c, 1990a)

Table 6-1. Remedial options potentially applicable to DNAPL Contamination sites (modified from USEPA, 1992).

METHOD	APPLICATION	PROCESS	EFFECTIVENESS / ADVANTAGES / LIMITATIONS	EXPERIENCE	REFERENCES
Steam and Hot Air Injection to Enhance Vacuum Extraction	Vacuum extraction in the unsaturated zone can be enhanced by steam and hot air injection.	Steam or hot air injected into or below the zone of soil contamination can improve the effectiveness of vacuum extraction systems. Heating and increased soil gas movement caused by steam and hot air injection raise the vaporization rate of volatile and some semi-volatile compounds. Additionally, contaminated soil water and low viscosity NAPLs can be physically displaced by the condensate that forms in front of the steam zone.	In general, steam or hot air injection will increase the effectiveness of a vacuum extraction system. Heating will reduce the viscosity and interfacial surface tension of residual or trapped DNAPL in the unsaturated zone, which may result in uncontrolled migration. Similarly, accumulation of DNAPL at the steam condensate front may result in uncontrolled downward or lateral migration of the DNAPL.	Successful laboratory and pilot-scale field studies have been conducted using steam and hot air stripping to enhance vacuum extraction recovery of solvents and petroleum contaminants from soil. Two different systems have been used: (1) a mobile unit consisting of a hollow stem auger rig outfitted for steam/air injection and vacuum extraction of vapors; and (2) a fixed system of injection and extraction wells.	Hunt et al. (1988b), Houthoofd et al. (1991), Johnson and Guffey (1990), Lord et al. (1987, 1988, 1989, and 1991), Udell and Stewart (1989 and 1990), USEPA (1990a and 1990b)
Radio Frequency Heating (RFH) to Enhance Vacuum Extraction	Vacuum extraction of chemicals that volatilize in the temperature range of 80° to 300° C. can be enhanced by using radio frequency heating of contaminated soil.	In situ radio frequency heating (RFH) involves heating soil with electromagnetic energy in the radio frequency band (typically 6.7 MHz to 2.5 GHz). Using a modified radio transmitter as a power source, energy is transmitted to the zone targeted for decontamination via electrodes placed in an array of boreholes. This energy heats the soil to temperatures between 150° and 300° C. Volatilized chemicals are recovered with soil gas for treatment by applying a vacuum to selected hollow electrodes. A rubber sheet barrier may be spread over the soil surface to provide thermal insulation and prevent fugitive emissions.	In general, RFH will increase the effectiveness of a vacuum extraction system. The technology is only applied to the unsaturated zone and its use is precluded where buried metal objects are present. As with vacuum extraction, this method is best suited to sites where volatile chemicals are present at shallow depth in homogeneous, coarse-grained soils. The uniformity of heating provided by RFH (which occurs due to dielectric heating mechanisms rather than the thermal conductivity of the soil), however, may result in more uniform decontamination than achieved using steam or hot air injection methods and may make this method more applicable to heterogeneous soils. Heating will reduce the viscosity and possibly the interfacial surface tension of residual or trapped DNAPL in the unsaturated zone, which may result in uncontrolled migration.	Several bench- and pilot-scale tests and limited field-scale testing have been conducted using RFH to remove 70% to 99% of various solvents, jet fuel, and PCBs from shallow soils. Although RFH continues to be in the pilot- and field-scale demonstration stage, at least one company has announced the availability of RFH on a commercial basis.	Dev (1986), Dev et al. (1988), Dev and Downey (1989), Houthoofd et al. (1991), Sims (1990), Sresty et al.(1986), USEPA (1990a,b)

Table 6-1. Remedial options potentially applicable to DNAPL Contamination sites (modified from USEPA, 1992).

METHOD	APPLICATION	PROCESS	EFFECTIVENESS / ADVANTAGES / LIMITATIONS	EXPERIENCE	REFERENCES
Bio-remediation	Bioremediation involves enhancement of natural processes to degrade hazardous chemicals in the subsurface.	Naturally occurring microbes can be used to degrade and/or detoxify hazardous chemicals in the subsurface. Bioremediation approaches include: (1) stimulation of biochemical mechanisms for degrading chemicals; (2) enhancement by delivery of exogenous acclimated or specialized microorganisms; (3) delivery of cell-free enzymes; and (4) vegetative uptake. Typically, oxygen and nutrients are delivered to the contamination zone via wells and/or drains to increase the rate of aerobic biodegradation.	Adequate characterization of the hydraulic conductivity distribution is necessary to achieve efficient delivery of oxygen, nutrients, and/or microbes within the contaminated zone. Many NAPLs are toxic to microbes and/or resistant to biodegradation. As a result, degradation may be limited to the periphery of NAPL contamination zones. Degradation may generate other undesirable chemicals. Bench and pilot studies are recommended. Bioremediation is typically used as a "polishing" step following the application of other chemical recovery and treatment processes.	Efforts to stimulate biodegradation have been employed at full field-scale at numerous contamination sites with varying degrees of success and documentation.	Downey and Elliot (1990), Hinchee (1989), Hinchee et al. (1990), Leuschner and Johnson (1990), Lokte (1984), Novak et al. (1984), Jhaveri and Mazzacca (1983), Lee et al. (1988), Lee and Ward (1984), Kuhn et al. (1985), Miller et al. (1990), Piontek et al. (1989), Raymond (1974), Sims (1988), Sims (1990), Suflita and Miller (1985), USEPA (1990a and 1990b), Vogel et al. (1987), Wilson et al. (1986), Wilson and Rees (1985), Yaniga and Mulry (1984)
Containment by Solidification, Stabilization, and/or In-Situ Vitrification	Solidification, stabilization, and in-situ vitrification are used to immobilize subsurface contaminants.	Waste solidification involves mixing cementing agents with soil to mechanically bind subsurface contaminants and thereby reduce their rate of release. Cementing agents (such as pozzolan-portland cement, lime-flyash pozzolan, and asphalt systems) can be combined with contaminated soils by injection, in-situ mechanical mixing, or above ground mechanical mixing. During waste stabilization, reagents are used to convert contaminants to their least toxic, soluble, or mobile form. In-situ vitrification utilizes an electrical network (125 or 138kV) to melt contaminated soils and sludges at temperatures of 1600 to 2000° C. Electricity is transmitted from a power source into contaminated soil via large electrodes. Organic contaminants are pyrolized and inorganic contaminants are incorporated within the vitrified mass. Vapors can be captured at the surface for treatment.	Obtaining complete and uniform mixing of the solidifying and/or stabilizing agents with the contaminated soil is a critical factor determining the success of solidification/stabilization systems. Successful application of these methods becomes more difficult with increasing depth. Costs are high, but may be competitive with other remedies. Bench studies and pilot field tests are generally necessary. Mixed, complex wastes present special challenges. Volatilization, mobilization, and migration of contaminants may be caused by these processes. The long-term stability and leaching characteristics of contaminated materials that have been solidified, stabilized or vitrified is unknown.	Several different solidification-stabilization processes have undergone pilot tests and full-scale field demonstrations. Six full-scale demonstrations of in-situ vitrification have been conducted at the DOE Hanford site, and more than 90 in-situ vitrification tests of various scales have been conducted on PCB wastes, and other solid combustibles and liquid chemicals.	Cullinane et al. (1986), Fitzpatrick et al. (1986), Sims (1990), USEPA (1990a and 1990b)

Table 6-1. Remedial options potentially applicable to DNAPL Contamination sites (modified from USEPA, 1992).

METHOD	APPLICATION	PROCESS	EFFECTIVENESS / ADVANTAGES / LIMITATIONS	EXPERIENCE	REFERENCES
Excavation	Excavation is used to remove contaminated materials from the subsurface.	Conventional excavating methods are used to remove contaminated materials from the subsurface for subsequent incineration, treatment and/or disposal.	Excavation can be a very effective site remedy where contaminant penetration is limited to shallow soils and where shallow contamination hot spots are identified. The cost and difficulty of excavation increases with the depth of contaminant migration and generally becomes prohibitive in bedrock. Additional concerns include potential fugitive dust, liquid, and gas emissions caused by excavating contaminated materials, and the possibility that DNAPL may have migrated beneath the excavation limit, thereby reducing the effectiveness of excavation as a remedy.	Excavation of contaminated materials is a widely-used remedy of proven value in appropriate situations.	

7 DNAPL SITE IDENTIFICATION AND INVESTIGATION IMPLICATIONS

Determining DNAPL presence should be a high priority at the onset of site investigation to guide the selection of site characterization methods. Knowledge or suspicion of DNAPL presence requires that special precautions be taken during field work to minimize the potential for inducing unwanted DNAPL migration. DNAPL presence may be inferred from information on DNAPL usage, release, or disposal at a site, and/or by examination and analysis of subsurface samples. However, due to limited and complex distributions of DNAPL at some sites, its occurrence may be difficult to detect, leading to inadequate site assessments and remedial designs.

Guides for evaluating the potential occurrence of DNAPL at contamination sites have recently been prepared by Newell and Ross (1992) and Cherry and Feenstra (1991). This chapter is derived from these and several related documents. A decision chart to evaluate the presence the DNAPL at a site is provided as Figure 7-1.

7.1 HISTORICAL SITE USE

As discussed in Chapter 3, DNAPL contamination is associated with industries and processes that utilize or generate DNAPL chemicals (such as halogenated solvents, creosote/coal tar, and PCB mixtures) and with waste disposal sites used by these businesses. Industries and industrial processes with a high probability of historical DNAPL release, and DNAPL chemicals which may contribute to contamination problems, are listed in Table 7-1.

Assessment of the potential for DNAPL contamination based on historical site use information involves careful examination of: landuse since site development; business operations and processes; types and volumes of chemicals used and generated; and the storage, handling, transport, distribution, generation, dispersal, and disposal of these chemicals and operation residues. Methods for conducting research on historic developments at suspected contamination sites are provided in environmental audit guidance documents (i.e., BNA, 1992; Wilson, 1990; Marburg Associates and Parkin, 1991). Site use information is available from numerous sources (Table 7-2). Suspect site areas often associated with contamination are listed in Table 7-3.

The potential for DNAPL contamination at a site based on historical information can be estimated using the decision chart (Figure 7-1) in conjunction with Table 7-1. Although this potential increases with the size and operating period of a facility, industrial process, or waste disposal practice (Newell and Ross, 1992), relatively small and short-term releases from pipeline leaks, overfilled tanks, or other sources can also create significant DNAPL contamination problems.

7.2 SITE CHARACTERIZATION DATA

Preexisting site characterization data is available at many sites being investigated. The potential for DNAPL contamination should be evaluated at the start and during the course of new field studies. As outlined in Figure 7-1, DNAPL presence can be: (1) determined directly by visual examination of subsurface samples; (2) inferred by interpretation of chemical analyses of subsurface samples; and/or (3) suspected based on interpretation of anomalous chemical distribution and hydrogeologic data.

7.2.1 Visual Determination of DNAPL Presence

Ideally, DNAPL presence can be identified by visual examination of soil, rock, and fluid samples. Direct visual detection may be difficult, however, where the DNAPL is colorless, present in low concentration, or distributed heterogeneously. Methods to visually detect DNAPL in subsurface samples are identified in Table 7-4 and discussed in Chapter 9.10.

7.2.2 Inferring DNAPL Presence Based on Chemical Analyses

Indirect methods for assessing the presence of DNAPL in the subsurface rely on comparing measured chemical concentrations to effective solubility limits for groundwater and to calculated equilibrium partitioning concentrations for soil and groundwater (Feenstra, 1990; Feenstra et al., 1991; Sitar et al., 1990; Mackay et al., 1991). Chemical concentrations and distributions indicative and/or suggestive of DNAPL presence are described in Table 7-4.

Where present as a separate phase, DNAPL compounds are generally detected at <10% of their aqueous solubility limit in groundwater. This is due to the effects of non-uniform groundwater flow, variable DNAPL distribution, the mixing of groundwater in a well, and the reduced effective solubility of individual compounds in a multi-liquid NAPL mixture (see Chapter 4, Chapter 9.10.2, and Worksheet 7-1). Typically, dissolved

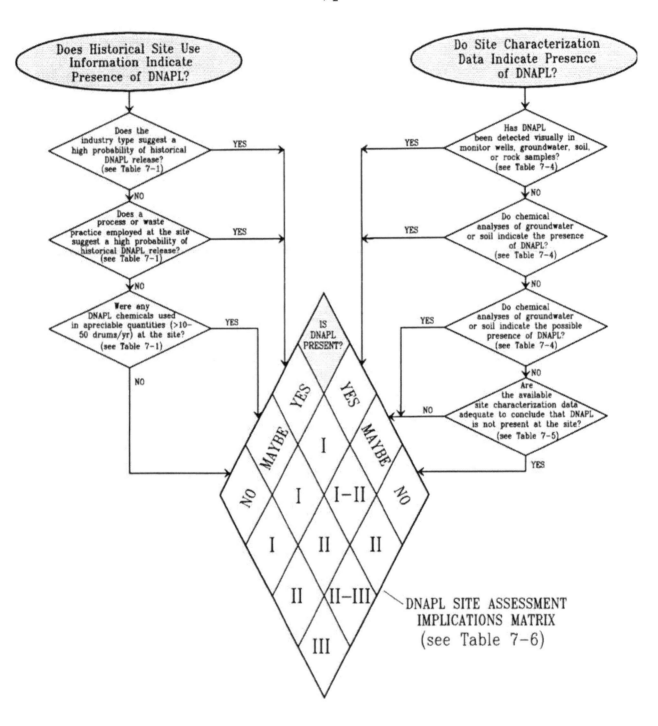

Figure 7-1. DNAPL occurrence decision chart and DNAPL site assessment implications Matrix (modified from Newell and Ross, 1992).

Table 7-1.   Industries and industrial processes using DNAPLs and some DNAPL chemicals (modified from Newell and Ross, 1992).

INDUSTRIES/BUSINESSES USING DNAPLS	PROCESSES INVOLVING DNAPLS	DNAPL CHEMICALS
Chemical manufacturing Solvent manufacturing, reprocessing, and/or packaging Commercial dry cleaning operations Electronic equipment manufacturing Computer component manufacturing Metal parts/products manufacturing Aircraft and automotive manufacturing, maintenance, and repair operations Machine shops and metal works Tool-and-die plants Musical instrument manufacturing Photographic film manufacturing and processing Plastics manufacturing Pharmaceutical manufacturing Illicit drug manufacturing Flame retardant materials manufacturing Refrigerants manufacturing Military equipment manufacturing and maintenance Manufacturing/processing of septic system/plumbing cleaners Manufacturing of typewriters, printers, and copiers Printing presses and publishing operations Textile processing, dying, and finishing operations Pesticide and herbicide manufacturing Wood preservation/treating plants Manufactured gas plants (prevalent between the mid-1800s to mid-1900s) Steel industry coking operations Asphalt processing and distribution plants Coal tar distillation plants Transformer/capacitor oil production, reprocessing, and disposal operations Waste disposal sites	Metal cleaning/degreasing Storage of solvents (and other DNAPLs) in drums in uncontained areas Tool-and-die operations Paint removing/stripping Metal machining Loading and unloading of solvents (and other DNAPLs) Solvent storage in underground and aboveground tanks Textile cleaning operations Dry plasma etching of semiconductor chips	Aniline Carbon tetrachloride Chlordane Chlorobenzene Chloroform 2-Chlorophenol Chlorotoluene Dibutyl phthalate 1,2-Dichlorobenzene 1,3-Dichlorobenzene 1,1-Dichloroethane 1,2-Dichloroethane 1,1-Dichloroethene trans-1,2-Dichloroethene cis-1,2-Dichloroethene 1,2-Dichloropropane cis-1,3-Dichloropropene trans-1,3-Dichloropropene Diethyl phthalate Dimethyl phthalate Ethylene dibromide Hexachlorobutadiene Hexachlorocyclopentadiene Malathion Methylene Chloride Nitrobenzene Parathion Polychlorinated Biphenyls 1,1,2,2-Tetrachloroethane Tetrachloroethene 1,2,4-Trichlorobenzene 1,1,1-Trichloroethane Trichloroethene 1,1,2-Trichlorofluoromethane 1,1,2-Trichlorotrifluoroethane Coal tar Creosote

Table 7-2. Sources and types of site history information.

Sources	Documents	Types of Available Information
Corporate Owner/ Operator Records	Account books (purchase journals, sales journals, cash journals, ledgers); warehouse receipts; inventory records; corporate archives; internal division annual reports; deeds; chemical process documents; waste generation and disposal documents; detailed site maps; chemical storage, distribution, transport, and utilization, documents; chemical spill and release incident reports; historic photographs; and environmental consulting reports.	Details regarding the storage, transport, use, generation, accidental release, and disposal of chemicals and wastes.
Municipal and/or County Offices	Tax Assessor records Fire Department records Public Works Department records Building Department records Utility Department records Sewer Department records Sanitation Department records Public Health Department records	Information on current and past owners; total area of parcel; site history; current and historical use of adjacent land; utilities, sewage and water supply maps and systems; historical maps; underground storage tank and waste discharge permits; chemical and hazardous waste inventories; maps of drainage features; building records; plot plans; waste disposal location maps; information on use, manufacture, storage, and discharge of hazardous materials and wastes; investigation and incident reports.
State Government	Waste Management Board Regional Water Quality Board Health Department	Information on active and inactive hazardous materials treatment, storage, and disposal sites; hazardous waste permits and registrations; notices of violation; regulatory and enforcement action documents; and reports on leaking storage tanks and contamination incidents.
Federal Government	USEPA records	Contamination site listings including the Comprehensive Environmental Response Compensation and Liability Information System (CERCLIS) list; the National Priority List of Superfund sites; and the RCRA TSD list; site investigation and incident reports; UST, CERCLA, and RCRA records.
	US Geological Survey records	Topographic and geologic maps; geologic/hydrogeologic reports; environmental contamination reports; aerial photographs.
	US Department of Commerce	Manufacturers census data with information on products shipped and materials consumed.
Universities, libraries, historical societies	Theses, archives, historical information	Information on site development, landuse, waste disposal, manufacturing activities, geology, hydrogeology, etc.
Key Personnel	Interviews	Information regarding all aspects of site history.
Miscella- neous	Aerial photographs: have been taken by private companies and public agencies every few years to several per year throughout much of the U.S. dating back to circa 1940. Listings of available public and private aerial photographs for given locations and periods are available from the National Cartographic Information Center in Reston, Virginia.	Information regarding site development, landuse, waste disposal practices, manufacturing activities, loading dock locations, pipeline and tank locations, settling or retention ponds, ponded fluid, stained soils, distressed vegetation, disturbed soils, etc.
	Maps: topographic, geologic, and hydrogeologic maps are available from the U.S. Geological Survey, State Geological Surveys, and University geology departments. Fire insurance maps dating back to the 1800s which depict manufacturing facilities and potential fire hazards such as tank locations are available for all regions of the U.S. from the Sanborn Map Company in Pelham, NY. Historic maps are available from local universities, historical societies, and libraries.	Information regarding site development, landuse, waste disposal locations, manufacturing activities, storage tank locations, etc.

Table 7-3. Industrial site areas frequently associated with contamination.

COMMON SUSPECT AREAS AT POSSIBLE DNAPL SITES
Floor drains
Sumps
Catch basins
Pits, ponds, lagoons, and other disposal areas
Septic tanks
Leach fields
French drains
Sewer systems
Process tanks
Wastewater tanks
Underground tank areas
Aboveground tank areas
Chemical storage areas
Chemical transfer areas
Pipelines
Waste storage areas
Loading dock areas
Work areas
Discolored soils
Discolored water
Stressed vegetation
Disturbed earth
Low-lying disturbed areas

Table 7-4. Determinant, inferential, and suggestive indications of DNAPL presence based on examination of subsurface samples and data (based on Newell and Ross, 1992; Cherry and Feenstra, 1991; and Cohen et al., 1992).

DETERMINING DNAPL PRESENCE BY VISUAL EXAMINATION OF SUBSURFACE SAMPLES	INFERRING DNAPL PRESENCE BY INTERPRETING CHEMICAL ANALYSES	SUSPECTING DNAPL PRESENCE BASED ON ANOMALOUS FIELD CONDITIONS
Methods to detect DNAPL in wells: • NAPL/water interface probe detection of immiscible phase at base of fluid column • Pumping from bottom of fluid column and inspecting retrieved sample • Retrieving a transparent, bottom-loading bailer from the bottom of a well and inspecting the fluid sample • Inspecting fluid retrieved from the bottom of a well using a mechanical discrete-depth sampler • Inspecting fluid retained on a weighted cotton string that was lowered down a well **Methods to enhance inspection of fluid samples for DNAPL presence:** • Centrifuge sample and look for phase separation • Add hydrophobic dye (such as Sudan IV or Red Oil) to sample, shake, and look for coloration of DNAPL fraction • Examine UV fluorescence of sample (many DNAPLs will fluoresce) • Assess density of NAPL relative to water (sinkers or floaters) by shaking solution or by using a syringe needle to inject NAPL globules into the water column **Methods to detect DNAPL in soil and rock samples** • Examine UV fluorescence of sample (many DNAPLs will fluoresce) • Add hydrophobic dye and water to soil sample in polybag or jar, shake, and examine for coloration of the NAPL fraction • Conduct a soil-water shake test without hydrophobic dye (can be effective for NAPLs that are neither colorless nor the color of the soil) • Centrifuge sample with water and look for phase separation • Perform a paint filter test, in which soil is placed in a filter funnel, water is added, and the filter is examined for separate phases	**Chemical analysis results from which DNAPL presence can be inferred (with more or less certainty depending on the strength of the overall data):** • Concentrations of DNAPL chemicals in groundwater are greater than 1% of the pure phase solubility or effective solubility (refer to Worksheet 7-1) • Concentrations of DNAPL chemicals on soils are greater than 10,000 mg/kg (equal to 1% soil mass) • Concentrations of DNAPL chemicals in groundwater calculated from water/soil partitioning relationships and soil samples are greater than pure phase solubility or effective solubility (refer to Worksheet 7-2) • Organic vapor concentrations detected in soil gas exceeds 100-1000 ppm	**Field conditions that suggest DNAPL presence:** • Concentrations of DNAPL chemicals increase with depth in a pattern that cannot be explained by advective transport • Concentrations of DNAPL chemicals increase up the hydraulic gradient from the contaminant release area (apparently due to contaminated soil gas migration and/or, DNAPL movement along capillary and/or permeability interfaces that slope counter to the hydraulic gradient) • Erratic patterns of dissolved concentrations of DNAPL chemicals in groundwater which are typical of DNAPL sites due to heterogeneity of (1) the DNAPL distribution, (2) the porous media, (3) well construction details, and (4) sampling protocols • Erratic, localized, very high contaminant concentrations in soil gas, particularly located just above the water table (where dense gas derived from DNAPL in the vadose zone will tend to accumulate) • Dissolved DNAPL chemical concentrations in recovered groundwater that decrease with time during a pump-and-treat operation, but then increase significantly after the pumps are turned off (although complexities of contaminant desorption, formation heterogeneity, and temporal and spatial variations of the contaminant source strength can produce similar results) • The presence of dissolved DNAPL chemicals in groundwater that is older than potential contaminant releases (using tritium analysis for age dating) suggests DNAPL migration (Uhlman, 1992) • Deterioration of wells and pumps (can be caused by DNAPL; i.e., chlorinated solvents degrade PVC)

Worksheet 7-1: Calculation of Effective Solubility (from Newell and Ross, 1992; after Shiu et al., 1988; and Feenstra et al., 1991)

For a single-component DNAPL, the pure-phase solubility of the organic constituent can be used to estimate the theoretical upper-level concentration of organics in aquifers or for performing dissolution calculations. For DNAPLs comprised of a mixture of chemicals, however, the effective solubility concept should be employed:

$$S^e_i = X_i S_i$$

where

S^e_i = the effective solubility (the theoretical upper-level dissolved-phase concentration of a constituent in groundwater in equilibrium with a mixed DNAPL; in mg/l)

X_i = the mole fraction of component i in the DNAPL mixture (obtained from a lab analysis of a DNAPL sample or estimated from waste characterization data)

S_i = the pure-phase solubility of compound i in mg/l (usually obtained from literature sources)

For example, if a laboratory analysis indicates that the mole fraction of trichloroethylene (TCE) in DNAPL is 0.10, then the effective solubility would be 110 mg/l. This is derived by multiplying the pure phase solubility of TCE by the TCE mole fraction:

$$1100 \text{ mg/l} * 0.10 = 110 \text{ mg/l.}$$

Effective solubilities can be calculated for all components in a DNAPL mixture. Nearly insoluble organics in the mixture (such as long-chained alkanes) will reduce the mole fraction and effective solubility of more soluble organics, but will contribute little dissolved-phase organics to groundwater.

Please note that this relationship is approximate and does not account for non-ideal behavior of mixtures, such as co-solvency, etc.

contaminant concentrations >1% of the aqueous solubility limit are highly suggestive of NAPL presence. Concentrations <1%, however, do not preclude the presence of NAPL.

In soil, contaminant concentrations in the percent range are generally indicative of NAPL presence. However, NAPL may also be present at much lower soil concentrations. Feenstra et al. (1991) detail an equilibrium partitioning method for assessing the presence of NAPL in soil samples based on determining total chemical concentrations, soil moisture content, porosity, organic carbon content, approximate composition of the possible NAPL, sorption parameters, and solubilities. This method is outlined in Chapter 9.10.2 and Worksheet 7-2.

7.2.3 Suspecting DNAPL Presence Based on Anomalous Conditions

Subsurface DNAPL can lead to anomalous chemical distributions at contamination sites. For example, dissolved chemical concentrations in horizontally-flowing groundwater may increase with depth beneath a waste site due to the density-driven downward movement of DNAPL. Anomalous field conditions suggestive of DNAPL presence are noted in Table 7-4, and characteristics of extensive field programs that can help indicate the absence of DNAPL are listed in Table 7-5.

7.3 IMPLICATIONS FOR SITE ASSESSMENT

If DNAPL presence is determined or suspected, special consideration must be given to (1) devising an effective site investigation strategy, and (2) preventing inducement of unwanted chemical migration during field activities. Implications of DNAPL presence on site assessment activities are highlighted in Table 7-6.

Worksheet 7-2: Method for Assessing Residual NAPL Based on Organic Chemical Concentrations in Soil Samples (from Newell and Ross, 1992; after Feenstra et al., 1991)

To estimate if NAPLs are present, a partitioning calculation based on chemical and physical analyses of soil samples from the saturated zone (from cores, excavation, etc.) can be applied. This method tests the assumption that all the organics in the subsurface are either dissolved in groundwater or adsorbed to soil (assuming dissolved-phase sorption, not the presence of NAPL). By using the concentration of organics on the soil and the partitioning calculation, a theoretical pore-water concentration of organics in groundwater is determined. If the theoretical pore-water concentration is greater than the estimated solubility of the organic constituent of interest, then NAPL may be present at the site. A worksheet for performing this calculation is presented below; see Feenstra et al. (1991) for the complete methodology.

Step 1: Calculate S^e_i, the effective solubility of organic constituent of interest. See Worksheet 7-1.

Step 2: Determine K_{oc}, the organic carbon-water partition coefficient from one of the following:

- Appendix A and associated references or

- Empirical relationships based on K_{ow}, the octanol-water partition coefficient, which also is found in Appendix A. For example, K_{oc} can be estimated from K_{ow} using the following expression developed for polyaromatic hydrocarbons:

$$\text{Log } K_{oc} = 1.0 * \text{Log } K_{ow} - 0.21$$

Step 3: Determine f_{oc}, the fraction of organic carbon on the soil, from a laboratory analysis of clean soils from the site. Values for f_{oc} typically range from 0.03 to 0.00017 mg/mg. Convert values reported in percent to mg/mg.

Step 4: Determine or estimate ρb, the dry bulk density of the soil, from a soils analysis. Typical values range from 1.8 to 2.1 g/ml(kg/l). Determine or estimate θw, the water-filled porosity.

Step 5: Determine Kd, the partition (or distribution) coefficient between the pore water (ground water) and the soil solids:

$$Kd = K_{oc} * f_{oc}$$

Step 6: Using Ct, the measured concentration of the organic compound in saturated soil in mg/kg, calculate the theoretical pore water concentration assuming no DNAPL (i.e., Cw in mg/l):

$$Cw = \frac{(Ct * \rho b)}{(Kd * \rho b + \theta w)}$$

Step 7: Compare Cw and S^e_i (from Step 1):

Cw > S^e_i suggests possible presence of DNAPL
Cw < S^e_i suggests possible absence of DNAPL

Table 7-5.   **Characteristics of extensive field programs that can help indicate the absence of DNAPL (modified from Newell and Ross, 1992).**

- Absence of conditions listed in Table 7-4 that are indicative or suggestive of DNAPL presence

- Numerous monitor wells, with screens in topographic lows on the surface of fine-grained, relatively impermeable units

- Multi-level fluid sampling capability

- Numerous organic chemical analyses of soil samples at different depths using GC or GC/MS methods

- Well-defined site stratigraphy, using numerous soil borings, a cone penetrometer survey, test pits, and/or geophysics

- Numerous subsurface explorations (test pits and/or borings) in the areas of contaminant release

- Data from pilot tests or "early action" projects that indicate the site responds as predicted by conventional solute transport relationships, rather than by responding as if additional sources of dissolved contaminants are present in the aquifer

Table 7-6. Implications of DNAPL presence on site assessment activities (see Figure 7-1) (modified from Newell and Ross, 1992).

CATEGORY	IMPLICATIONS FOR SITE ASSESSMENT
I CONFIRMED OR HIGH POTENTIAL FOR DNAPL AT SITE	• The risk of spreading contaminants increases with the proximity to a potential DNAPL zone. Special precautions should be taken to ensure that drilling does not create pathways for continued vertical migration of free-phase DNAPLs. In DNAPL zones, drilling should be suspended when a low-permeability unit or DNAPL is first encountered. Wells should be installed with short screens (≤ 10 feet). If required, deeper drilling through known DNAPL zones should be conducted only by using double or triple-cased wells to prevent downward migration of DNAPL. As some DNAPLs can penetrate fractures as narrow as 10 microns, special care must be taken during all grouting, cementing, and well sealing activities conducted in DNAPL zones. • In some hydrogeologic settings, such as fractured crystalline rock, it is impossible to drill through DNAPL with existing technology without causing vertical migration of DNAPL down the borehole, even when double or triple casing is employed. • The subsurface DNAPL distribution is difficult to delineate accurately at some sites. DNAPL migrates preferentially through selected pathways (fractures, sand layers, etc.) and is affected by small-scale stratigraphic changes. Therefore, the ultimate path taken by DNAPL can be very difficult to characterize and predict. • In most cases, fine-grained aquitards (such as clay or silt units) should be assumed to permit downward migration of DNAPL through fractures unless proven otherwise in the field. At some sites, it can be exceptionally difficult to prove otherwise even with intensive site investigation. • Drilling in areas known to be DNAPL-free should be performed before drilling in DNAPL zones in order to form a reliable conceptual model of site hydrogeology, stratigraphy, and potential DNAPL pathways. In areas where it is difficult to form a reliable conceptual model, an "outside-in" strategy may be appropriate: drilling in DNAPL zones is avoided or minimized in favor of delineating the outside dissolved-phase plume. Many fractured rock settings may require this approach to avoid opening further pathways for DNAPL migration during site assessment.
II MODERATE POTENTIAL FOR DNAPL AT SITE	• Due to the potential risk for exacerbating groundwater contamination problems during drilling through DNAPL zones, the precautions described for Category I should be considered during site assessment. Further work should focus on determining if the site is a DNAPL site.
III LOW POTENTIAL FOR DNAPL AT SITE	• DNAPL is not likely to be a problem during site characterization, and special DNAPL precautions are probably not needed. Floating, less-dense-than-water nonaqueous phase liquids (LNAPLs), sorption, and other factors can complicate site assessment and remediation activities, however.

8 NONINVASIVE CHARACTERIZATION METHODS

Noninvasive methods can often be used during the early phases of field work to optimize the cost-effectiveness of a DNAPL site characterization program. Specifically, surface geophysical surveys, soil gas analysis, and photointerpretation can facilitate characterization of contaminant source areas, geologic controls on contaminant movement (i.e., stratigraphy and utilities), and the extent of subsurface contamination. Conceptual model refinements derived using these methods reduce the risk of spreading contaminants during subsequent invasive field work.

Measurements made at or just below ground surface using geophysical and soil gas surveying cost much less than invasive methods of subsurface data acquisition such as drilling, monitor well installation, sampling, and chemical analyses. Although less costly, subsurface data acquired indirectly using noninvasive methods are also less definitive than data acquired directly. As a result, invasive techniques (i.e., borings and monitor wells) are usually needed to confirm interpretations derived using noninvasive methods.

The applicability of noninvasive methods must be evaluated on a site-specific basis. Advantages and limitations of surface geophysical methods, soil gas analysis, and aerial photograph interpretation are discussed in Chapters 8.1, 8.2, and 8.3, respectively.

8.1 SURFACE GEOPHYSICS

Several surface geophysical survey techniques have been used with varied success to enhance contamination site characterization since the late 1960s. Surface methods utilized most commonly with proven effectiveness for specific applications include: ground-penetrating radar, electromagnetic (EM) conductivity, electrical resistivity, seismic, magnetic, and metal detection. At contamination sites, geophysical surveys are usually conducted to: (1) assess stratigraphic and hydrogeologic conditions; (2) detect and map electrically conductive contaminants; (3) locate and delineate buried wastes and utilities; (4) optimize the location and spacings of test pits, borings, and wells; and (5) facilitate interpolation of subsurface conditions among boring locations.

8.1.1 Surface Geophysical Methods and Costs

The application of surface geophysical methods to groundwater and contamination site investigations is the subject of several reviews (Zohdy et al., 1974; Benson et al., 1982; Rehm et al., 1985; GRI, 1987; USEPA, 1987; Benson, 1988, 1991) and many published case studies (see references in Table 8-1). An expert system for selecting geophysical methods based on site-specific conditions and investigative objectives is provided by Olhoeft (1992). Detailed procedures for instrument operation are provided in product manuals. Capabilities, operating principles, and limitations of the most common surface geophysical survey methods are described in Table 8-1 and include the following.

- *Ground-penetrating radar* (GPR) is used to measure changes in dielectric properties of subsurface materials by transmitting high-frequency electromagnetic waves into the subsurface and continuously monitoring their reflection from interfaces between materials with different dielectric properties. GPR is primarily used to help delineate site stratigraphy, buried wastes, and utilities.

- *Electromagnetic conductivity* methods measure the bulk electrical conductance of the subsurface by recording changes in the magnitude of electromagnetic currents that are induced in the ground. The measured currents are proportional to the bulk electrical conductivity of the subsurface and can be interpreted to infer lateral stratigraphic variations, and the presence of conductive contaminants, buried wastes, and utilities.

- *Electrical resistivity* methods measure the bulk electrical resistance (the reciprocal of conductance) of the subsurface directly by transmission of current between electrodes implanted at ground surface. Electrical resistivity can be used to aid determination of site stratigraphy, water-table depth, conductive contaminant plumes, and buried wastes.

- *Seismic refraction and reflection* methods use geophones implanted in the ground surface and seismographs to measure and record the subsurface transmission of sound waves generated using a hammer blow or explosive device at a point source. Seismic waves are reflected and refracted as they pass through media with different elastic properties enabling interpretation of geologic layering and waste zone geometry based on analysis of wave arrival times.

Table 8-1. Summary of various surface geophysical survey methods (modified from Benson, 1991; Gretsky et al., 1990; O'Brien and Gere, 1988).

METHOD AND TARGETS	OPERATING PRINCIPLES/DESCRIPTION	OPERATING PARAMETERS, ADVANTAGES, DISADVANTAGES	REFERENCES
GROUND-PENETRATING RADAR • stratigraphy (layering and lateral variations) • water table in coarse media • metallic and nonmetallic buried drums, tanks, and pipes • bedrock surface • fracture zones • very limited capacity to delineate NAPL presence	High-frequency electromagnetic (<100 to 1000 MHz radio) waves transmitted from a radar antenna at the ground surface are reflected from interfaces between subsurface materials with contrasting dielectric properties back to a receiving antenna. The reflections are amplified, processed, and displayed in real-time on a recorder and/or color video monitor. Such contrasts are due to changing clay content, fluid content, dielectric constants, porosity, fracture density, bedding, cementation, the presence of manmade objects, the bedrock surface, etc. The depth of penetration (and measurement of reflected waves) is proportional to the receiving antenna response time.	• Provides a continuous visual profile of shallow subsurface objects, structure, and lithology in real time. • The graphic output can often be interpreted in the field; thereby facilitating direction of the survey. • Traverse rates range from 600 to 6000 ft/hr for detailed studies and up to 10 mph for low-resolution reconnaissance work. • Depth of penetration is site-specific; typically between 6 and 30 ft. Radar penetration increases in coarse, dry, sandy, or massive rock; and decreases with increasing clay content, fluid content, and fluid conductivity. • Approximate and relative depths are calculated using simple interpretative methods and assumptions. Depth calibrations, however, require careful onsite work and are often nonlinear. • Depending on the frequency used, GPR provides very high resolution from an inch to a few feet. The survey can be optimized to local conditions by changing antennas (frequency). High frequency provides the best resolution; lower frequency provides deeper penetration. • GPR can be used in fresh water and through ice to obtain profiles of sediment depth. • Access is limited due to bulkiness of GPR equipment. • Limited use during wet weather. • A variety of processing options can be used to enhance data interpretation and presentation. • Quantitative interpretation is difficult. The data can be affected by various sources of system noise.	Benson (1991); Benson and Glaccum (1979); Benson et al. (1982); Benson and Yuhr (1987); Koerner et al. (1981); Olhoeft (1984; 1986); Redman et al. (1991); Wright et al. (1984)
ELECTRO-MAGNETIC CONDUCTIVITY • detection and mapping of conductive contaminant plumes • stratigraphy (layering and lateral variations) • waste disposal areas containing metallic and nonmetallic buried drums • USTs and buried pipes • fracture zones • very limited capacity to delineate NAPL presence	Electromagnetic methods measure changes in the bulk subsurface electrical conductivity (also referred to as terrain conductivity). A transmitter induces circular eddy current loops in the subsurface. The magnitude of the induced current is proportional to and altered by the terrain conductivity in the vicinity of the loop. Each current loop generates a magnetic field which is proportional to the current magnitude. A portion of the generated magnetic field is measured by the EM receiver and results in an output voltage that is proportional to terrain conductivity. Terrain conductivity generally increases with increasing (1) subsurface fluid conductivity, (2) clay content, (3) fluid content, and, (4) porosity (including fracture porosity). Interpretation is typically qualitative based on consideration of spatial variability, anomalies, and the aforementioned factors.	• Profiling or vertical sounding data can be acquired from various depths between 2 and 200 ft by combining measurements from various common EM systems and by varying the coil orientation and/or spacing between the EM transmitter and receiver. Compared to the resistivity method, however, EM has reduced vertical sounding resolution due to the limited number of transmitter-receiver spacings available. • The influence of subsurface materials on the measured EM conductivity decreases with depth; a confounding factor that must be considered when interpreting EM data. • Continuous EM profiling can be obtained from 2.5 to 50 ft providing increased survey speed, density, and resolution. Data can be recorded on an analog strip chart or digital data logger. • Spatial variability and anomalies can be caused by several factors, thereby confounding unique interpretation of conductivity measurements. • Various objects emit noise and interfere with EM surveys including natural atmospheric noise, power lines, buried metal objects, radio transmitters, buried pipes and cables, fences, vehicles, and buildings. • Limited use in wet weather. • Frequency-domain EM systems measure changes in continuously transmitted currents. Time-domain EM systems measure changes in cyclically induced currents. Time-domain systems offer enhanced vertical sounding capability to depths of 150 to >1000 ft. • Frequency-domain EM systems measure in-phase and out-of-phase components of EM conductivity. The in-phase component responds to magnetic susceptibility and can be used to detect metals. The out-of-phase component measures electrical conductivity.	Benson et al. (1982); Davis (1991); Grady and Haeni (1984); Greenhouse and Slaine (1983); Greenhouse and Monier-Williams (1985); Griffith and King (1969); Ladwig (1983); McNeill (1980); Slaine and Greenhouse (1982); Rumbaugh et al. (1987); Stewart (1982); Telford et al. (1982); Zohdy et al. (1974)

Table 8-1. Summary of various surface geophysical survey methods (modified from Benson, 1991; Gretsky et al, 1990; O'Brien and Gere, 1988).

METHOD AND TARGETS	OPERATING PRINCIPLES/DESCRIPTION	OPERATING PARAMETERS, ADVANTAGES, DISADVANTAGES	REFERENCES
ELECTRICAL RESISTIVITY • detection and mapping of conductive contaminant plumes • stratigraphy (layering and lateral variations) • waste disposal areas containing metallic and nonmetallic buried drums • fracture zones • very limited capacity to delineate NAPL presence	Electrical resistivity methods measure the bulk electrical conductivity of the subsurface in a manner different from the EM Conductivity method. Electrical current is transmitted into the ground from a pair of surface electrodes and the voltage drop due to bulk subsurface resistivity is measured at the surface between a second pair of electrodes. The depth of measurement increases with, but is generally less than, the electrode spacing. Of the various electrode spacing geometries used, the Wenner array, with four electrodes equally spaced along a line and the two transmitting electrodes at each end, is the simplest. As the reciprocal of conductivity, electrical resistivity generally decreases with increasing (1) subsurface fluid conductivity, (2) clay content, (3) fluid content, and, (4) porosity (including fracture porosity).	• Survey speed is slower than with EM methods because of need with resistivity to drive electrodes into the ground. Station measurements are made; continuous measurements are not possible. • Profiling or vertical sounding data can be acquired from various electrode spacings by using different electrode spacings. Because the current electrodes can be spaced at any distance (assuming site access), resistivity is capable of providing better vertical resolution of subsurface conductivity than EM methods (which rely on fixed intercoil lengths). • The vertical sounding technique, however, requires that subsurface conditions be relatively consistent laterally. • Results are amenable to qualitative interpretation based on observed spatial variability and anomalies. However, quantitative interpretation of stratigraphic layering can be made based on vertical soundings. • Spatial variability and anomalies can be caused by several factors, thereby confounding unique interpretation of resistivity measurements. • The influence of subsurface materials on the measured resistivity decreases with depth; a confounding factor that must be considered when interpreting EM data. • Resistivity is affected less than EM methods by noise associated with power lines, buried metal objects such as pipes and cables, fences, vehicles, and buildings. Thus, resistivity can be used to provide reliable measurements in some locations near metal objects where EM is of little use. • Limited use in wet weather.	Benson et al. (1982); Cartwright and McComas (1968); Griffith and King (1969); Mooney (1975,1980); Orellana and Mooney (1966); Rodgers and Kean (1980); Stollar and Roux (1975); Sweeney (1984); Telford et al. (1982); USEPA (1978); Urish (1983); Warner (1969); Zohdy et al. (1974)
SEISMIC REFRACTION AND REFLECTION • bedrock surface • depth to water table • locate fractures, faults, and buried bedrock channels • characterize rock type and degree of weathering • stratigraphy (layering and lateral variations) • depth of landfills, trenches, and disturbed zones	The subsurface transmission of seismic waves emitted from a point source are measured using geophones implanted in the ground along a straight line and recorded digitally by a seismograph. The seismic source may be the impact of a sledge or mechanical hammer on a steel plate, or an explosive device. Compressional, shear, and surface seismic waves radiate from the energy source. The compressional waves are refracted and reflected as they pass through media with different seismic velocities (densities). Refracted and/or reflected compressional wave travel times associated with different source-to-geophone wave paths are interpreted using analytical models to determine the depth to a one or more geologic units.	• Seismic refraction is generally used for shallow investigations (<200 ft depth). A refraction survey may require a maximum source-to-geophone distance of up to five times the depth of investigation. • Given sufficient velocity contrast between adjacent horizontal layers, as many as three or four layers can be delineated using seismic refraction. A lower velocity layer under a higher velocity layer and thin layers, however, cannot be resolved using the refraction method. • Geophone spacing can be varied from a few to hundreds of feet depending on the desired measurement depth and resolution. For shallow investigations, 12 or 24 geophones may be positioned at equal spacings as close as 5 to 10 ft and seismic waves may be initiated separately at each end of the line. • Seismic reflection can be used for much deeper investigations (to >1000 ft). It is similar to GPR in that the depth of measurement is as a function of wave reflection travel time. • Seismic data is collected as station measurements and surveying is slow compared to continuous measurement methods. • Data interpretation is confounded by heterogeneous subsurface conditions. • Use may be limited during wet and very cold weather. • Seismic methods are subject to interference from vibration noise associated with various natural and cultural sources (i.e., walking, machinery, and vehicles).	Benson et al. (1982); Griffith and King (1969); Haeni (1986); Hunter et al. (1982); Lankston and Lankston (1983); Mooney (1980); Palmer (1980); Redpath (1973); Steeples (1984); Sverdrup (1986); Telford et al. (1982); Zohdy et al. (1974)

Table 8-1. Summary of various surface geophysical survey methods (modified from Benson, 1991; Gretsky et al, 1990; O'Brien and Gere, 1988).

METHOD AND TARGETS	OPERATING PRINCIPLES/DESCRIPTION	OPERATING PARAMETERS, ADVANTAGES, DISADVANTAGES	REFERENCES
MAGNETICS • buried ferrous metal objects (drums, pipelines, USTs, etc.) • waste zones containing ferrous metal	Magnetometers are used to measure the strength of the earth's magnetic field and respond to ferrous metals perturb the earth's natural magnetic field. Two common types of magnetometers are available: the total field magnetometer and the gradiometer. Typically, a hand-held magnetometer is used to measure total magnetic field intensity along a grid, allowing detection of anomalies associated with shallow buried ferrous metal objects. A nearby base station can be used to record background diurnal variations in the earth's magnetic field.	• Magnetometers respond only to ferrous metals (iron or steel). • Magnetometers provide greater depth range than metal detectors: single drums and drum masses can be detected at depths to 20 and 60 ft, respectively, using a total field magnetometer. • The total field magnetometer response is proportional to the ferrous target mass and inversely proportional to cube of the distance to the target. Gradient measurements using a gradiometer are inversely proportional to the fourth power of the distance to the target and thus minimize interferences but are less sensitive than total field measurements. • Magnetometers can be used to provide continuous or station measurements, and can be mounted on vehicles for coverage of large sites. • Magnetometer response is subject to interference noise from many sources including steel fences, buildings, vehicles, iron debris, utilities, ferrous soil minerals, vehicles, etc. • Data interpretation may be confounded by heterogeneous subsurface conditions and/or the presence of iron-rich geologic media (such as greensands and red hematitic soils). • Magnetometry may also be used to study regional geologic conditions, and occasionally to map the bedrock surface.	Breiner (1973); Gilkeson et al. (1986); Telford et al. (1982); Zohdy et al. (1974)
METAL DETECTORS • shallow buried metal objects (drums, pipelines, USTs, etc.) • shallow waste zones containing metal	Metal detectors sense ferrous and nonferrous metals. The area of detection is typically 1 to 3 ft (equal to the detector coil size or spacing). The depth of detection is commonly limited to less than 10 ft because the response is proportional to the target cross-section and inversely proportional to the sixth power of the distance to the target.	• Metal detectors are routinely used to locate buried cables and pipes. • Quart-size metal objects and drum masses can be detected to depths of 3 and 10 - 25 ft, respectively. • Buried and above-ground metal objects (cars, fences, buildings, etc.) can interfere with measurements.	

- *Magnetometer* surveys are used at contamination sites to measure the perturbation to the earth's magnetic field caused by buried ferrous metal objects such as steel drums, ferrous metal waste in landfills, and iron pipes.

- *Metal detectors* are used to sense ferrous and nonferrous metals at shallow depths, thereby allowing detection of shallow metal pipelines, drums, and waste zones containing metal.

The success of a noninvasive geophysical survey depends on several factors. Reasonable survey objectives should be developed based on consideration of the site conceptual model, characterization requirements, and available methods. Generally, discrete (station) or continuous survey measurements are made along transect or grid lines. Station and transect line spacings and orientation should be commensurate with the expected size and geometry of the subsurface targets (i.e., buried channels, contamination plumes, drums, etc.). As shown in Figure 8-1, the enhanced resolution provided by continuous measurements compared to station measurements may be critical to accurate data interpretation. For methods lacking continuous measurement capability, resolution generally increases with decreased station spacing. Success requires that the survey method be capable of measuring the subsurface properties (and anomalies) of interest, and the existence of sufficient contrast in the measured subsurface properties. Finally, the survey should be conducted by an experienced field crew using functional instrumentation and interpreted by personnel knowledgeable about both geophysical survey data analysis and site conditions.

Surface geophysical equipment is available for rent or purchase from several distributors in North America that advertise in technical journals such as *Ground Water Monitoring Review*. It will generally be cost-effective, however, to contract an experienced firm (or individual) to design, conduct, and interpret surface geophysical surveys at contamination sites. Estimated equipment rental, equipment purchase, and survey contract prices in 1992 dollars based on quotes from distributors and geophysical surveyors are given in Table 8-2. These prices do not include mobilization or shipping fees.

8.1.2 Surface Geophysical Survey Applications

Geophysical surveys can provide relatively quick and inexpensive information on subsurface conditions over a wide area. Inferences derived therefrom during the early phase of a DNAPL site investigation can be used to optimize the cost-effectiveness of subsequent invasive field work. Drilling locations can be selected to test the refined site conceptual model and examine geophysical anomalies. Geophysical data can facilitate interpolation of limited well or boring data and reduce the number of wells/borings needed to adequately characterize a site. Similarly, sequential measurements can be made along transects between wells to augment a detection monitoring program. The utility of alternative methods for various site characterization applications is rated in Table 8-3 based upon experience at a large number of contamination sites (Benson, 1988).

8.1.2.1 Assessing Geologic Conditions

A more detailed summary of the applicability of different surface geophysical methods for assessing stratigraphic and hydrogeologic conditions is given in Table 8-4. For survey speed and resolution, ground-penetrating radar and EM conductivity are probably the two best techniques for mapping lateral variations in soil and rock (Benson, 1991). Radar performance, however, is highly site-specific and typically limited to depths less than 30 ft. High-resolution seismic methods can provide detailed bedrock surface profiles and are sometimes used to provide vertical information at depths below effective radar penetration. Examples of stratigraphic interpretations derived from surface geophysical data are given in Figure 8-2.

8.1.2.2 Detecting Buried Wastes and Utilities

The feasibility of using different surface geophysical survey methods to locate buried wastes and utilities is summarized in Table 8-5. Ground-penetrating radar and EM-conductivity are recommended for detecting non-metallic buried waste; and EM-conductivity, magnetometers, and metal detectors are well-suited for detecting metallic waste. Metal detector use is typically limited to a depth of less than 10 ft, and magnetometers respond only to ferrous metals. The in-phase component of EM-conductivity responds to magnetic susceptibility and should be used to detect ferrous metals. Seismic, electrical resistivity, and magnetic methods can be used to

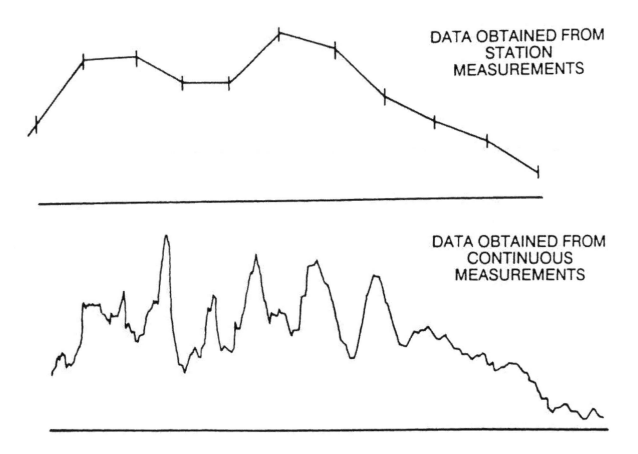

DATA OBTAINED FROM
STATION
MEASUREMENTS

DATA OBTAINED FROM
CONTINUOUS
MEASUREMENTS

Figure 8-1. Comparison of station and continuous surface EM conductivity measurements made along the same transect using an EM-34 with a 10 m coil spacing (from Benson, 1991). The electrical conductivity peaks are due to fractures in gypsum bedrock.

Table 8-2. Surface geophysics equipment rental, equipment purchase, and contract surveying estimated prices in 1992 dollars not including mobilization or shipping fees.

Method	Rental Price	Sale Price	Contractor Price
Ground Penetrating Radar	GSSI SIR-3 with one antenna: $100/d, $700/wk, $3000/mo GSSI SIR-10 with one antenna: $195/d, $1365/wk, $5850/mo	GSSI SIR-3 with one antenna: $20,000-25,000 GSSI SIR-10 with one antenna: $40,000-$47,000	Daily charge for equipment and field crew is $2000-$3500. This price includes subsequent data analysis and reporting. Typical daily coverage is 5000 ft of continuous measurement.
EM Conductivity	Geonics EM31: $39-65/d, $273-450/wk, $1170-$1360/mo Geonics EM34-3: $55/d, $385-$740/wk, $1650-$2240/mo Geonics EM34-3XL: $66/d, $462-$825/wk, $1980-$2480/mo Digital dataloggers for Geonics meters: $150-$260/wk Geonics Protem EM47: $125/d, $875/wk, $3750/mo	*Frequency Domain EM Systems* Geonics EM31 (20 ft penetration): $12,985 Geonics EM34-3 (variable depth to 200 ft): $18,900 Geonics EM34-3XL (variable depth to 200 ft): $21,285 Geonics EM38 (shallow penetration to 5 ft): $6250 Datalogger: $4000-$4800 *Time-Domain EM Systems* Protem 47/P (Profiling for depths to 330 ft): $39,240 Protem 47/S (Soundings 15 to 500 ft): $41,800	Daily charge for equipment and field crew using Geonics EM Conductivity instruments is $2000-$3000 for 8-hrs field work. This price includes subsequent data analysis and reporting. Typical daily profiling survey coverages are: 1000 station measurements or several thousand feet of continuous measurements using the EM31; and 100 to 250 station measurements using the EM34. Typical coverage for Time-Domain soundings is 6 to 12 soundings per day.
Electrical Resistivity	ABEM Terrameter SAS-300B/C: $25-$45/d, $175-$315/wk, $750-$1350/mo ABEM SAS 2000 Booster: $19-$25/d, $133-$175/wk, $570-$700/mo	Soiltest Stratameter R-50 (200-500 ft operating depth): $5000 Bison "BOSS" offset sounding system Model 2365: $5995 Bison signal averaging earth resistivity receiver/transmitter: $9995 ABEM Terrameter SAS-300C with cable and four electrodes: $12,790 ABEM SAS 2000 Booster: $7295	Daily charge for equipment and field crew is $2000-$3000 for 8-hrs field work. This price includes subsequent data analysis and reporting. Typical daily soundings survey coverage is 6 to 12 soundings per day with 10 to 15 electrode array spacings per sounding.
Seismic	GeoMetrics ES1225F (12-channel with filters) with phones: $53/d, $371/wk, $1590/mo GeoMetrics ES2401&8 Hz phones (24-channel): $128/d, $896/wk, $3840/mo Bison 9000 series DIFP Seismograph: 48-channel, $200/d; 24-channel, $175/d; 12-channel, $150/d Bison 7000 series DIFP seismograph: 24-channel, $129/d; 12-channel, $80/d Bison 12-channel 5000 series DIFP seismograph: $70/d ABEM Miniloc 36 channels with geophones and cable: $70/d, $490/wk, $2100/mo ABEM Terraloc Mark 6 24-channel seismograph: $125/d, $875/wk, $3750/mo	Bison digital instantaneous floating point (DIFP) signal stacking seismograph with standard accessories (various models) 12-channel: $12,500-42,500 24-channel: $25,200-$55,600 48-channel: $58,700-$85,500 96-channel: $143,800 120-channel: $173,200. ABEM Miniloc refraction seismograph with geophones and cable: $18,000 ABEM Terraloc Mark 6 seismograph 24 channels: $41,000 48 channels: $48,000	Daily charge for equipment and field crew is $3200-$5000 for 8-hrs field work. This price includes subsequent data analysis and reporting. Typical daily seismic (refraction or reflection) survey coverage includes 4 to 6 spreads per day using 12 or 24 geophones (or approximately 1500 ft of survey line).
Magnetom-meters	GeoMetrics G816 portable magnetometer: $10/d, $70/wk, $300/mo GeoMetrics G856AX extended memoray magnetometer: $14/d, $98/wk, $420/mo GEM GSM-19G: $40-$50/d, $280-350/wk, $1200-$1500/mo	GEM GSM-19 field unit/base station: $10,000 GSM-19 gradiometer: $5,000 GSM-19 VLF: $6,000 GSM-19 "Walking Mag" option: $3500 GEM GSM-8 Magnetometer: $5000	Daily charge for equipment and field crew is $2000-$3000 for 8-hrs field work. This price includes subsequent data analysis and reporting, and (for at least one contractor) includes the simultaneous operation of two magnetometers. Typical daily survey coverage includes approx. 750-1000 station measurements per instrument per day.

Table 8-3. Applications of selected surface geophysical survey methods (modified from Benson, 1988).

APPLICATION	GPR	EM COND.	ELEC. RES.	SEISMIC	METAL DET.	MAG-NETICS
Evaluation of natural geologic and hydrologic conditions						
Depth and thickness of soil and rock layers and vertical variations	1a	2	1	1	NA	NA
Mapping lateral variations in soil and rock (fractures, karst features, etc.)	1a	1	2	2 (refr.) 1 (refl.)	NA	NA
Depth of water table	3	2	1		NA	NA
Evaluation of subsurface contamination and post-closure monitoring						
Inorganics (high TDS and electrically conductive)						
Early warning contaminant detection	3	1	2	NA	NA	NA
Detailed lateral mapping	3	1	2	NA	NA	NA
Vertical extent	3	2	1	NA	NA	NA
Changes of plume with time (flow direction and rate)	3	1	2	NA	NA	NA
Post cleanup/closure monitoring	3	1	2	NA	NA	NA
Organics (typically nonconductive)						
Early warning contaminant detection	3	3	3	NA	NA	NA
Detailed lateral mapping	2a	2	3	NA	NA	NA
Vertical extent	2a	3	2	NA	NA	NA
Changes of plume with time (flow direction and rate)	3	3	3	NA	NA	NA
Post clean-up/closure monitoring	3	3	3	NA	NA	NA
Location of buried wastes and trench boundaries						
Bulk waste trenches without metal	1	1	2	3	NA	NA
Bulk waste trenches with metal	1	1	2	3	1a	1b
Depth of trenches and landfill	2	3	2	2	NA	NA
Detection of 55-gallon steel drums	2a	2	NA	NA	1a	1
Estimates of depth and quantity of 55-gallon steel drums	2a	3	3	NA	2	1
Location of utilities						
Buried pipes and tanks	1	1c	NA	NA	1c	1b
Potential pathways of contaminant migration via conduits and permeable trench backfill	1	2	NA	NA	2	2
Abandoned wells with metal casing	3	NA	NA	NA	2	1b

Notes: 1 = Primary choice under most field conditions   a = shallow
          2 = Secondary choice under most field conditions   b = Assumes ferrous metals to be present
          3 = Limited field application under most field conditions   c = Assumes metals to be present
          NA = Not applicable

This table is intended as a general guide. The application ratings given are based upon actual experience at a large number of sites. The rating system is based upon the ability of each method to produce results under general field conditions when compared to other methods applied to the same task. One must consider site-specific conditions before recommending an optimum approach. Site-specific conditions may dictate the choice of a method rated 2 or 3 in preference to a method rated 1.

Table 8-4. Surface geophysical methods for evaluating natural hydrogeologic conditions (modified from Benson, 1991).

Method	General Application	Continuous Measurements?	Depth of Penetration	Major Limitations
Ground-penetrating radar	Highest resolution of any method for profiling and mapping	Yes	Typically less than 30 ft; to 100 ft under ideal conditions	Penetration limited by increasing clay content, fluid content, and fluid conductivity
EM Conductivity (Frequency Domain)	Very rapid profiling and mapping	Yes (to 50 ft)	To 200 ft	Affected by cultural features including metal fences, pipes, buildings, and vehicles
EM Conductivity (Time Domain)	Soundings	No	To >1000 ft	Cannot be used to provide measurements shallower than about 150 ft
Electrical Resistivity	Soundings or profiling and mapping	No	No limit, but commonly used to a depth of a <300 ft	Requires good ground contact and long electrode arrays; integrates a large volume of subsurface; affected by cultural features including metal fences, pipes, buildings, and vehicles
Seismic Refraction	Profiling and mapping soil and rock	No	No limit, but commonly used to <300 ft	Requires considerable energy for deeper surveys; sensitive to ground vibrations
Seismic Reflection	Profiling and mapping soil and rock	No	To >1000 ft	Very slow surveying; requires extensive data reduction; sensitive to ground vibrations
Magnetics	Profiling and mapping soil and rock	Yes	No limit, but commonly used to <300 ft	Only applicable in certain rock environments; limited by cultural ferrous metal features

Note: Actual results depend on site-specific conditions. In some applications, an alternate method may provide better results.

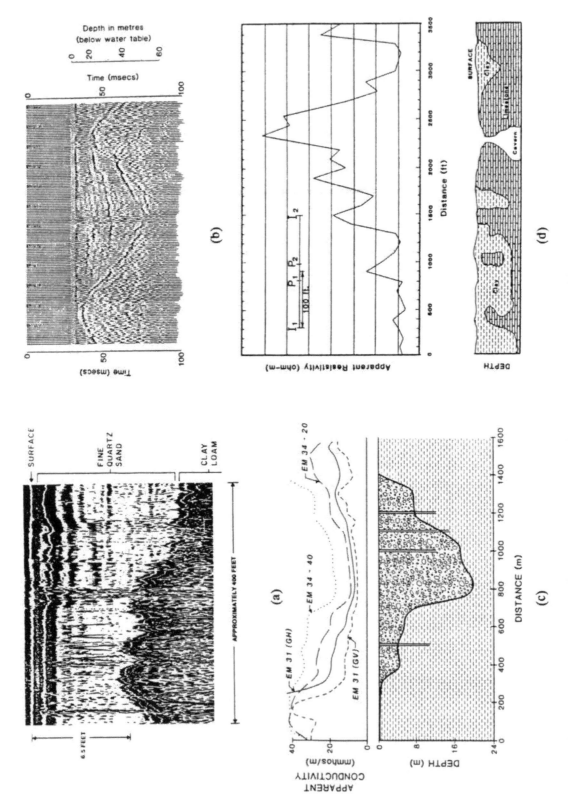

Figure 8-2.   Examples of stratigraphic interpretations using surface geophysical surveys: (a) ground-penetrating radar (from Benson et al., 1982); (b) delineating of a bedrock channel by seismic reflection (from Benson, 1991); (c) relationship of EM conductivity data and a sand/gravel channel (from Hoekstra and Hoekstra, 1991); and (d) electrical resistivity profile of karst terrain (from Hoekstra and Hoekstra, 1991).

Table 8-5. Surface geophysical methods for locating and mapping buried wastes and utilities (modified from Benson, 1991).

Method	Bulk Wastes Without Metals	Bulk Wastes With Metals	55-Gallon Drums	Pipes and Tanks
Ground-penetrating radar	Very effective in coarse, dry, sandy soil or massive rock; sometimes effective to obtain shallow boundaries in soils with poor GPR penetration	Very effective in media with good GPR penetration; sometimes effective to obtain shallow boundaries in soils with poor GPR penetration	Good if soil conditions permit sufficient GPR penetration; may provide drum depth	Very good for metal and non-metal tanks and pipes if soil conditions permit sufficient GPR penetration; may provide object depth
EM Conductivity	Excellent to depths less than 20 ft	Excellent to depths less than 20 ft	Very good; may identify single drum to 6-8 ft	Very good for metal tanks
Electrical Resistivity	Good; soundings may provide depth	Good; soundings may provide depth	Not applicable	Not applicable
Seismic Refraction	Fair; may provide depth	Fair; may provide depth	Not applicable	Not applicable
Magnetometry	Not applicable	Very good for ferrous metal only; deeper sensing than metal detectors	Very good for ferrous metal only; deeper sensing than metal detectors	Very good for ferrous metal only; deeper sensing than metal detectors
Metal Detection	Not applicable	Very good for shallow ferrous and non-ferrous metals	Very good for shallow ferrous and non-ferrous metals	Very good for shallow ferrous and non-ferrous metals

Note: Actual results depend on site-specific conditions. In some applications, an alternate method may provide better results.

estimate the depth to the base of a landfill. Examples of geophysical survey measurements over buried wastes and pipes are shown in Figure 8-3.

8.1.2.3 Detecting Conductive Contaminant Plumes

The application of surface geophysical survey methods to map conductive contaminant plumes is summarized in Table 8-6. A two-fold strategy is generally employed to map conductive contaminant plumes: (1) potential preferential pathways for contaminant transport such as buried channels, fracture zones, or sand lenses, are delineated using applicable methods (Table 8-4); and, (2) conductive contaminants (i.e., dissolved ions in groundwater) are mapping using electrical resistivity or EM conductivity. The application of electrical methods to map contaminated groundwater was initiated in the late 1960s (e.g., Cartwright and McComas, 1968; Warner, 1969) and popularized during the 1980s (see references in Table 8-1). Examples of contaminant plume detection using electrical resistivity and EM conductivity are shown in Figure 8-4.

8.1.2.4 Detecting DNAPL Contamination

Subsurface DNAPL is generally a poor target for conventional geophysical methods. Although ground-penetrating radar, EM conductivity, and complex resistivity have been used to infer NAPL presence at a very limited number of sites (e.g., Olhoeft, 1986; Davis, 1991), direct detection and mapping of non-conductive subsurface DNAPL using surface geophysical techniques is an unclear, and apparently limited, emerging technology (WCGR, 1991; USEPA, 1992).

Ground-penetrating radar was used to monitor the infiltration of tetrachloroethene following a controlled spill under relatively ideal conditions at the Borden, Ontario DNAPL research site (Annan et al., 1991; Redman et al., 1991; WCGR, 1991). Subtle changes in ground-penetrating radar reflectivity patterns at different times before and after release of the tetrachloroethene appear to document its downward migration and pooling on a capillary barrier. At a coal tar spill site, Davis (1991) inferred the presence of large volumes of subsurface coal tar DNAPL based on low conductivity anomalies detected using EM surveys. Others, however, have concluded that surface geophysical methods have very limited capacity to detect NAPL (Pitchford et al.,

1989) and that their use for this purpose should be considered with caution (Benson, 1991).

The value of surface geophysics at most DNAPL sites will be to aid characterization of waste disposal areas, stratigraphic conditions, and potential routes of DNAPL migration. Objectives and methods for using surface geophysical surveys to evaluate DNAPL site contamination are described in Table 8-7. The use of geophysical surveying for direct detection of NAPL is currently limited by a lack of: (1) demonstrable methods, (2) documented successes, and (3) environmental geophysicists trained in these techniques (USEPA, 1992).

8.2 SOIL GAS ANALYSIS

Soil gas analysis became a popular screening tool for detecting volatile organic chemicals in the vadose zone at contamination sites during the 1980s (Devitt et al., 1987; Marrin and Thompson, 1987; Thompson and Marrin, 1987; Kerfoot, 1987, 1988; Marrin, 1988; Marrin and Kerfoot, 1988; Tillman et al., 1989a, b). The American Society for Testing and Materials (ASTM) is in the process of developing a standard guide for soil gas monitoring in the vadose zone. Soil gas surveys generate extensive chemical distribution data quickly at a fraction of the cost of conventional invasive methods and offer the benefits of real-time field data. Consideration should be given to its use during the early phases of site investigation to assist delineation of DNAPL in the vadose zone, contaminant source areas, contaminated shallow groundwater, and contaminated soil gas; and, thereby, guide subsequent invasive field work.

8.2.1 Soil Gas Transport and Detection Factors

Many DNAPLs, including most halogenated solvents, have high vapor pressures and will volatilize in the vadose zone to form a vapor plume around a DNAPL source as described in Chapters 4.8 and 5. Volatile organic compounds (VOCs) dissolved in groundwater can also volatilize at the capillary fringe into soil gas. Although this latter process is poorly understood (Barber et al., 1990), it may be enhanced by water-table fluctuations (Lappala and Thompson, 1983). Equilibrium vapor concentrations are probably never attained in the field, however, due to rapid vapor diffusion above the water table (Marrin and Thompson, 1984).

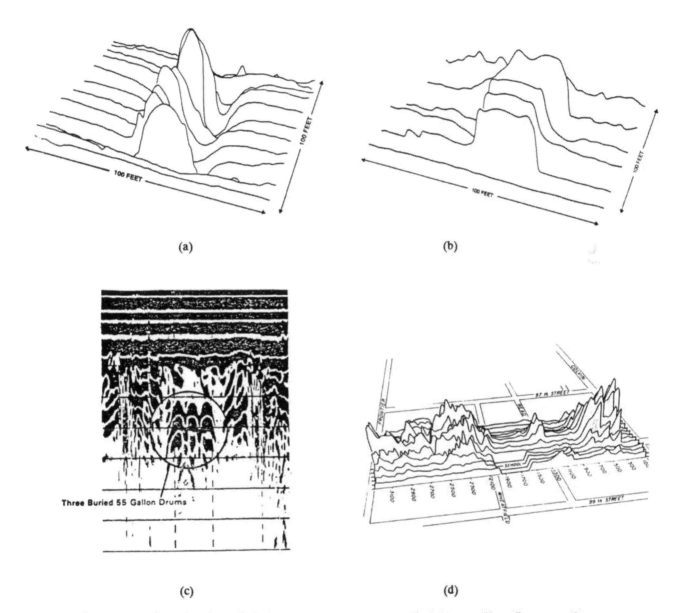

(a)

(b)

Three Buried 55 Gallon Drums

(c)

(d)

Figure 8-3. Examples of geophysical survey measurements over buried wastes (from Benson et al., 1982; Technos, 1980): (a) a gradiometer magnetometer survey over metal drums buried in a trench measuring approximately 20 ft by 100 ft by 6 ft deep; (b) a metal detector survey of the same trench; (c) a ground-penetrating radar image showing three buried drums; and (d) shallow EM conductivity survey data at the Love Canal landfill showing the presence of large concentrations of conductive materials and buried iron objects (i.e., drums) associated with chemical waste disposal areas at each end of the landfill.

Table 8-6. Surface geophysical methods for mapping conductive contaminant plumes (modified from Benson, 1991).

Mapping Permeable Pathways, Bedrock Channels, etc.

The fundamental approach to evaluating the direction of groundwater flow and the possible extent of a contaminant plume is by determining the hydrogeologic characteristics of the site (refer to Table 8-4).

Mapping Conductive Inorganic or Mixed Inorganic-Organic Contaminants

Conductive contaminants (inorganics or organic-inorganic mixtures) can be mapped using electrical methods (electrical resistivity and EM conductivity) and sometimes using ground-penetrating radar when the ionic strength of the contaminated fluids sufficiently exceeds that of background fluid. The higher specific conductance of the contaminated pore fluid acts as a tracer which can be mapped using electrical surface geophysical methods.

Note: Actual results depend on site-specific conditions. In some applications, alternate methods may provide better results.

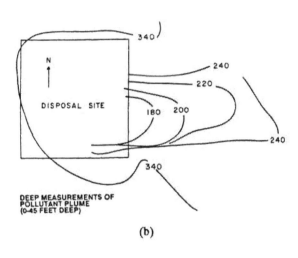

(a)

(b)

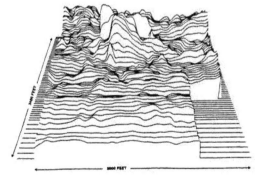

(c)

Figure 8-4. Examples of conductive plume detection using: (a) shallow and (b) deep electrical
resistivity surveys at an approximately 1 square mile landfill with values given in ohm-feet;
and (c) a continuous EM conductivity survey showing a large inorganic plume (center
rear) (from Benson, 1991).

Table 8-7. Surface geophysical methods for evaluating DNAPL site contaminations.

Objective	Methods
Delineate limits of waste disposal areas	Following review of historical documentation, interviews, aerial photographs, and available site data, consider use of GPR, EM conductivity, magnetometer, and metal detection surveys (see Table 8-5).
Delineate buried utility corridors	Following review of historical documentation, utility records, interviews, aerial photographs, and available site data, consider use of GPR, EM conductivity, magnetometer, and metal detection surveys (see Table 8-5).
Map stratigraphy, particularly permeable pathways and the surfaces of fine-grained capillary barrier layers and bedrock, to determine potential routes of DNAPL migration and stratigraphic traps	Following review of aerial photographs and available site, local, and regional hydrogeologic data, consider use of GPR, EM conductivity, electrical resistivity, and seismic surveys (see Table 8-4).
Delineate conductive inorganic contaminant plumes that may be associated with DNAPL contamination	Following review of available data on waste disposal practices, chemical migration, and hydrogeologic conditions, consider use of EM conductivity, electrical resistivity surveys (see Table 8-6).
Delineate geophysical anomalies that may result from an accumulation of DNAPL	Consider using GPR to infer DNAPL accumulations at shallow depths in low conductivity soils because DNAPL presence will probably alter the dielectric properties of the subsurface (Olhoeft, 1986; WCGR, 1991). Also consider using electrical methods to infer DNAPL accumulation based on the presence of low conductivity anomalies (Davis, 1991). Note that the use of surface geophysical methods for direct detection of DNAPL presence is an emerging, but limited technology, that may not be cost-effective. Very few geophysicists are experienced in the application of surface survey methods for DNAPL detection. These applications, therefore, should be treated with caution.
Note: Actual survey results depend on site-specific conditions. At some sites, an alternate method may provide better results.	

Experiments conducted at the Borden, Ontario DNAPL research site suggest that soil gas contamination will usually be dominated by volatilization and vapor phase transport from contaminant sources in the vadose zone rather than from groundwater, and that the upward transport of VOCs to the vadose zone from groundwater is probably limited to dissolved contaminants that are very near to the water table (Hughes et al., 1990a; Rivett and Cherry, 1991; Rivett et al., 1991; Chapter 8.2.4). Rivett and Cherry (1991) attribute the limited upward diffusion of groundwater contaminants to the low vertical transverse dispersivities (mm range) which are observed in tracer studies. Soil gas contamination, therefore, is not a reliable indicator of the distribution of DNAPL or groundwater contamination at depth below the water table.

VOCs in soil gas diffuse due to the chemical concentration gradient. Modeling analyses indicate that contaminated vapors can diffuse tens of yards or more from a DNAPL source in the vadose zone within weeks to months (Mendoza and McAlary, 1990; Mendoza and Frind, 1990a,b) and that transport velocities can be reduced significantly by vapor-phase retardation (Chapter 5.3.20). Field experiments involving trichloroethene (TCE) vapor transport in a sand formation confirm the modeling study findings (Hughes et al., 1990a). These experiments demonstrate the formation of significant groundwater plumes from a solvent vapor source in a thin (< 12 ft thick) vadose zone over a period of weeks.

Volatilization of contaminants with high molecular weights and saturated vapor concentrations (Table 4-6) can engender density-driven gas migration in media with high gas phase permeability (i.e., $k_g > 1 \times 10^{-11}$ m^2 in uniform media) (Falta et al., 1989; Sleep and Sykes, 1989; Mendoza and Frind, 1990b; Mendoza and McAlary, 1990). Density-driven gas flow causes VOCs to sink and move outward above (and dissolve in) the saturated zone. This relative density effect should decrease with decreasing VOC concentrations and increasing distance from a contaminant source.

Although advective processes due to density effects or high vapor pressure gradients may influence VOC migration, gaseous diffusion is considered the predominant vapor transport mechanism at most sites (Marrin and Kerfoot, 1988). At steady-state, solute flux is proportional to the air-filled porosity, the VOC diffusion coefficient, and the gas-phase concentration gradient.

Vapor transport can cause shallow groundwater contamination in directions opposite to groundwater and/or DNAPL flow. The resulting groundwater contamination plumes can have high dissolved chemical concentrations, but tend to be very thin in vertical extent, and occur close to the water table.

Subsurface geologic heterogeneities, soil porosity, moisture conditions, VOC source concentrations, and sorption equilibria can significantly affect VOC gradients in soil gas (Marrin and Thompson, 1987). For example, false negative interpretations may result from the presence of vapor barriers (perched groundwater, clay lenses, or irrigated soils) below the gas probe intake. Thus, sample locations and depths influence the measured vapor concentrations. Profiles of soil gas concentrations under a variety of field conditions are illustrated in Figure 8-5.

Several chemical characteristics indicate whether a measurable vapor concentration can be detected (Devitt et al., 1987; and Marrin, 1988). Ideally, compounds such as VOCs monitored using soil gas analysis will: (1) be subject to little retardation in groundwater; (2) partition significantly from water to soil gas (Henry's Law constant >0.0005 atm-m^3/mole); (3) have sufficient vapor pressure to diffuse significantly upward in the vadose zone (>0.0013 atm @ 20° C); (4) be persistent; and (5) be susceptible to detection and quantitation by affordable analytical techniques. Vapor pressures, solubilities, and Henry's Law constants for DNAPL chemicals are plotted in Figure 8-6 and given in Appendix A.

A generalized flowchart for conducting a soil-gas survey is provided in Figure 8-7. Under typical conditions, 20 to 40 soil gas grab samples can be acquired per day for analysis. The cost to obtain and analyze soil gas samples using lab-grade gas chromatography with appropriate detectors typically ranges between $115 and $195 per sample in 1992 dollars. This cost includes report preparation, but not mobilization fees. Methods of soil gas sampling and analysis are discussed below.

8.2.2 Soil Gas Sampling Methods

Soil gas sampling methods have been reviewed by Devitt et al. (1987), Marrin and Kerfoot (1988), GRI (1987), and Tillman et al. (1989b). Essentially, there are two basic types of sample acquisition: (1) grab sampling and (2) passive sampling.

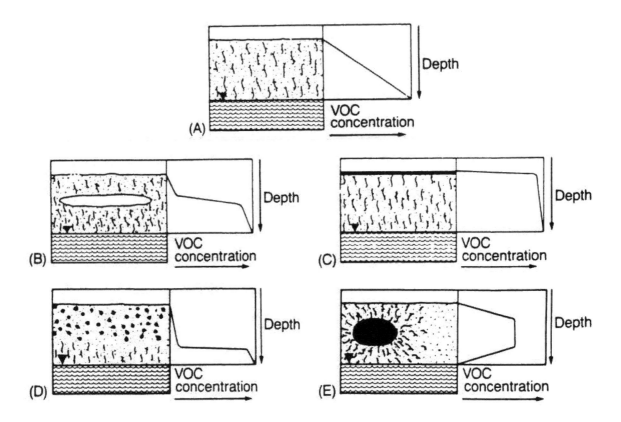

(A) Homogeneous porous material with sufficient air-filled porosity
(B) Impermeable subsurface layer (e.g., clay or perched water)
(C) Impermeable surface layer (e.g., pavement)
(D) Zone of high microbiological activity (circles and wavy
    lines indicate different compounds)
(E) VOC source in the vadose zone

Figure 8-5.   Soil gas concentration profiles under various field conditions (reprinted with permission ACS, 1988).

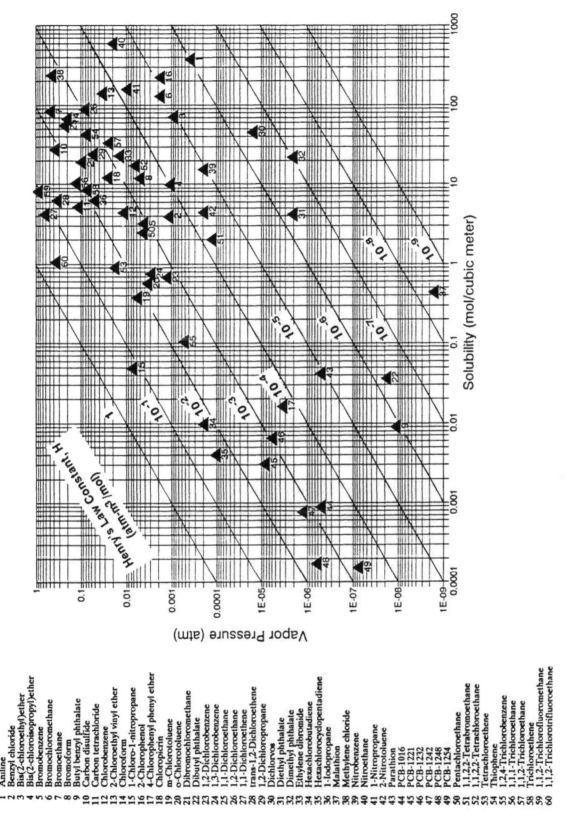

Figure 8-6. Solubility, vapor pressure, and Henry's Law Constants for selected DNAPLs (refer to Appendix A).

1 Aniline
2 Benzyl chloride
3 Bis(2-chloroethyl)ether
4 Bis(2-chloroisopropyl)ether
5 Bromobenzene
6 Bromochloromethane
7 Bromoethane
8 Bromoform
9 Butyl benzyl phthalate
10 Carbon disulfide
11 Carbon tetrachloride
12 Chlorobenzene
13 2-Chloroethyl vinyl ether
14 Chloroform
15 1-Chloro-1-nitropropane
16 2-Chlorophenol
17 4-Chlorophenyl phenyl ether
18 Chloropicrin
19 m-Chlorotoluene
20 o-Chlorotoluene
21 Dibromochloromethane
22 Dibutyl phthalate
23 1,2-Dichlorobenzene
24 1,3-Dichlorobenzene
25 1,1-Dichloroethane
26 1,2-Dichloroethane
27 1,1-Dichloroethene
28 trans-1,2-Dichloroethene
29 1,2-Dichloropropane
30 Dichlorvos
31 Diethyl phthalate
32 Dimethyl phthalate
33 Ethylene dibromide
34 Hexachlorobutadiene
35 Hexachlorocyclopentadiene
36 1-Iodopropane
37 Malathion
38 Methylene chloride
39 Nitrobenzene
40 Nitroethane
41 1-Nitropropane
42 2-Nitrotoluene
43 Parathion
44 PCB-1016
45 PCB-1221
46 PCB-1232
47 PCB-1242
48 PCB-1248
49 PCB-1254
50 Pentachloroethane
51 1,1,2,2-Tetrabromoethane
52 1,1,2,2-Tetrachloroethane
53 Tetrachloroethene
54 Thiophene
55 1,2,4-Trichlorobenzene
56 1,1,1-Trichloroethane
57 1,1,2-Trichloroethane
58 Trichloroethene
59 1,1,2-Trichlorofluoromethane
60 1,1,2-Trichlorotrifluoroethane

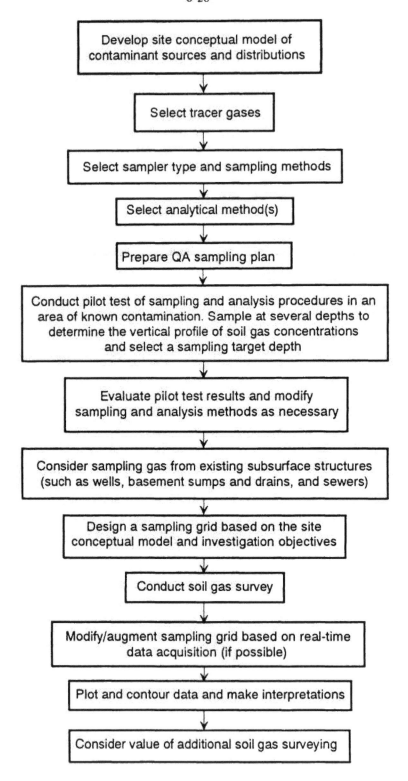

Figure 8-7. Flowchart for conducting a soil-gas survey.

Grab sampling typically involves driving a small-volume hollow probe with a conical tip to a depth of 3 to 10 ft, pumping soil gas from the probe, and collecting a sample from the moving soil gas. Samples are usually taken from a depth of at least 3 ft to diminish the effects of surface contamination, changes in barometric pressure and temperature, rainfall, and air pollution (Tillman et al., 1989b). The transient effects related to weather can be minimized by conducting the soil gas survey in the shortest time possible. Probes can be inserted to greater depths from the bottom of a hollow-stem auger hole. Vertical profiles of soil gas concentrations can be developed by taking samples from different depths at the same location.

Although there are many different probe designs (Devitt et al., 1987), small-diameter hollow steel probes with conical tips and openings just above the tip to allow soil gas entry (Figure 8-8) are used most frequently to minimize soil disturbance and the mixing of air with soil gas. Before insertion, the probes are cleaned and purged using an inert gas or filtered air. The hollow probes can be pushed into the ground hydraulically or pneumatically, hammered in using a variety of manual or mechanical hammers, or inserted into holes created by a hand-held rotary percussion drill or a drive rod.

A small volume of soil gas (e.g., 3 to 10 liters) is withdrawn using a pump to purge the sampling system, and the sample is taken for immediate onsite analysis, or encapsulated (preferably in glass or stainless steel containers) for subsequent laboratory analysis. Samples taken for VOC determination should be analyzed within two days of collection. Continuous or excessive pumping should be avoided; it may dilute soil gas with surface air or distort the actual soil gas concentration patterns. After sampling has been completed, the probes can be removed hydraulically or by use of a jacking device.

Grab sampling of hydrophobic compounds can also be accomplished by pumping soil gas through an adsorbent collection medium such as charcoal or a carbonized molecular sieve adsorbent (USEPA-ERT, 1988a,b). Instructions regarding pumping rates and durations are provided along with commercially available sorbent traps. Generally, trapped soil gas contaminants are desorbed thermally or using a solvent prior to chemical analysis.

Primary advantages of dynamic grab sampling include: its low cost relative to drilling; the quick acquisition and analysis of samples (e.g., 40 to 70 samples per day reported by Tillman et al., 1989b); its noninvasive nature;

and its documented utility. Onsite analysis of grab samples also allows efficient field direction of the soil gas survey. Problems may include: the potential collection of unrepresentative samples due to excessive pumping, disturbance to the vadose zone, and air leakage into the sampling apparatus; plugging of sampling syringes by pieces of the rubber septum and of the probe screens with wet cohesive soils; and misinterpretation of results due to complex chemical or hydrogeologic conditions (Hughes et al., 1990b; Devitt et al., 1987).

Passive sampling provides an integrated measure of VOC concentrations over the duration of sample collection; and thereby may overcome short-term perturbations due to variable weather conditions or other factors. Typically, a sorbent material such as activated carbon is placed below ground within a hollow probe, sampling chamber, or can for an extended period to trap VOCs that diffuse through soil gas (Figure 8-9). The sampling duration can be varied to promote accumulation of a detectable quantity of trapped contaminant. The sorbent is then retrieved and submitted for analysis. Passive soil gas sampling is used infrequently at contamination sites. Among other reasons, this is because: (1) the long sampling times involved are inconvenient; (2) the volume of soil gas sampled is not measured making determination of soil gas concentrations impossible; and, (3) chemical degradation may affect sampling results due to the length of the sampling period.

To further delineate the nature and extent of subsurface contamination, gas samples derived from the headspace of capped wells and jars containing drill cuttings, basement sumps and drains, sewer lines, etc. can also be analyzed using the following methods. It should be noted that basements and utility corridors can act as vapor sinks, inducing the transport of contaminated soil gas from source areas.

8.2.3 Soil Gas Analytical Methods

Selection of an analytical method for soil gas depends on the sensitivity, selectively, and immediacy of analytical results required for a particular survey. Devitt et al. (1987) provide a detailed comparative review of alternative methods of soil gas analysis, including: portable VOC analyzers (i.e., Flame Ionization Detectors, FIDs, Photoionization Detectors, PIDs, and IR analyzers); portable gas chromotographs (GCs) with various detectors; and laboratory grade GCs with various detectors which can be installed in a mobile laboratory.

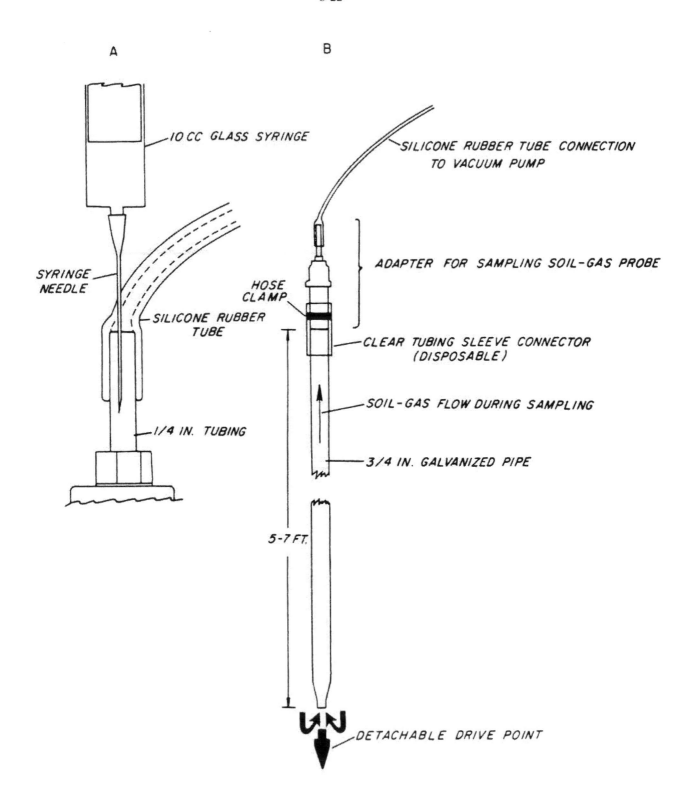

Figure 8-8. Soil gas probe sampling apparatus: (a) close-up view of syringe sampling through evacuation tube; and (b) hollow soil gas probe with sampling adapter (from Thompson and Marrin, 1987).

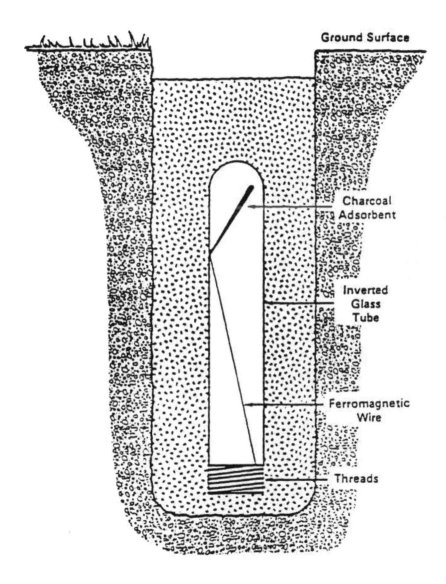

Figure 8-9. Passive soil gas sampling apparatus (from Kerfoot and Barrows, 1987).

Following chromatographic separation, soil gas samples can be analyzed using a (Devitt et al., 1987):

- Flame ionization detector (FID) for analysis of nearly all organic compounds;

- Photoionization detector (PID) for measuring aromatic hydrocarbons concentrations;

- Electron capture detector (ECD) for selective measurement of halogenated organic compound concentrations;

- Hall electrolytic conductivity detector (HECD) for specific determination of halogenated compounds, nitrogen species, and sulfur species; and,

- flame photometric detector (FPD) for determination of sulfur and phosphorous compounds.

Typical analytes and products detectable by soil gas surveys are listed in Table 8-8. The advantages and limitations of specific analytical methods are described in Table 8-9.

8.2.4 Use of Soil Gas Analysis at DNAPL Sites

Several published case studies are available that describe the application of soil gas survey techniques to delineate DNAPL chemical contamination (Glaccum et al., 1983; Voorhees et al., 1984; Spittler et al., 1985; Wittmann et al., 1985; Devitt et al., 1987; Marrin and Thompson, 1987; Thompson and Marrin, 1987; Kerfoot, 1987; Newman et al., 1988; Shangraw et al., 1988; Marrin and Kerfoot, 1988; Bishop et al., 1990; Hughes et al., 1990a; Rivett and Cherry, 1991). Studies during the 1980s generally indicated the utility of soil gas surveying for delineating VOC source areas and VOC-contaminated groundwater (e.g., Marrin and Thompson, 1987; Thompson and Marrin, 1987). Examples of the correlation between halogenated solvent concentrations in soil gas and dissolved in groundwater at three different sites is shown in Figure 8-10.

More recently, controlled and highly-documented field experiments were conducted at the Borden, Ontario DNAPL research site to investigate the behavior and distributions of TCE in soil gas caused by (1) vapor transport from a DNAPL source in the vadose zone and (2) dissolved transport with groundwater from a DNAPL source below the water table (Hughes et al., 1990a; Rivett

and Cherry, 1991). As shown in Figures 8-11 and 8-12, the extent and magnitude of soil gas contamination derived from the vadose zone source was much greater than that derived from the groundwater source.

TCE concentrations in soil gas associated with the plume of contaminated groundwater having TCE concentrations greater than 1000 ug/L less than 5 ft below the water table were generally <1 ug/L in a limited zone around the source (Figures 8-11a and 8-12a). Rivett and Cherry (1991) believe that the soil gas contamination at the groundwater source site actually derived from accidental TCE spillage above the water table during the source emplacement, rather than from upward diffusion of TCE from contaminated groundwater. The vadose zone source produced TCE concentrations in soil gas and groundwater over a much wider area than the groundwater source site. The TCE in groundwater derived from the vadose zone source, however, was less concentrated than at the groundwater source site and was restricted to the upper 5 ft of the saturated zone.

Rivett and Cherry (1991) provide the following interpretation of the observed TCE distributions.

"High vapor concentrations are transported laterally by diffusion in the vadose zone from the residual. Groundwater underlying the vapor plume is contaminated by partitioning from the vapor. In addition, during recharge events infiltration is contaminated after passage through the vadose zone residual or vapor and transported to groundwater. Any rise in the water table would be into a zone containing contaminated soil gas and thus groundwater contamination would follow. The wide shallow groundwater plume observed at Site B is a result of the above processes and is entirely derived from source materials in the vadose zone. . . . Due to lateral transport of vapors in the vadose zone, the groundwater plume produced will be much wider than the source of residual NAPL. Thus, although similar source widths were used at Sites A and B, the groundwater plume derived from vapor contact at site B will be much wider than the plume at Site A that is only subject to transverse dispersion in the groundwater zone, which is a weak process (Sudicky, 1986; Sudicky and Huyakorn, 1991) . . . The process of upward transport of VOCs [from groundwater to soil gas] is probably only really important for the partitioning of very shallow water table

Table 8-8. Typical analytes and products detectable by soil gas surveys (modified from Tillman et al., 1990a).

ANALYTES	PRODUCTS
Acetone Benzene 1-Butane n-Butane Carbon Tetrachloride Chloroform 1,1-Dichloroethane (1,1-DCA) 1,1-Dichloroethene (1,1-DCE) trans-1,2-Dichloroethene (1,2-DCE) Ethane Ethylbenzene Isopropyl Ether (DIPE) Methane Methylene Chloride (Dichloromethane) Methyl Ethyl Ketone (MEK) Methyl-Isobutyl Ketone (MIBK) Methyl-Tert Butyl Ether (MTBE) Propane Tetrachloroethene (PCE) Toluene 1,1,1-Trichloroethane (1,1,1-TCA) 1,1,2-Trichloroethane (1,1,2-TCA) Trichloroethene (TCE) 1,1,2-Trichlorotrifluoroethane (Freon-113) Xylenes	Cleaning Fluids Coal Tar Creosote Degreasers Diesel Fuel Gasoline Heating Oil Jet Fuel Solvents Turpentine

Table 8-9. Advantages and limitations of several soil gas analytical methods (modified from Devitt et al., 1987).

METHOD	ADVANTAGES	LIMITATIONS
Portable VOC Detectors	• Easy to transport to the field • Minimum operator training required • Elimination of sample collection steps minimizes uncertainties and expense of sample collection, storage and transport • Immediate analysis provides guidance for additional sampling	• Limited sensitivity due to lack of an analyte concentration step; typical detection limit is 1 ppmv • Limited selectivity and interference problems because of the lack of a separation step; variable response to different compounds • Limited accuracy because of the inability to calibrate adequately for chemical mixtures • Relatively large sample volumes required
Portable FID Analyzer (e.g., Century OVA)	• Flame ionization detectors use a hydrogen-fed flame to ionize organic gases and generate a current that is proportional to concentration • Capable of detecting all organic compounds • Sensitivity is to <1 ppmv (methane) without the GC option • Less variability in instrument response to different organic compounds than with FIDs • Less susceptible to than PIDs to interference by high humidity • FIDs can be operated in a survey mode or, with appropriate attachments, in a GC mode • The GC option can be used with proper standards to separate, identify, and quantitate individual organic compounds in soil gas to the ppt or low ppb level	• High humidity will reduce the relative response • Successful use of the GC mode requires significant operator training and experience, and equipment maintenance • Approximate cost of Century System OVA-128 is $6,800.
Portable PID Analyzer (e.g., HNU, Photovac TIP)	• An ultraviolet light is used to ionize gas or vapor molecules which are then collected and produce a current which is proportional to concentration • Capable of monitoring many organic and some inorganic gases and vapors • Sensitivity to 0.1 ppmv (benzene); detection range is approximately 0.1 to 2000 ppmv • No fuel or flame required • With the 10.2 eV lamp, alcohols, halogenated alkanes, and most inorganic gases have no response; alkanes have little or no response; and alkenes, aromatics, organosulfur compounds, and carbonyl compounds have high response • With the 11.7eV lamp, all organic compounds except methane produce instrument response	• The energy generated by the lamp must exceed the ionization potential of the target chemical to permit detection • The instrument's response to different chemicals varies • UV lamp intensity declines slowly with age, but can be compensated for during instrument calibration • Airborne dust can interfere with UV light transmission and reduce readings • High humidity can condense on the UV lamp and reduce the light transmission, and also can decrease the ionization of chemicals, thereby reducing readings • PIDs cannot detect light hydrocarbons, such as methane • Approximate cost of a HNU meter is $4300 in 1992 dollars.
Portable Hot-Wire Detector (Bacharach TLV Sniffer)	• Vapors are catalytically combusted using a hot-wire • Similar advantages to FIDs • Sensitive to approximately 2 ppm; detection range is approximately 2 to 10,000 ppmv	• Similar disadvantages as FIDs • Approximate cost of a Bacharach TLV Sniffer is $2000 in 1992 dollars.

Table 8-9. Advantages and limitations of several soil gas analytical methods (modified from Devitt et al., 1987).

METHOD	ADVANTAGES	LIMITATIONS
Portable and Field GC Units with various detectors	• GC units can be used to separate a mixture of gaseous compounds prior to detection; • Several portable VOC analyzers have GC options, including the Photovac PID and the Century OVA FID, allowing individual compounds to be monitored and quantitated to the ppb range • These instruments are easy to transport • Portable GCs are most effective on samples with large concentrations of easy-to-separate VOCs • Field GC units contain temperature controlled ovens and a variety of injectors and detectors facilitating better precision and accuracy	• No temperature control for the portable GC column hinders high resolution compound separation • Reproducible retention times are difficult to replicate in the field using portable GCs due to temperature variation • Estimated cost in 1992 dollars for GC units without a detector is from $2500 to >$30,000.
Lab-grade GC Units with various detectors or MS	• Fast analytical response with better-controlled analytical conditions; can typically process 50-70 samples per day • Produces highest quality data • Provides significantly lower detection limits for a wider range of compounds than possible using a portable GC • Detection limits can be in the ppt range • Can be mounted in a mobile van	• Requires significant operator training • Lab-grade GC units are expensive

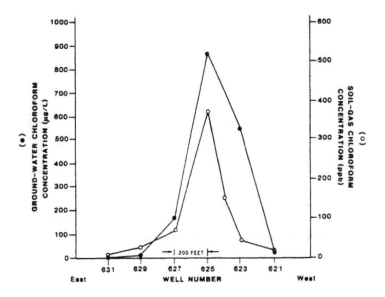

(a)

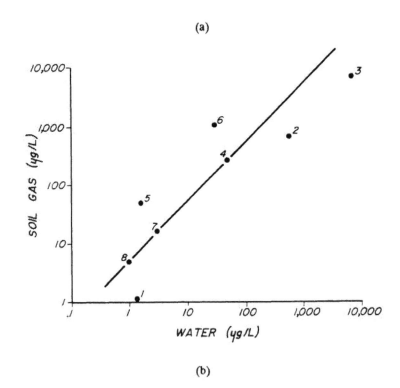

(b)

Figure 8-10. Correlations between halogenated solvent concentrations in shallow soil gas sampled between 3 and 8 ft below ground and underlying groundwater: (a) chloroform along a transect perpendicular to the direction of groundwater flow at an industrial site in Nevada (reprinted with permission from ACS, 1987); and, (b) 1,1,2-trichlorotrifluoroethane (Freon 113) at an industrial site in California (from Thompson and Marrin, 1987).

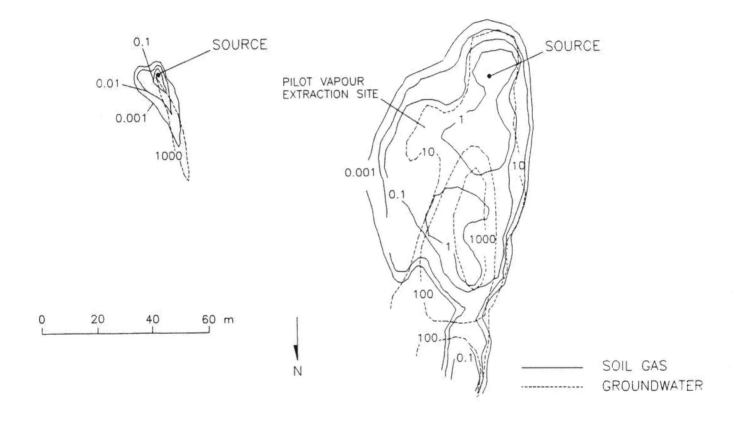

Figure 8-11. Aerial extent of soil gas and groundwater contamination derived from TCE emplaced below the water table (Site A) and in the vadose zone (Site B) (from Rivett and Cherry, 1991). All values in ug/L. Refer to Rivett and Cherry (1991) for details.

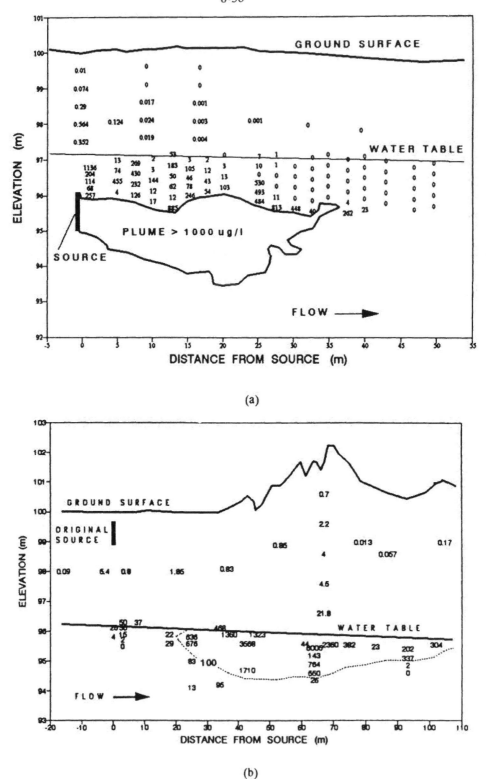

Figure 8-12. Longitudinal profiles showing the extent of soil gas and groundwater contamination parallel to the direction of flow and through the source areas derived from TCE emplaced (a) below the water table (Site A) and (b) in the vadose zone (Site B) (from Rivett and Cherry, 1991). All values in ug/L. Refer to Rivett and Cherry (1991) for details.

contamination to the soil gas. These experiments and real site soil gas surveys confirm transfer is occurring, however it may be expected that in non-arid climatic regions, after sufficient time and travel distance, continued recharge onto the shallow plume may prevent partitioning to the soil gas."

Rivett and Cherry describe several conceptual models of soil gas contamination that may arise from different DNAPL release scenarios. These models are provided in Table 8-10 and illustrate that soil gas surveying can be generally useful to delineate VOC contamination in the vadose zone and in shallow groundwater, but may fail to discern areas of deeper groundwater contamination that are not coincident with shallow soil gas contamination due to vapor transport.

Finally, although soil gas surveying has been used successfully to assist delineation of vadose zone and shallow groundwater contamination at many sites, it can provide misleading results if subsurface conditions are not understood adequately. Thus, interpretations of regarding the subsurface VOC distribution derived from a soil gas survey must be confirmed by analysis of soil and fluid samples taken at depth.

8.3 AERIAL PHOTOGRAPH INTERPRETATION

Aerial photographs should be acquired during the initial phases of a site characterization study to facilitate analysis of waste disposal practices and locations, drainage patterns, geologic conditions, signs of vegetative stress, and other factors relevant to contamination site assessment. Additionally, aerial photograph fracture trace analysis should be considered at sites where bedrock contamination is a concern.

8.3.1 Photointerpretation of Site Conditions

Government agencies have made extensive use of aerial photographs since the 1930s for the study of natural resources (Stoner and Baumgardner, 1979). For example, the U.S. Soil Conservation Service uses aerial photographs for soil mapping; the U.S. Department of Agriculture analyzes photos to check farmer compliance with government programs; and the U.S. Geological Survey uses photos to map and interpret geologic conditions. During contamination site investigations, the U.S. Environmental Protection Agency and others analyze

historic aerial photographs to document waste disposal practices and locations, geologic conditions, drainage patterns, pooled fluids, site development including excavations for pipelines and underground storage tanks, signs of vegetative stress, soil staining, and other factors relevant to assessing subsurface chemical migration (Phillipson and Sangrey, 1977).

Most photointerpretation involves qualitative stereoscopic analysis of a series of historic aerial photographs available for a particular site. For example, selected waste disposal features at a manufacturing site that were noted on aerial photographs taken during 1965 and 1966 are shown in Figure 8-13. Historical topographic maps and/or topographic profiles can also be derived from aerial photographs by use of analytical stereoplotting instrumentation. Such quantitative analysis facilitates examination of drainage patterns, land recontouring associated with waste disposal, waste volumes, etc.

Conventional aerial photography has been produced on behalf of government agencies for the entire United States. Much of the photography is vertical black and white coverage of moderate scale, typically about 1:20,000. Aerial photographs and remote sensing images are available through several agencies (Table 8-11). In particular, the National Cartographic Information Center (NCIC) in Reston, Virginia catalogs and disseminates information about aerial photographs and satellite images available from public and private sources. NCIC will provide a listing of available aerial photographs for any location in the United States and order forms for their purchase. Historic aerial photographs taken every few years dating back to the 1940s are available for many parts of the United States.

8.3.2 Fracture Trace Analysis

Fracture trace analysis involves stereoscopic study of aerial photographs to identify surface expressions of vertical or nearly-vertical subsurface zones of fracture concentrations (Figure 8-14). In fractured rock terrain, particularly in karst areas, groundwater flow and chemical transport are usually concentrated in fractures.

Lattman (1958) defined a photogeologic fracture trace as "a natural linear feature consisting of topographic (including straight stream segments), vegetation, or soil tonal alignments, visible primarily on aerial photographs, and expressed continuously for less than one mile . . . [that are not] related to outcrop pattern of tilted beds,

Table 8-10. Conceptual models (longitudinal sections) of soil gas and groundwater contamination resulting from NAPL releases (modified from Rivett and Cherry, 1991).

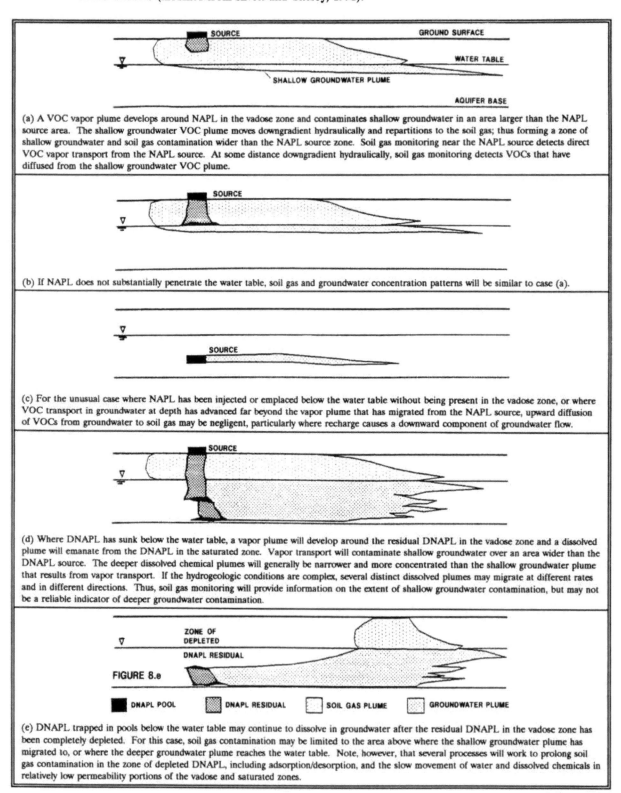

(a) A VOC vapor plume develops around NAPL in the vadose zone and contaminates shallow groundwater in an area larger than the NAPL source area. The shallow groundwater VOC plume moves downgradient hydraulically and repartitions to the soil gas; thus forming a zone of shallow groundwater and soil gas contamination wider than the NAPL source zone. Soil gas monitoring near the NAPL source detects direct VOC vapor transport from the NAPL source. At some distance downgradient hydraulically, soil gas monitoring detects VOCs that have diffused from the shallow groundwater VOC plume.

(b) If NAPL does not substantially penetrate the water table, soil gas and groundwater concentration patterns will be similar to case (a).

(c) For the unusual case where NAPL has been injected or emplaced below the water table without being present in the vadose zone, or where VOC transport in groundwater at depth has advanced far beyond the vapor plume that has migrated from the NAPL source, upward diffusion of VOCs from groundwater to soil gas may be negligent, particularly where recharge causes a downward component of groundwater flow.

(d) Where DNAPL has sunk below the water table, a vapor plume will develop around the residual DNAPL in the vadose zone and a dissolved plume will emanate from the DNAPL in the saturated zone. Vapor transport will contaminate shallow groundwater over an area wider than the DNAPL source. The deeper dissolved chemical plumes will generally be narrower and more concentrated than the shallow groundwater plume that results from vapor transport. If the hydrogeologic conditions are complex, several distinct dissolved plumes may migrate at different rates and in different directions. Thus, soil gas monitoring will provide information on the extent of shallow groundwater contamination, but may not be a reliable indicator of deeper groundwater contamination.

(e) DNAPL trapped in pools below the water table may continue to dissolve in groundwater after the residual DNAPL in the vadose zone has been completely depleted. For this case, soil gas contamination may be limited to the area above where the shallow groundwater plume has migrated to, or where the deeper groundwater plume reaches the water table. Note, however, that several processes will work to prolong soil gas contamination in the zone of depleted DNAPL, including adsorption/desorption, and the slow movement of water and dissolved chemicals in relatively low permeability portions of the vadose and saturated zones.

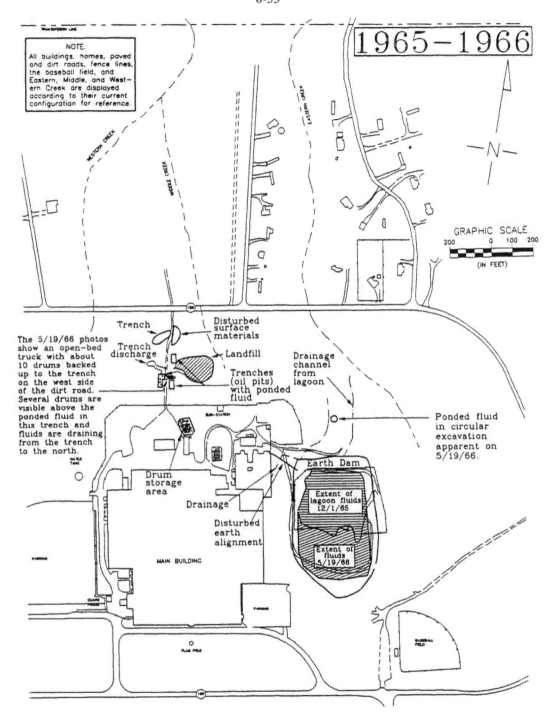

Figure 8-13. Selected waste disposal features identified at a manufacturing site using aerial photographs taken between 1965 and 1966.

Table 8-11. Sources of aerial photographs and related information.

National Cartographic Information Center
U.S. Geological Survey National Center
Reston, Virginia 22092

U.S. Department of the Interior
Geological Survey
EROS Data Center
Sioux Falls, South Dakota 57198

Western Aerial Photograph Laboratory
Administrative Services Division
ASCS-USDA
2505 Parley's Way
Salt Lake City, Utah 84109

The National Climatic Center
National Oceanic and Atmospheric Administration
Federal Building
Asheville, North Carolina 28801

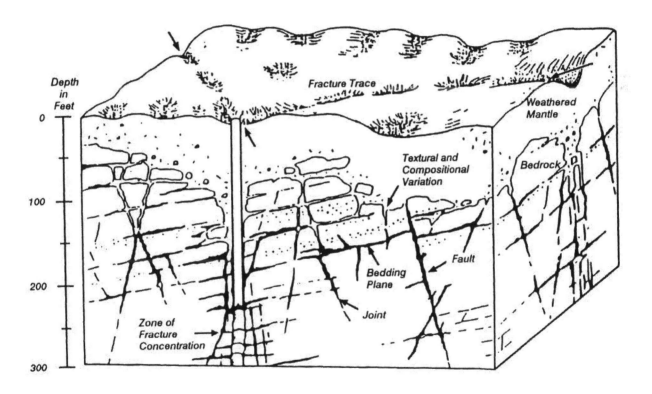

Figure 8-14. Relationship between fracture traces and zones of subsurface fracture concentration (from Lattman and Parizek, 1964).

lineation and foliation, and stratigraphic contacts." A lineament is a similar feature that is expressed for more than one mile in length. Numerous papers document the application of fracture trace analysis to hydrogeologic investigations (e.g., Lattman and Parizek, 1964; Siddiqui and Parizek, 1971; Parizek, 1976; Parizek, 1987). A manual on the principles and techniques of fracture trace analysis was prepared by Meiser and Earl (1982).

Fracture trace analysis is widely used to site productive wells and guide the placement of bedrock monitor wells at contamination sites. The significance of preferential groundwater flow along fracture zones for monitoring contaminant migration is illustrated in Figure 8-15. Extensive contaminant migration occur undetected if monitor wells are located in low permeability rock between zones of fracture concentration. A case study involving fracture trace analysis to define and remediate TCE contamination in fractured bedrock was described by Schuller et al. (1982).

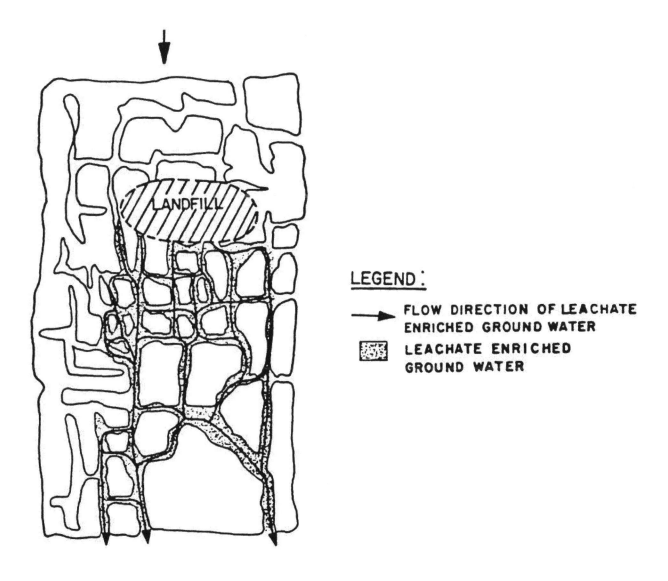

Figure 8-15. Preferential migration of contaminants in fracture zones can bypass a detection monitoring system (from USEPA, 1980).

9 INVASIVE METHODS

Following development of the site conceptual model (Chapter 5) based on available information (Chapter 7) and noninvasive field methods (Chapter 8), invasive techniques will generally be required to advance site characterization and enable the conduct of risk and remedy assessments. Various means of subsurface exploration are utilized to directly observe and measure subsurface materials and conditions. Generally, these invasive activities include: (a) drilling and test pit excavation to characterize subsurface solids and liquids; and (b) monitor well installation to sample fluids, and to conduct fluid level surveys, hydraulic tests, and borehole geophysical surveys. Invasive site characterization methods are described in numerous texts (Table 2-1; e.g., USEPA, 1991a). Their application to DNAPL site investigation is discussed in this Chapter.

9.1 UTILITY OF INVASIVE TECHNIQUES

Invasive field methods should be used in a phased manner to test and advance the site conceptual model based on careful consideration of site-specific conditions and DNAPL transport processes. Although the methods selected for invasive study are site-specific, their application will generally be made to (Chapter 6; Figure 6-1):

- delineate DNAPL source (entry) areas (Figure 9-1);

- define the stratigraphic, lithologic, structural, and/or hydraulic controls on the movement and distribution of DNAPL, contaminated groundwater, and contaminated soil gas;

- characterize the fluid and fluid-media properties that affect DNAPL migration and the feasibility of alternative remedies;

- estimate or determine the nature and extent of contamination, and the rates and directions of contaminant transport;

- evaluate exposure pathways; and

- design monitoring and remedial systems.

9.2 RISKS AND RISK-MITIGATION STRATEGIES

The risk of enlarging the zone of chemical contamination by use of invasive methods is an important consideration that must be evaluated during a DNAPL site investigation. DNAPL transport caused by site characterization activities may: (1) heighten the risk to receptors, (2) increase the difficulty and cost of site remediation, and/or (3) generate misleading data, leading to development of a flawed conceptual model, and flawed assessments of risk and remedy (USEPA, 1992).

Drilling, well installation, and pumping activities typically present the greatest risk of promoting DNAPL migration during site investigation. Drilling and well installations may create vertical pathways for DNAPL movement. In the absence of adequate sampling and monitoring as drilling progresses, it is possible to drill through a DNAPL zone without detecting the presence of DNAPL (USEPA, 1992). Increased hydraulic gradients caused by pumping may mobilize stagnant DNAPL. Precautions should be taken to minimize these risks. If the risks cannot be adequately minimized, alternate methods should be used, if possible, to achieve the characterization objective; or the objective should be waived. It is important to consider the question, "Is this data really necessary?"

Conventional drilling methods have a high potential for promoting downward DNAPL migration (USEPA, 1992). Special care should be exercised, in particular, to avoid causing the downward movement of mobile, perched DNAPL or DNAPL-contaminated soil that may result from drilling through a barrier layer. Similarly, DNAPL may sink preferentially along the inside or outside of a well. Specific conditions which may result in downward DNAPL migration include:

- an open borehole during drilling and prior to well installation;

- an unsealed or inadequately sealed borehole;

- a well screen that spans a barrier layer and connects an overlying zone with perched DNAPL to a lower transmissive zone;

- an inadequately sealed well annulus that allows DNAPL to migrate through: (a) the well-grout interface, (b) the grout, (c) the grout-formation interface, or (d) vertically-connected fractures in the disturbed zone adjacent to the well; and,

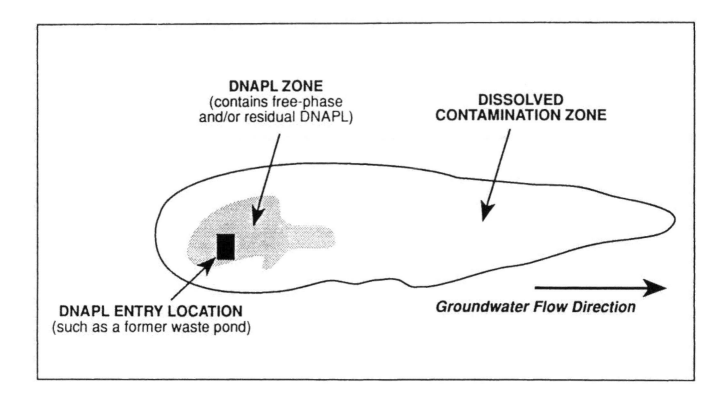

Figure 9-1. Defined areas at a DNAPL site (from USEPA, 1992).

- structural degradation of bentonite or grout sealant, or well casing, due to chemical effects of DNAPL or the groundwater environment.

To minimize the risk of inducing DNAPL migration as a result of drilling, site investigators should:

- avoid unnecessary drilling within the DNAPL zone;

- minimize the time during which a boring is open;

- minimize the length of hole which is open at any time;

- use telescoped casing drilling techniques to isolate shallow contaminated zones from deeper zones;

- utilize the site conceptual model (knowledge of site stratigraphy and DNAPL distribution), and carefully examine subsurface materials brought to the surface as drilling progresses, to avoid drilling through a barrier layer beneath DNAPL (i.e., stop drilling above or at the top of the barrier layer);

- consider using a dense drilling mud (i.e., with barium sulfate additives, also known as barite) to prevent DNAPL from sinking down the borehole during drilling;

- select optimum well materials and grouting methods based on consideration of site-specific chemical compatibility; and,

- if the long-term integrity of a particular grout sealant is questionable, consider placing alternating layers of different grout types and sealing the entire distance between the well screen and surface to minimize the potential for vertical migration of DNAPL.

Pumping from beneath or adjacent to the DNAPL zone can induce downward or lateral movement of DNAPL, particularly in fractured media due to the development of relatively high fluid velocities. In general, groundwater should not be pumped from an uncontaminated aquifer directly beneath a capillary barrier and overlying DNAPL zone. The risk of mobilizing stagnant DNAPL by pumping groundwater can be assessed using equations provided in Chapter 5.3.

The risk of causing DNAPL migration generally increases where there is fractured media, heterogeneous strata, multiple release locations, large DNAPL release volumes, and barrier layers that are subtle (e.g., a thin silt layer

beneath sand) rather than obvious (e.g., a thick soft clay layer beneath sand). At many sites, the DNAPL zone can be adequately characterized by limiting drilling to shallow depths. Characterization of deeper units can be accomplished by deeper borings and wells beyond the edge of the DNAPL zone.

Two basic strategies for invasive site characterization referred to as the outside-in and inside-out approaches are described as follows by USEPA (1992).

"The outside-in strategy of conducting initial invasive characterization outside suspected DNAPL areas and working toward the source has the advantage of allowing acquisition of significant geologic and hydrogeologic data at relatively low risk. These data can be used to refine the conceptual model and guide additional investigations. However, it should be noted that uncertainty exists in determining the extent of the suspected DNAPL area at any site. One disadvantage to the outside-in approach is that much additional time and expense may be incurred during the study if characterization is started too far from the suspected DNAPL zone. At some sites, it may be appropriate to avoid drilling directly within areas of known or suspected DNAPL contamination and focus on characterizing dissolved contaminant plumes migrating from source areas. However, drilling in suspected DNAPL areas may be required to provide the necessary information for site characterization and remedial design (e.g., assessing the presence and locations of DNAPLs and for application of in situ restoration technologies).

The inside-out approach of initially drilling within areas suspected to be the most contaminated and subsequently drilling in more remote areas to define the extent of contamination has been the traditional approach for many site investigations. One potential advantage of this method is that fewer boreholes may be required to determine the extent of contamination when investigation is started within an area of known contamination and progresses outward. The obvious disadvantage to this strategy is the increased potential for providing pathways for rapid downward migration of DNAPLs. The increased risks of remobilizing certain high mobility DNAPLs (e.g., chlorinated solvents) may render the inside-out approach the least desirable strategy at certain sites. This strategy appears to be most applicable at DNAPL sites where the immiscible fluids are relatively immobile (e.g., many creosote and coal tar sites). The relative time and cost advantages of the inside-out approach may be offset by additional investigation and remediation costs incurred if DNAPLs are mobilized to greater depths.

The choice of characterization strategy depends on the conceptual model of the site and the physical properties of the contaminants. Immiscible liquids with relatively high densities and low viscosities (e.g., chlorinated solvents) will be relatively more mobile than less dense, more viscous liquids (e.g., creosote or coal tar). More mobile liquids represent greater risk for spreading contamination to deeper zones during site characterization. In addition, these contaminants may have migrated considerably farther from the DNAPL entry location than less dense, more viscous liquids. The use of site history and noninvasive techniques may provide useful information for guiding invasive study. However, it will generally not be possible to determine the extent of immiscible liquids prior to invasive study. Locations for invasive study must be chosen using the best available site specific information and knowledge of DNAPL contaminant transport principles. Regardless of whether the outside-in or inside-out approach is chosen, characterization should proceed from shallow depths to greater depths. In this manner, more information is acquired concerning shallow geologic features and contamination and the risk of mobilizing DNAPLs to greater depths during drilling is reduced."

9.3 SAMPLING UNCONSOLIDATED MEDIA

Subsurface explorations (i.e., test pits and borings) permit direct measurement of stratigraphic, hydrogeologic, and contaminant conditions. Soil and waste zone sampling is conducted to pursue the site characterization goals outlined in Figure 6-1. Overburden sampling locations are selected to test and verify the site conceptual model. The numbers and locations of borings and samples required during a site characterization study depend on site-specific conditions and objectives. Estimated costs of drilling and test pit excavation in 1987 dollars are provided in Table 9-1.

The value of subsurface exploration relies, in large part, on the types and quality of measurements made and recorded during the investigation. Information which should be considered for logging is listed in Table 9-2. In addition to describing soil characteristics such as texture,

consistency, and moisture state, it is important to document visual and olfactory (if consistent with site safety plans) observations, and vapor monitoring detections, of contaminant presence. Inner sections of retrieved samples should be inspected to avoid misinterpreting core surfaces that may be contaminated during sampling. Particular care should be taken to examine soil fractures, ped faces, and macropores for NAPL presence. Methods for visual detection of NAPL presence are described in Chapter 9.10.

9.3.1 Excavations (Test Pits and Trenches)

Excavating test pits and/or trenches can be a very rapid and cost effective means to:

- characterize the nature and continuity of shallow overburden stratigraphy, including the macropore distribution;

- identify, delineate, and characterize waste disposal and grossly contaminated areas;

- help determine the horizontal and vertical extent of shallow contamination;

- locate and examine buried structures, tanks, pipelines, etc. and their bedding/backfill that may act as contaminant reservoirs or preferential pathways; and,

- acquire samples for chemical and physical analyses.

Guidelines for test pit and trench excavation, sampling, and backfilling are provided by USEPA (1987) and USDI (1974).

The main advantage of excavation for site characterization is that it provides an opportunity to examine a large, continuous subsurface section. As a result, excavations can reveal conditions such as subtle or complex stratigraphic relations, soil fracture patterns, heterogeneous NAPL distributions, and irregular disposal areas, that can be difficult to characterize by examining drill cuttings or samples. The potential risk of causing DNAPL migration is limited by the relatively shallow depth of excavation.

Limitations and/or disadvantages of excavation as a site characterization tool include: the limited depth of exploration using a backhoe; the diminished view of excavation sidewalls with depth due to shadows, viewing

Table 9-1. Drilling and excavation costs in April, 1987 dollars (from GRI, 1987).

ITEM	HIGH COST	LOW COST	MEAN COST
Drilling Soil Borings (3¼")	$39/ft	$18/ft	$28/ft
Rock Coring	$50/ft	$40/ft	$44/ft
Stainless Steel Screen (2", installed)	$375/5 ft	$175/5 ft	$252/5 ft
Stainless Steel Riser Pipe (2", installed)	$37/ft	$11/ft	$21/ft
PVC Screen (2", installed)	$50/5 ft	$35/5 ft	$43/5 ft
PVC Riser Pipe (2", installed)	$8/5 ft	$5/5 ft	$6/5 ft
Protective Casing	$150/each	$90/each	$113/each
Shelby Tube Samples (3")	$125/each	$40/each	$85/each
Water Truck Rental			$400/day
Steam Cleaner Rental	$125/day	$60/day	$85/day
Steam Cleaning Time	$140/hr	$112/hr	$125/hr
Stand By Time	$140/hr	$112/hr	$125/hr
Drilling in Level C Protection (Add)	$125/hr	$35/hr	$87/hr
Mobilization and Demobilization (200 miles)	$1250	$900	$1075
Test Pit Excavation Small Rubber Tired Backhoe and Operator			$70 - $110/hr
Large, Track-Mounted Backhoe (2 yd³ shovel) and Operator			$100 - $170/hr
Mobilization and Demobilization			$50 - $100/hr

Table 9-2. Information to be considered for inclusion in a drill or test pit log (modified from USEPA, 1987; Aller et al., 1989).

General:
- Project name/number
- Hole name/number
- Date started and finished
- Hole location; map and elevation

- Weather conditions
- Rig type, bit size/auger size
- Classification system used (e.g., Unified Soil Classification)

- Geologist's name
- Driller's name
- Sheet number

Information Columns:
- Depth
- Sample location/number
- Low counts and advance rate

- Percent sample recovery
- Narrative description
- Depth to saturation

- Well construction details
- Other remarks

Narrative Geologic Description:
- Soil/rock type
- Soil/rock texture and structure
- Color (Munsell) and stain
- Petrology and mineralogy
- Friability
- Moisture content (dry, moist, wet)
- Degree of weathering
- Presence of carbonate
- Fractures, joints (orientation, size, and spacing)

- Bedding nature and spacing
- Soil gradation or plasticity
- Discontinuities descriptions
- Water-bearing zones
- Formation strike and dip
- Fossils
- Depositional structures
- Organic content
- Solution cavities
- Rock core total breakage and breaks/ft

- Particle roundness or angularity
- Estimate of density of granular soil or consistency of cohesive soil (usually based on standard penetration test)
- Slickensides
- Roots, rootholes
- Residual or relict structure
- Buried horizons
- Disturbed earth, waste materials
- Rock Quality Designation (RQD)

Sampling Information:
- Types of sampler(s) used
- Diameter and length of sampler(s)
- Number of each sample
- Start and finish depth of each sample
- Percent sample recovery

- Split-spoon sampling
  + size and weight of drive hammer
  + number of blows required to penetrate each 6-inch interval
  + free fall distance used to drive sampler
- Thin-walled sampling
  + ease or difficulty pushing sampler
  + psi required to push sampler

- Rock coring
  + core barrel drill bit design
  + penetration rate
  + fluid gain or loss

Drilling Observations:
- Loss of circulation
- Advance rates
- Rig chatter
- Water-levels

- Changes in drilling method/equipment
- Drilling difficulties
- Amount of water yield/loss during drilling at different depths

- Caving/hole stability
- Amount of air used; air pressure
- Running sands
- Amounts and types of drilling fluids used

Well Construction Details:
- Well Design:
  + casing length, schedule, and diameter
  + joint type
  + screen length, schedule, and diameter
  + screen slot size
  + percent open area in screen
  + filter pack depth interval
  + elevations of top of casing, bottom and top of protective casing, ground surface, bottom of borehole, bottom and top of well screen, annular seal and grout intervals, etc.
  + well location coordinates and map
  + other backfill materials and intervals

- Materials:
  + casing and screen
  + filter pack (i.e., grain size analysis)
  + seal and physical form
  + slurry or grout mix
- Installation:
  + drilling method
  + drilling fluids
  + source of water
  + timing
  + method of sealant/grout emplacement
  + volumes of all materials used

- Development:
  + time and date
  + water level elevation before after development
  + development method
  + time spent developing well
  + volume of fluid removed
  + volume of fluid added
  + clarity of water and sediment before and after development
  + amount of sediment at well bottom
  + pH, specific conductance, and temperature readings

Other Remarks:
- Chemical odors
- Sample fluorescence
- NAPL sheens or presence
- HNU or OVA readings

- Sample shipping reference
- Equipment failures
- Deviations from drilling protocols
- Photograph cross-reference

- Air-monitoring data
- Hydrophobic dye test results
- Equipment decontamination procedures

angle, and distance (binoculars can enhance sidewall viewing); potential sidewall stability problems, of particular concern near structures, utilities, and roads; potential airborne release of contaminated vapors and dust; potential creation of a preferential pathway for contaminant transport along a trench; potential increased waste handling requirements; and potential subsidence problems after the excavation has been backfilled.

9.3.2 Drilling Methods

Borings and monitor wells are installed to evaluate subsurface stratigraphic, hydrogeologic, and contaminant conditions. Selection of drilling locations, depths, and methods is based on available information regarding site conditions. The potential for causing DNAPL migration by drilling through a barrier layer should be considered before and during drilling and hence minimized (Chapter 9.2).

Drilling methods applicable to contamination site investigations are documented by Aller et al. (1989), Driscoll (1986), Davis et al. (1991), GeoTrans (1989), Hackett (1987, 1988), Clark (1988a), GRI (1987), USEPA (1987), Rehm et al. (1985), Barcelona et al. (1985), Cambell and Lehr (1984), Acker (1974), USDI (1974), Morgan (1992), and ASTM (1990a). Additionally, the adaptation and use of cone penetrometers for delineating contaminated groundwater and soil during the past several years has been described by Robertson and Campanella (1986), Lurk et al. (1990), Smolley and Kappmeyer (1991), Christy and Spradlin (1992), and Chiang et al. (1992). These methods are described briefly, and their applications and limitations are noted, in Table 9-3. Diagrams depicting several drilling methods are shown in Figure 9-2.

Selecting a drilling method generally involves a trade-off between the advantages and limitations of the different techniques. Due to the risks associated with drilling at DNAPL sites, special consideration should be given to drilling methods which allow for: (1) continuous, high-quality sampling to facilitate identification of DNAPL presence and potential capillary barriers, and (2) highly-controlled well construction and hole abandonment.

Drilling in unconsolidated media at DNAPL sites is most commonly done using hollow-stem augers with split-spoon sampling. Riggs and Hatheway (1988) estimate that greater than 90 percent of all monitor wells in unconsolidated media in North America are installed using hollow-stem auger rigs. A detailed summary of drilling, sampling, and well construction methods using hollow-stem auger drill rigs is provided by Hackett (1987, 1988). Despite the advantages of drilling with hollow-stem augers, DNAPL can flow down through the disturbed zone along the outside of the augers and/or possibly enter the hollow stem auger through joints and then sink to the bottom of the boring (WCGR, 1991).

Some potential for causing vertical DNAPL migration is associated with all drilling methods. When drilling at DNAPL sites, this potential can be minimized by careful consideration of drilling strategies (Chapter 9.2).

Boreholes which are abandoned (not used for well construction) should be properly sealed to prevent vertical fluid movement in the borehole. Typically, this involves pumping a grout mixture through a tremie pipe and filling the hole from the bottom up to prevent segregation, dilution, and bridging of the sealant (Aller et al., 1989). Compositions and characteristics of various grouts are summarized in Table 9-4. Several investigators have reported that organic immiscible fluids cause permeability enhancement in bentonite clay which is frequently used as a sealant or grout additive for well construction and borehole abandonment (Anderson et al., 1981; Palombo and Jacobs, 1982; Brown et al., 1984; Evans et al., 1985; Abdul et al., 1990). Hydrophobic organic liquids tend to shrink the bentonite swelling clay structure and thereby produce fractures. Where DNAPLs are present, the use of bentonite, may be inappropriate (except perhaps as a minor-component grout additive). One strategy to limit vertical migration is to emplace layers of different sealants to fill the borehole (or well annulus). Thus, degradation of one sealant type will have a limited deleterious effect.

9.3.3 Sampling and Examination Methods

Soil or waste samples brought to the surface are typically examined to log some of the characteristics listed in Table 9-2. Drill cuttings or core material can be screened for volatile organic contaminant presence using portable instruments such as an organic vapor analyzer (OVA) and a HNU meter. Methods for visual detection of NAPL presence in soils are described in Chapter 9.10. Samples are also taken for detailed chemical analyses and to characterize media properties (i.e., wettability, capillary pressure and relative permeability curves, porosity, etc.; Chapter 10).

Table 9-3. Drilling methods, applications, and limitations (modified from Aller et al., 1989; GRI, 1987; Rehm et al., 1985; USEPA, 1987).

METHOD	APPLICATIONS/ADVANTAGES	LIMITATIONS
HAND AUGERS -- A hand auger is advanced by turning it into the soil until the bucket or screw is filled. The auger is then removed from the hole. The sample is dislodged from the auger, and drilling continues. Motorized units are also available.	• Shallow soil investigations (0 to 15 ft) • Soil samples collected from the auger cutting edge • Water-bearing zone identification • Contamination presence examination; sample analysis • Shallow, small diameter well installation • Experienced user can identify stratigraphic interfaces by penetration resistance differences as well as sample inspection • Highly mobile, and can be used in confined spaces • Various types (i.e., bucket, screw, etc.) and sizes (typically 1 to 9 inches in diameter) • Inexpensive to purchase	• Limited to very shallow depths (typically < 15 ft) • Unable to penetrate extremely dense or rocky or gravelly soil • Borehole stability may be difficult to maintain, particularly beneath the water table • Potential for vertical cross-contamination • Labor intensive
SOLID-FLIGHT AUGERS -- A cutter head ($\geq$ 2-inch diameter) is attached to multiple auger flights. As the augers are rotated by a rotary drive head and forced down by either a hydraulic pulldown or a feed device, cuttings are rotated up to ground surface by moving along the continuous flighting.	• Shallow soils investigations (< 100 ft) • Soil samples are collected from the auger flights or using split-spoon or thin-walled samplers if the hole will not cave upon retrieval of the augers • Vadose zone monitoring wells • Monitor wells in saturated, stable soils • Identification of depth to bedrock • Fast and mobile; can be used with small rigs • Holes up to 3-ft diameter • No fluids required • Simple to decontaminate	• Low-quality soil samples unless split-spoon or thin-wall samples are taken • Soil sample data limited to areas and depths where stable soils are predominant • Unable to install monitor wells in most unconsolidated aquifers because of borehole caving upon auger removal • Difficult penetration in loose boulders, cobbles, and other material that might lock up auger • Monitor well diameter limited by auger diameter • Cannot penetrate consolidated materials • Potential for vertical cross-contamination
HOLLOW-STEM AUGERS -- Hollow-stem augering is done in a similar manner to solid-flight augering. Small-diameter drill rods and samplers can be lowered through the hollow augers for sampling. If necessary, sediment within the hollow stem can be cleaned out prior to inserting a sampler. Wells can be completed below the water table using the augers as temporary casing.	• All types of soil investigations to <100 ft below ground • Permits high-quality soil sampling with split-spoon or thin-wall samplers • Water-quality sampling • Monitor well installation in all unconsolidated formation • Can serve as a temporary casing for coring rock • Can be used in stable formations to set surface casing • Can be used with small rigs in confined spaces • Does not require drilling fluids	• Difficulty in preserving sample integrity in heaving (running sand) formations • If water or drilling mud is used to control heaving will invade the formation • Potential for cross-contamination of aquifers where annular space not positively controlled by water or drilling mud or surface casing • Limited auger diameter limits casing size (typical augers are: 6¼-in OD with 3¼-in ID, and 12-in OD with 6-in ID) • Smearing of clays may seal off interval to be monitored

Table 9-3. Drilling methods, applications, and limitations (modified from Aller et al., 1989; GRI, 1987; Rehm et al., 1985; USEPA, 1987).

METHOD	APPLICATIONS/ADVANTAGES	LIMITATIONS
DIRECT MUD ROTARY -- Drilling fluid is pumped down the drill rods and through a bit attached to the bottom of the rods. The fluid circulates up the annular space bringing cuttings to the surface. At the surface, drilling fluid and cuttings are discharged into a baffled sedimentation tank, pond, or pit. The tank effluent overflows into a suction pit where drilling fluid is recirculated back through the drill rods. The drill stem is rotated at the surface by top head or rotary table drives and down pressure is provided by pull-down devices or drill collars.	• Rapid drilling of clay, silt, and reasonably compacted sand and gravel to great depth (>700 ft) • Allows split-spoon and thin-wall sampling in unconsolidated materials • Allows drilling and core-sampling in consolidated rock • Abundant and flexible range of tool sizes and depth capabilities • Sophisticated drilling and mud programs available • Geophysical borehole logs	• Difficult to remove drilling mud and wall cake from outer perimeter of filter pack during development • Bentonite or other drilling fluid additives may influence quality of ground-water samples • Potential for vertical cross-contamination • Circulated cutting samples are of poor quality; difficult to determine sample depth • Split-spoon and thin-wall samplers are expensive and of questionable cost effectiveness at depths > 150 ft • Wireline coring techniques for sampling both unconsolidated and consolidated formations often not available locally • Drilling fluid invasion of permeable zones may compromise integrity of subsequent monitor well samples • Difficult to decontaminate pumps
AIR ROTARY -- Air rotary drilling is similar to mud rotary drilling except that air is the circulation medium. Compressed air injected through the drill rods circulates cuttings and groundwater up the annulus to the surface. Typically, rotary drill bits are used in sedimentary rocks and down-hole hammer bits are used in harder igneous and metamorphic rocks. Monitor wells can be completed as open hole intervals beneath telescoped casings.	• Rapid drilling of semi-consolidated and consolidated rock to great depth (>700 ft) • Good quality/reliable formation samples (particularly if small quantities of drilling fluid are used) because casing prevents mixture of cuttings from bottom of hole with collapsed material from above • Allows for core-sampling of rock • Equipment generally available • Allows easy and quick identification of lithologic changes • Allows identification of most water-bearing zones • Allows estimation of yields in strong water-producing zones with short "down time"	• Surface casing frequently required to protect top of hole from caving • Drilling restricted to semi-consolidated and consolidated formations • Samples reliable, but occur as small chips that may be difficult to interpret • Drying effect of air may mask lower yield water producing zones • Air stream requires contaminant filtration • Air may modify chemical or biological conditions; recovery time is uncertain • Potential for vertical cross-contamination • Potential exists for hydrocarbon contamination from air compressor or down-hole hammer bit oils
AIR ROTARY WITH CASING DRIVER -- This method uses a casing driver to allow air rotary drilling through unstable unconsolidated materials. Typically, the drill bit is extended 6 to 12 inches ahead of the casing, the casing is driven down, and then the drill bit is used to clean material from within the casing.	• Rapid drilling of unconsolidated sands, silts, and clays • Drilling in alluvial material (including boulder formations) • Casing supports borehole, thereby maintaining borehole integrity and reducing potential for vertical cross-contamination • Eliminates circulation problems common with direct mud rotary method • Good formation samples because casing (outer wall) prevents mixture of caving materials with cuttings from bottom of hole • Minimal formation damage as casing pulled back (smearing of silts and clays can be anticipated)	• Thin, low pressure water-bearing zones easily overlooked if drilling not stopped at appropriate places to observe whether or not water levels are recovering • Samples pulverized as in all rotary drilling • Air may modify chemical or biological conditions; recovery time is uncertain

Table 9-3. Drilling methods, applications, and limitations (modified from Aller et al., 1989; GRI, 1987; Rehm et al., 1985; USEPA, 1987).

METHOD	APPLICATIONS/ADVANTAGES	LIMITATIONS
DUAL-WALL REVERSE ROTARY -- Circulating fluid (air or water) is injected through the annulus between the outer casing and drill pipe, flows into the drill pipe through the bit, and carries cuttings to the surface through the drill pipe. Similar to rotary drilling with the casing driver, the outer pipe stabilizes the borehole and reduces cross-contamination of fluids and cuttings. Various bits can be used with this method.	• Very rapid drilling through both unconsolidated and consolidated formations • Allows continuous sampling in all types of formations • Very good representative samples can be obtained with reduced risk of contamination of sample and/or water-bearing zone • Allows for rock coring • In stable formations, wells with diameters as large as 6 inches can be installed in open hole completions	• Limited borehole size that limits diameter of monitor wells • In unstable formations, well diameters are limited to approximately 4 inches • Equipment available more common in the southwest U.S. than elsewhere • Air may modify chemical or biological conditions; recovery time is uncertain • Unable to install filter pack unless completed open hole
CABLE TOOL DRILLING -- A drill bit is attached to the bottom of a weighted drill stem that is attached to a cable. The cable and drill stem are suspended from the drill rig mast. The bit is alternatively raised and lowered into the formation. Cuttings are periodically removed using a bailer. Casing must be added as drilling proceeds through unstable formations.	• Drilling in all types of geologic formations • Almost any depth and diameter range • Ease of monitor well installation • Ease and practicality of well development • Excellent samples of coarse-grained media can be obtained • Potential for vertical cross-contamination is reduced because casing is advanced with boring • Simple equipment and operation	• Drilling is slow, and frequently not cost-effective as a result • Heaving of unconsolidated materials must be controlled • Equipment availability more common in central, north central, and northeast sections of the U.S.
ROCK CORING -- A carbide or diamond-tipped bit is attached to the bottom of a hollow core barrel. As the bit cuts deeper, the rock sample moves up into the core tube. With a double-wall core barrel, drilling fluid circulates between the two walls and does not contact the core, allowing better recovery. Clean water is usually the drilling fluid. Standard core tubes are attached to the bottom of a drill rod and the entire string of rods must be removed after each core run. With wireline coring, an inner core barrel is withdrawn through the drill string using an overshot device that is lowered on a wireline into the drill string.	• Provides high-quality, undisturbed core samples of stiff to hard clays and rock • Holes can be drilled at any angle • Can detect location and nature of rock fractures • Can use core holes to run a complete suite of geophysical logs • Variety of core sizes available • Core holes can be utilized for hydraulic tests and monitor well completion • Can be adapted to a variety of drill rig types and operations	• Relatively expensive and slow rate of penetration • Can lose a large quantity of drilling water into permeable formations • Potential for vertical cross-contamination

Table 9-3. Drilling methods, applications, and limitations (modified from Aller et al., 1989; GRI, 1987; Rehm et al., 1985; USEPA, 1987).

METHOD	APPLICATIONS/ADVANTAGES	LIMITATIONS
CONE PENETROMETER -- Hydraulic rams are used to push a narrow rod (e.g., 1.5-inch diameter) with a conical point into the ground at a steady rate. Electronic sensors attached to the test probe measure tip penetration resistance, probe side resistance, inclination and pore pressure. Sensors have also been developed to measure subsurface electrical conductivity, radioactivity, and optical properties (fluorescence and reflectance). Cone penetrometer tests (CPT) are generally performed using a special rig and a computerized data collection, analysis, and display system. To facilitate interpretation of CPT data from numerous tests, CPT data from at least one test per site should be compared to a log of continuously sampled soil at an adjacent location. References: Robertson and Campanella (1986), Lurk et al. (1990), Smolley and Kappmeyer (1991), Christy and Spradlin (1992), Edge and Cordry (1989), and, Chiang et al. (1992).	• Efficient tool for stratigraphic logging of soft soils • Measurement of some soil/fluid properties (e.g., tip penetration resistance, probe side fraction, pore pressure, electrical conductivity, radioactivity, fluorescence), with proper instrumentation, can be obtained continuously rather than at intervals; thus improving the detectability of thin layers (i.e., subtle DNAPL capillary barriers) and contaminants • There are virtually no cuttings brought to the ground surface, thus eliminating the need to handle cuttings • Process presents a reduced potential for vertical cross-contamination if the openings are sealed with grout from the bottom up upon rod removal • Porous probe samplers can be used to collect groundwater samples with minimal loss of volatile compounds • Soil gas sampling can be conducted • Fluid sampling from discrete intervals can be conducted using special tools (e.g., the Hydropunch™ manufactured by Q.E.D. Environmental Systems of Ann Arbor, Michigan)	• Unable to penetrate dense geologic conditions (i.e., hard clays, boulders, etc.) • Limited depth capability (depends on • Soil samples cannot be collected for examination or chemical analyses, unless special equipment is utilized • Only very limited quantities of groundwater can be sampled • Limited well construction capability • Limited availability

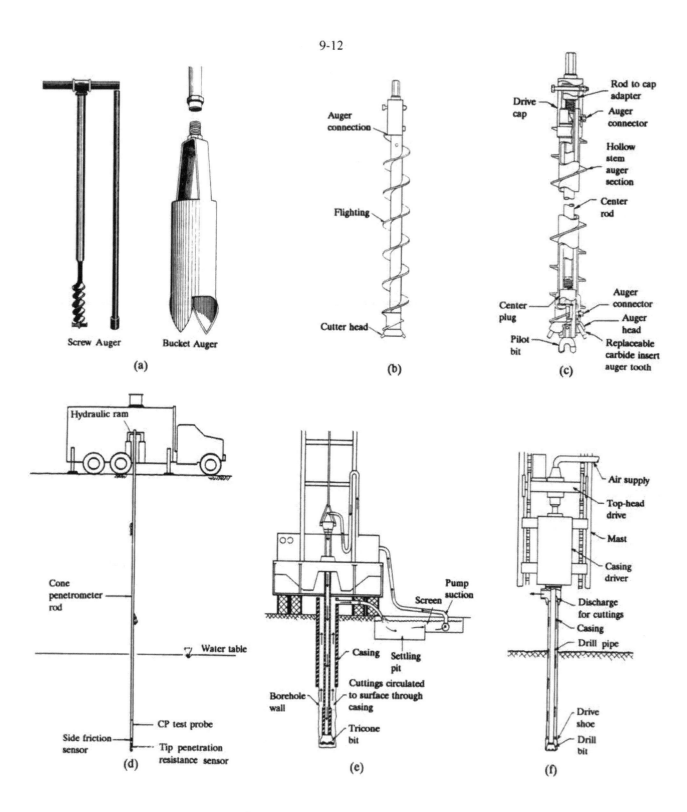

Figure 9-2. Schematic diagrams of several boring methods: (a) screw and bucket augers, (b) solid stem auger, (c) hollow-stem auger, (d) cone penetrometer test probe, (e) mud rotary, and (f) air rotary with a casing driver (reprinted with permission, EPRI, 1985).

Table 9-4. Borehole and well annulus grout types and considerations (modified from Aller et al., 1989; Edil et al., 1992).

BENTONITE AND BENTONITE-CEMENT GROUTS: Bentonite is a hydrous aluminum silicate comprised primarily of montmorillonite clay. The volume of hydrated bentonite in water is typically 10 to 15 times greater than that of dry bentonite because water is incorporated within the expanding clay lattice. The low permeability and expansion of bentonite in water are desirable properties for sealing abandoned boreholes and well annular spaces. Bentonite grouts are best prepared using mechanical mixers and should be pumped under pressure in place from the base of the interval to be grouted through a tremie pipe. Bentonite grouts should be mixed in batches so that they can be pumped before becoming too viscous. Bentonite grout should not be placed in the vadose zone because it will dry, shrink, and fracture. Bentonite grout may also shrink and fracture in the presence of hydrophobic NAPLs. Several available bentonite grout types are described below.

Bentonite Slurry Grout is commonly prepared by mixing dry bentonite powder in fresh water at a ratio of 15 lbs of bentonite to 7 gallons of water to make 1 ft^3 of slurry. Thick slurries may gel prematurely and be impossible to emplace. Due to their low solids content, bentonite slurries tend to settle as liquid bleeds off, requiring the emplacement of more slurry.

Quick-Gel® Bentonite Drilling Mud Grout is slurry of sodium bentonite and water that is marketed primarily as a drilling mud. Grouts of varying viscosity and strength can be obtained by mixing different proportions of Quick Gel®, water, and sand. Slurries containing sand appear more stable than pure Quick Gel®. Edil et al. (1992) found that Quick Gel® slurries of different sand content and viscosity form poorer annulus seals than neat cement, cement-bentonite, and Benseal®-bentonite slurry grouts.

Volclay® Bentonite Powder Grout is a commercial bentonite-based clay grout that is formulated for sealing boreholes and well annular spaces. Edil et al. (1992) mixed 2.1 lbs of Volclay® per gallon of water and added 2 lbs of magnesium oxide powder as a setting inhibitor to each 50 lbs of Volclay® slurry. They determined that Volclay® grout has a stiff gel structure which adheres to PVC but not steel well casing; and that it is not as effective a well sealant as neat cement, cement-bentonite, and Benseal®-bentonite slurry grouts.

Benseal® - Bentonite Slurry Grout is a mixture of Benseal®, a granular nondrilling mud grade bentonite developed for use in sealing and grouting well casings, and bentonite powder with water. Edil et al. (1992) mixed 30 lbs of Natural Gel® (a natural, unaltered bentonite powder) with 100 gallons of water, and then used a venturi pump to mix in 125 lbs of Benseal® to the slurry. They found that this grout adheres to steel and PVC casing, has low permeability, good swelling characteristics, and flexibility, and is an excellent sealant.

Bentonite-Cement Grout is a slurry incorporating 5 to 6 gallons of water and 2 to 6 lbs of bentonite powder for each 94 lbs (1 ft^3) of Portland cement. Bentonite improves the workability of the cement slurry, reduces slurry density, and reduces grout shrinkage during setting. Edil et al. (1992) found the addition of 5 lbs of bentonite per 94 lbs of cement forms a rigid well annulus seal with low permeability and high durability; and that the grout adheres to steel casing, but appeared to allow some infiltration along the grout-PVC casing interface.

Bentonite Pellets can be used to seal borehole or well annulus intervals. Wet pellets, however, tend to stick to well casing and borehole walls, and bridge high above their intended placement depth. A tamper can be used to break up bridges, but this technique becomes ineffective at depths greater than approximately 20 ft. Pellets can be frozen using refrigeration or liquid nitrogen to increase their fall distance.

PORTLAND CEMENT: Neat cement is a mixture of Portland cement (ASTM C-150) and water in the proportion of 5 to 6 gallons of clean water per bag (94 lbs or 1 ft^3) of cement. Five types of Portland cement are produced: Type I for general use; Type II for moderate sulfate resistance of moderate heat of hydration; Type III for high early strength; Type IV for low heat of hydration; and Type V for high sulfate resistance. Type I is most widely-used in well construction or hole abandonment. A typical 14 lb/gallon neat cement slurry with a mixed volume of 1½ ft^3 will have a set volume of 1.2 ft^3, reflecting a 17% shrinkage. The setting time ranges from 48 to 72 hrs depending primarily on water content.

Common additives include: (1) 2 to 6% bentonite to reduce shrinkage, improve workability, reduce density, and produce a lower cost per volume of grout; (2) 1 to 3% calcium chloride to accelerate the setting time and thereby create higher early strength, of particular value in cold climates; (3) 3 to 6% gypsum to produce a quick-setting very hard cement that expands upon setting; (4) <1% aluminum powder to produce a quick-setting strong cement that expands upon setting; (5) 10 to 20% flyash to increase sulfate resistance and provide early compressive strength; (6) hydroxylated carboxylic acid to retard setting time and improve workability without compromising set strength; and (7) diatomaceous earth to reduce slurry density, increase water demand and thickening time, and reduce set strength.

Edil et al. (1992) found neat cement grout forms a rigid seal with low permeability and high durability that adheres fairly well to steel and PVC casing. Kurt and Johnson (1982), however, report that neat cement annular seals are subject to channeling between the casing and grout due to temperature changes during curing, swelling and shrinkage during curing, and poor bonding between the ground and casing. Cement shrinkage can produce fractures, thereby degrading the integrity of the grout seal. Cement slurries can infiltrate the well sandpack, particularly if well development occurs prior to when the cement has completely set. Thus, a minimum of 1 to 2 ft of filter pack is usually extended in the annulus above the top of the well screen. The high heat of cement hydration can compromise the integrity of thermoplastic casing. Cement is a highly alkaline substance with a pH that ranges from 10 to 12. This can alter groundwater pH.

Unconsolidated media sampling methods are described by Davis et al. (1991), Aller et al. (1989), Clark (1988a), GRI (1987), USEPA (1987), Driscoll (1986), Rehm et al. (1985), Acker (1974), USDI (1974), ASTM (1990a), Zapico et al. (1987), Clark (1988b), Ostendorf et al. (1991), McElwee et al. (1991), Starr and Ingleton (1992), and Christy and Spradlin (1992). High-quality soil sampler method descriptions, advantages, and limitations are provided in Table 9-5. Diagrams depicting several samplers are shown in Figure 9-3.

Soil samples submitted for chemical analysis are frequently collected using split-spoon samplers. Typically, soil from the inner portion of the core is removed from the spoon and placed in a sample jar containing air. This procedure may result in significant loss of volatile organic compounds (VOCs) from the soil sample during storage and handling (WCGR, 1991). For example, an analytical laboratory log of a soil sample taken this way at the Love Canal DNAPL site in Niagara Falls, New York (New York State Department of Health, 1980) notes "oily film found in desiccation crack although it evaporated quickly."

The loss of VOCs can be reduced by storing samples in VOA vials, and minimized by placing samples in jars containing methanol followed by analysis of the solvent (WCGR, 1991). Care must be taken to avoid exposure of the methanol to contaminated materials (other than the soil sample) or air. Trichloroethene concentrations detected in soils that were stored in wide-mouth jars with air and in jars with methanol are compared in Table 9-6 (WCGR, 1991). The data show that storing soil in a jar with air may result in a significant underestimation of VOC concentrations. Volatile loss can also be minimized by sealing the ends of core barrel samplers and submitting sealed samples directly to the laboratory (Cherry and Feenstra, 1991). Finally, samples should be kept refrigerated, analyzed quickly, and not agitated to limit volatile loss.

9.4 ROCK SAMPLING

Bedrock drilling and sampling are conducted to characterize the extent of contamination in the subsurface. Drilling in rock is primarily accomplished using rotary and coring methods (Table 9-3; Figure 9-2). Drilling method references are given in Chapter 9.3. Bedrock characteristics which should be considered for logging are listed in Table 9-2.

Drilling in rock typically poses a significantly greater risk of promoting vertical DNAPL movement than drilling in unconsolidated media. This is due to the brittle and heterogeneous, fractured nature of rocks. Fracture networks in rock are usually ill-defined. Similarly, the distribution of DNAPL in rock fractures is typically difficult to predict or characterize (Figure 5-8a). Factors to consider when evaluating the distribution of DNAPL in rock fractures include: (1) fracture orientations, spacings, and apertures, (2) DNAPL density, viscosity, and interfacial tension, (3) DNAPL release volume, (4) the DNAPL pool height driving density flow, and (5) hydraulic gradients. When drilling in rock, DNAPL can enter and exit the borehole unpredictably via fractures and drilling may create or widen fractures in the near-well environment. Where possible, drilling through bedrock contaminated with DNAPL should be avoided. Where drilling is necessary, the risk minimization strategies listed in Chapter 9.2 should be implemented as practicable.

An example protocol for characterizing fractured bedrock at DNAPL sites is that utilized at the S-Area Landfill CERCLA site in Niagara Falls, New York (Conestoga-Rovers and Associates, 1986). The drilling sequence for coring, hydraulic testing, and grouting the Lockport Dolomite is illustrated in Figure 9-4 and adapted with slight modification from Conestoga-Rovers and Associates (1986) below.

(1) At each bedrock survey well location, the overburden will be penetrated using a hollow-stem auger drill rig as follows:

(a) Drilling at each location will advance to the bedrock/overburden interface taking continuous split-spoon samples of the overburden materials. The augers will penetrate to refusal on competent bedrock. Eight-inch ID augers will be required for this operation. Once the augers are seated on top of the rock, the auger plug will be advanced an additional 6 inches into the rock to create a pilot hole for the 6-inch diameter permanent steel casing.

(b) A clean six-inch diameter permanent steel casing (¼ to ½ inch wall thickness) will be inserted through the augers (or through a temporary casing) and will be seated and properly sealed into the top of the bedrock so as to ensure that no overburden groundwater has access to the borehole. Seating and sealing of permanent casing into the top of the bedrock will be

Table 9-5. Soil sampler descriptions, advantages, and limitations (modified from Acker, 1974; Rehm et al., 1985; Aller et al., 1989).

METHOD DESCRIPTION	ADVANTAGES	LIMITATIONS
SPLIT-SPOON (SPLIT-BARREL) SAMPLERS The Standard Penetration Test procedure for driving a split-spoon sampler to obtain a representative soil sample and a measure of soil penetration resistance is described by ASTM Test Method D1586-84. The split-spoon sampler is 18 to 30 inches long with a 1½-inch ID and made of steel. It is attached to the end of drill rods, lowered (typically through a hollow-stem auger) to the bottom of the borehole which must be clean, and then hammered into the undisturbed soil by dropping a 140-lb weight a distance of 30 inches onto an anvil that transmits the impact to the drill rods. The number of blows required to drive the sampler each 6-inch interval is counted to determine penetration resistance. Continuous or noncontinuous samples can be taken, and various other split-barrel diameter sizes are available. These samplers can also be pushed into the ground rather than hammered.	• High quality samples can be evaluated for mineralogical, stratigraphic, physical, and chemical properties • Steel, brass or plastic liners can be used with split-spoon samplers so that samples can be sealed to minimize changes in sample chemical and physical conditions prior to delivery to a laboratory • Relatively inexpensive • Widely available	• Hammering creates a stress that can consolidate and alter the sample • Sample transfer from the split spoon can result in disaggregation of cohesionless soil • Sample handling exposes soil to atmosphere and may result in loss of volatile chemicals • Cannot penetrate rock, cobbles, and some gravels • Poor recovery of some loose or flowing cohesionless samples (although sample retainers can be used to minimize this problem)
THIN-WALL (SHELBY) OPEN-TUBE SAMPLERS Open-tube thin-wall samplers consist of a connector head and a 30 or 36 inch long thin-wall steel, aluminum, brass, or stainless steel tube which is sharpened at the cutting edge. The wall thickness should be less than 2½% of the tube outer diameter, which is commonly 2 or 3 inches. The sampler is attached via its connector head to the end of drill rods, lowered (typically through a hollow-stem auger) to the bottom of the borehole which must be clean, and then pushed down into the undisturbed soil using the hydraulic or mechanical pulldown of the drill rig. This procedure is described by ASTM Method D1587-83. The Central Mining Company (CME) recently developed a 5-ft long continuous thin-wall sampling system. The tube is kept in place by a latching mechanism that allows the sample to be retracted by wireline when full and replaced with an empty tube.	• Provides undisturbed samples in stiff, cohesive soils and representative samples in soft to medium cohesive soils • High quality samples can be evaluated for mineralogical, stratigraphic, physical, and chemical properties • Sample can be preserved and stored within the sample tube by sealing its ends, thereby minimizing sample disturbance prior to lab analysis • Widely available • Relatively inexpensive	• The sampler should be at least six times the diameter of the longest particle size to minimize disturbance of the sample • Large gravel or cobbles can disturb the finer grained soil within which they are embedded and/or can damage the sampler walls • Due to thin wall and limited structural strength, the sampler cannot be easily pushed into dense or consolidated materials • Generally not effective for cohesionless soils
THIN-WALL PISTON CORE SAMPLERS These samplers consist of a thin-wall tube, with an internal piston, and mechanisms to control movement between the piston and tube. Thin-wall piston samplers are typically set up and pushed into the ground in the same manner as thin-wall open-tube samplers. The internal pistons generate a vacuum on the sample as the sampler is withdrawn from the hole. Starr and Ingleton (1992) recently developed a drive point piston sampler to collect high quality core samples of sands, silts, and clays without drilling fluids or a drilling rig to a depth of approximately 30 ft.	• Provides undisturbed samples in cohesive soils, silts, and sands above or below the water table • Vacuum enables recovery of cohesionless soils • High quality samples can be evaluated for mineralogical, stratigraphic, physical, and chemical properties • Sample can be preserved and stored within the sample tube by sealing its ends, thereby minimizing sample disturbance prior to lab analysis	• As with open-tube sampler, large particles may disturb sample or damage sampler walls and the sampler cannot be easily pushed into dense or consolidated materials • If used with a clam shell fitted auger head, only 1 sample can be obtained per borehole because the clam shell will not close after being opened; continuous sampling not possible • Some piston samplers require use of drilling fluid for hydrostatic control • Not as widely available as split-spoon or open-tube samplers • Relatively expensive
CORE BARREL SAMPLERS (see ROCK CORING description in Table 9-3)	See ROCK CORING advantages in Table 9-3	See ROCK CORING limitations in Table 9-3

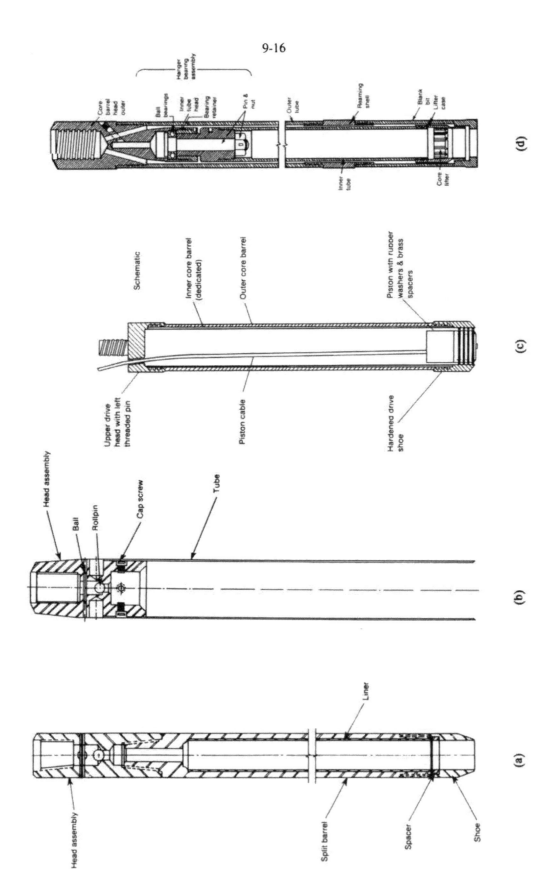

Figure 9-3.   Schematic diagrams of a (a) split-spoon sampler, (b) thin-wall open-tube sampler, and (c) thin-wall piston sampler used to obtain undisturbed soil samples; and of a (d) double-tube core barrel used to obtain rock core (modified from Aller et al, 1989).

Table 9-6. Comparison of trichloroethene (TCE) concentrations determined after storing soil samples in jars containing air versus methanol; showing apparent volatilization loss of TCE from soil placed in jars containing air (from WCGR, 1991).

SPLIT-SPOON SAMPLE PLACED IN WIDE-MOUTH JAR CONTAINING AIR AND THEN SUB-SAMPLED, EXTRACTED, AND ANALYZED		SPLIT-SPOON SAMPLE PLACED IN WIDE-MOUTH JAR CONTAINING METHANOL TO PRESERVE VOLATILES AND SUBSEQUENT ANALYSIS OF THE SOLVENT	
Sample Depth (ft BGS)	TCE Concentration (mg/Kg)	Sample Depth (ft BGS)	TCE Concentration (mg/Kg)
5.0 - 7.0	2.2	7.0	3,100
20.0-20.5	9.2	20.0	420
30.0-30.5	<0.022	30.0	210

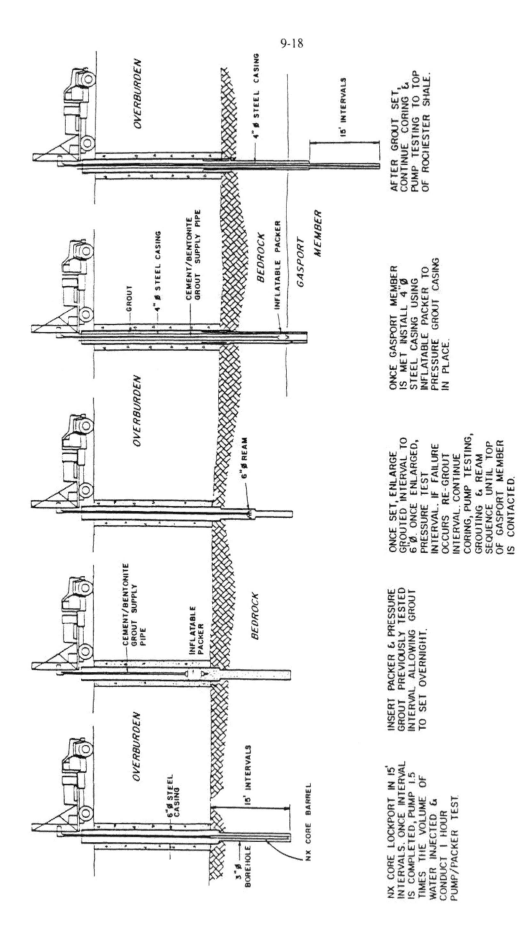

Figure 9-4. The drilling sequence for coring, hydraulic testing, and grouting through the Lockport Dolomite utilized at the Occidental Chemical Corporation S-Area DNAPL site in Niagara Falls, New York (from Conestoga-Rovers and Associates, 1986).

completed as shown in Figure 9-5 and described below:

(i) Seat the casing into the top of rock taking care to center the casing.

(ii) Insert an inflatable packer through the 6-inch diameter casing to within 12 inches of the bottom of the hole and pump grout under pressure through the packer via the drill rod and into the annular space between the casing and the auger/temporary casing inside wall. Continue pumping grout until it is observed at ground surface. As grouting progresses, remove the augers (or temporary casing) from the permanent casing, and add grout as required. All grout will be placed using positive placement procedures. Upon completion of grouting, the casing will be tapped firmly into the pilot hole using the 140 lb hammer.

(iii) The materials required to grout wells shall consist of Portland cement and clean water. Hydrated lime may be added to facilitate placing of the grout mixture. Bentonite will be necessary to reduce shrinkage. All additives will be added in accordance with ASTM standards. Cement grout placed to seal the annular space between permanent well casing and an oversized hole shall be mixed in the proportion of not less than five or more than six gallons of water to one 94-lb bag of cement. Hydrated lime (ten percent by volume) may be needed to facilitate pumping and bentonite (3% by volume) will be needed to prevent shrinking.

(iv) The packer will be released and the grout will be allowed to set for 48 hours before drilling continues.

(v) The grout seal will be tested by conducting a hydraulic test modified from USDI (1974, p.575). After drilling through the grout until rock is encountered, the casing is filled with water. The water-level decline, ΔH, from the top of the casing is measured over a five minute period. The seal shall

be regrouted if the water level declines by more than that calculated by

$$\Delta H = (5.5 \, K \, H \, t) \, / \, (\pi \, r) \qquad (9\text{-}1)$$

where H is the head of water applied (top of casing minus initial water-level), r is the casing radius, t is the test duration, and K is the allowable grout hydraulic conductivity (1 X 10-5 cm/sec). Any consistent units may be used.

(c) All drilling rods used in the overburden materials will be cleaned prior to the bedrock coring operation to prevent cross-contamination from the overburden into bedrock.

(2) Drilling of the Lockport Dolomite will be carried out in 15-ft increments to facilitate testing of the bedrock formation. The drilling sequence will be completed as shown in Figure 9-4 and described below.

(a) Collect continuous NX rock core (3-inch OD hole, 2.2-inch OD core) of the top 15-ft bedrock interval. The drill water will be spiked with an approved tracer (see 2c).

(b) Insert a high volume pump (10 gpm minimum) to remove 1 to 1½ times the volume of water introduced to the formation during the coring operation.

(c) Insert a packer/pump assembly as shown in Figure 9-6. Inflate the packer and pump test the 15-ft bedrock interval for one hour. Collect a groundwater sample for chemical analysis if interval yields ≥ 0.3 gpm during the last 15 minutes of the test. The sample will also be analyzed for the drill water tracer to account for dilution effects in the analytical results.

(d) Remove the packer/pump assembly and insert a packer system 2 ft above the top of the tested bedrock interval and pressure grout the interval. Grouting pressure will not be allowed to exceed 0.7 psi per foot of overlying earth. The cement will be quick set cement (type 3 with accelerator) and allowed to set overnight. In cases where an interval yielded < 0.3 gpm during the last 15 minutes of the 1 hour pump test, grouting will be optional. Control samples of the grout will be observed to document the quality of the set.

9-20

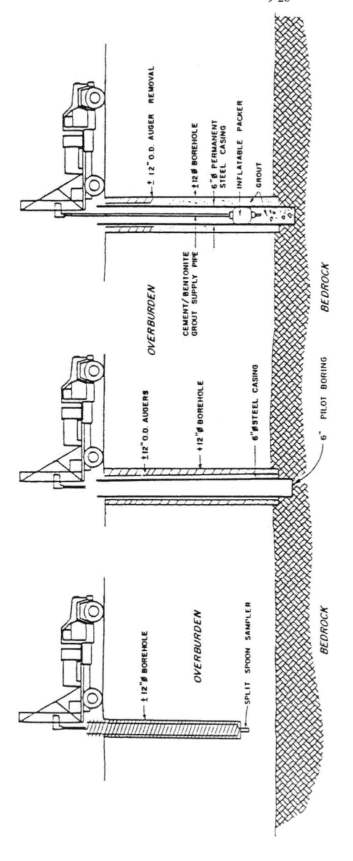

Auger and continuous split spoon to top of competent bedrock.

Once to top of competent bedrock, install 6"-diameter steel casing inside augers to top of bedrock.

Install inflatable packer inside 6"-diameter steel casing (within 12" of bottom of hole), inflate, and commence pumping grout through the packer, forcing grout to ground surface. Remove augers adding grout as required to maintain continous grout envelope. Once all augers have been removed, seat 6"-diameter casing into rock by driving with 140 lb. hammer until refusal.

Figure 9-5. Overburden casing installation procedure utilized prior to drilling into the Lockport Dolomite at the Occidental Chemical Corporation S-Area DNAPL site in Niagara Falls, New York (from Conestoga-Rovers and Associates, 1986).

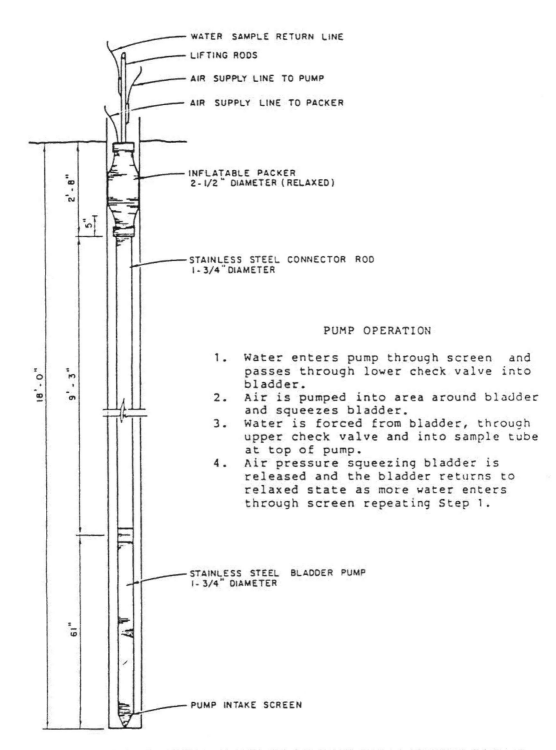

WATER SAMPLE RETURN LINE
LIFTING RODS
AIR SUPPLY LINE TO PUMP
AIR SUPPLY LINE TO PACKER

INFLATABLE PACKER
2-1/2" DIAMETER (RELAXED)

STAINLESS STEEL CONNECTOR ROD
1-3/4" DIAMETER

PUMP OPERATION

1. Water enters pump through screen and passes through lower check valve into bladder.
2. Air is pumped into area around bladder and squeezes bladder.
3. Water is forced from bladder, through upper check valve and into sample tube at top of pump.
4. Air pressure squeezing bladder is released and the bladder returns to relaxed state as more water enters through screen repeating Step 1.

STAINLESS STEEL BLADDER PUMP
1-3/4" DIAMETER

PUMP INTAKE SCREEN

18'-0"
2'-8"
5"
9'-3"
61"

Figure 9-6. Typical packer/pump assembly used for bedrock characterization at the Occidental Chemical Corporation S-Area DNAPL site in Niagara Falls, New York (from Conestoga-Rovers and Associates, 1986).

After the set time, the grout will be drilled out using 6-inch diameter enlarging tools and flushed for one minute or until clear water return is observed. The 15-ft interval will be pressure tested. The test will consist of installing and inflating a packer at the top of the previously tested interval and applying a constant head of water on the sealed interval. The water loss must be less than that calculated by (USDI, 1974, p.576)

$$Q = (2 \pi K L H) / \ln(L/r) \qquad (9-2)$$

where Q is the constant rate of flow into the hole, K is the maximum allowable hydraulic conductivity (1×10^{-5} cm/sec), L is the length of the test interval, H is the differential head of water, and r is the radius of the tested interval. If the interval fails this test, it will be regrouted and retested.

(e) Remove the pressure testing assembly and core an additional 15 ft of bedrock.

(f) Packer/pump test this 15-ft bedrock interval. Continue coring, pump testing, groundwater sampling, grouting, and pressure testing in 15-ft intervals through the upper more permeable portion of the Lockport Dolomite down to the top of the Gasport Member of the formation.

(g) Once the top of the Gasport Member has been contacted, a 4-inch diameter black steel casing will be inserted into the borehole and grouted into place using the procedure outlined for overburden casing installation.

(h) Once the grout has set, the coring and testing will continue in 15-ft increments to the top of the Rochester Shale Formation. When available, all hydraulic test and chemical analysis data will be evaluated to select the appropriate interval in the lower Lockport Dolomite for completing a permanent monitor well.

(i) If NAPL is encountered in the borehole, a decision will be made to:

(i) continue to test and grout in accordance with the plan;

(ii) grout the borehole and attempt to complete the testing in a nearby boring; or,

(iii) grout the borehole and assume that NAPL extends to the top of the Rochester Shale Formation.

Results of the bedrock drilling program at the S-Area site (Conestoga Rovers and Associates, 1988) were presented, in part, on geologic cross sections as exemplified in Figure 9-7. Although the drilling and testing protocols developed for S-Area and other chemical waste sites in Niagara Falls were designed to minimize the potential for cross-contamination, the extent to which DNAPL migration might have been facilitated by drilling is unknown.

If DNAPL is observed in fractures of cores obtained during drilling, samples can be placed in jars and submitted to a laboratory for extraction and analysis. For determination of non-volatile DNAPL components, WCGR (1991) suggests using a solvent-soaked gauze pad to wipe the NAPL-contaminated fracture surface and then submitting the gauze pad to a laboratory for extraction and analysis.

9.5 WELL CONSTRUCTION

Monitor wells are installed to characterize immiscible fluid distributions, flow directions and rates, groundwater quality, and media hydraulic properties. Pertinent data are acquired by conducting fluid thickness and elevation surveys, fluid sampling surveys, hydraulic tests, and borehole geophysical surveys. The locations and design of monitor wells are selected based on consideration of the site conceptual model and specific data collection objectives.

Well construction methods and concerns are discussed in numerous references, including Nielsen and Schalla (1991), Aller et al. (1989), Barcelona et al. (1983, 1985), USEPA (1986, 1987, 1991a), ASTM (1990b), Driscoll (1986), GeoTrans (1989), Cambell and Lehr (1984), and NWWA (1981). These documents provide detailed information on: drilling techniques, well design, well materials (casing, screen, sandpack, sealant, and grout), well construction/ installation methods, and well development. Details of multiple-level monitoring systems that can be installed in a single drill hole are discussed by Ridgeway and Larssen (1990), Cherry and

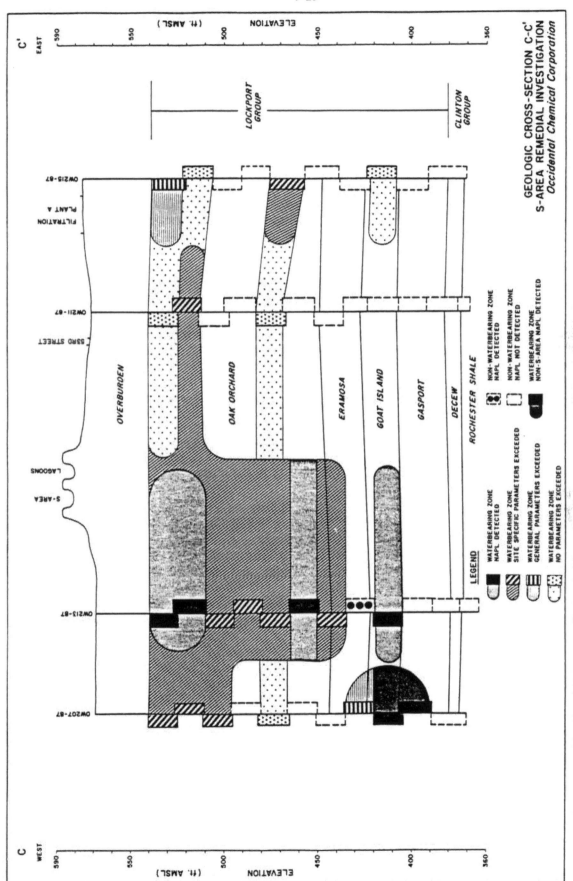

Figure 9-7. Results of the Lockport Dolomite characterization program at the S-Area DNAPL site reflect heterogeneous subsurface conditions (from Conestoga Rovers and Associates, 1988). The non-S-Area DNAPL detected at depth in Well OW207-87 has different chemical and physical properties than S-Area DNAPL, and is believed to derive from another portion of the Occidental Chemical Corporation plant site (Conestoga-Rovers and Associates, 1988).

Johnson (1982), Black et al. (1986), Korte and Kearl (1991), and Welch and Lee (1987).

The design and construction of wells at DNAPL sites require special consideration of (1) the effect of well design and location on immiscible fluid movement and distribution in the well and near-well environment; (2) the compatibility of well materials with NAPLs and dissolved chemicals; and (3) well development options.

Inadequate well design can increase the potential for causing vertical DNAPL migration and misinterpretation of fluid elevation and thickness measurements. Although much theoretical and experimental research has been conducted to examine the relationship between LNAPL presence in wells and formations (Mercer and Cohen, 1990; van Dam, 1967; Zilliox and Muntzer, 1975; Schiegg, 1985; Abdul et al., 1989; Fiedler, 1989; Farr et al., 1990; Kemblowski and Chiang, 1990; Lenhard and Parker, 1990; Hampton, 1988; Hampton and Miller, 1988; de Pastrovich et al., 1979), relatively little work has focused on interpreting measurements of DNAPL thickness and elevation in wells. Adams and Hampton (1990) conducted physical model experiments to evaluate the effects of capillarity on DNAPL thickness in wells and adjacent sands. Based on their experiments, field experience, and the principles of DNAPL movement, it is apparent that several factors may cause the elevation and thickness of DNAPL in a well to differ from that in formation and/or lead to vertical DNAPL migration. These factors include the following.

- If the well screen or casing extends below the top of a DNAPL barrier layer, the measured DNAPL thickness may exceed that in the formation by the length of the well below the barrier layer surface (Mercer and Cohen, 1990) as shown in Figure 9-8.

- If the well bottom is set above the top of a DNAPL barrier layer, the DNAPL thickness in the well may be less than the formation thickness as shown in Figure 9-9.

- If the well connects a DNAPL pool above a barrier layer to a deeper permeable formation, the DNAPL elevation and thickness in the well are likely to be erroneous and the well will cause DNAPL to short-circuit the barrier layer and contaminate the lower permeable formation as shown in Figure 9-10. The height of the DNAPL column at the well bottom will tend to equal or be less than the critical DNAPL height, z_n, required to overcome the capillary resistance offered by the sandpack and/or formation (Adams and Hampton, 1990; Chapter 5.1).

- DNAPL which enters a coarse sandpack may sink to the bottom of the sandpack rather than flow through the well screen (Figure 9-11). Small quantities of DNAPL may elude detection by sinking down the sandpack and accumulating below the base of the well screen.

- Similarly, if the bottom of the well screen is set above the bottom of the sand pack and there is no casing beneath the screen, small quantities of DNAPL may elude detection by sinking out the base of the screen and into the underlying sandpack as shown in Figure 9-12.

- Sandpacks generally should be coarser than the surrounding media, however, to ensure that mobile DNAPL can enter the well. Screen or sandpack openings that are too small may act as a capillary barrier to DNAPL flow (Figure 9-13).

- If the well screen is located entirely within a DNAPL pool and water is pumped from the well, DNAPL will upcone in the well to maintain hydrostatic equilibrium causing the DNAPL thickness in the well to exceed that in the formation as shown in Figure 9-14 (Huling and Weaver, 1991; Villaume, 1985).

- The elevation of DNAPL in a well may exceed that in the adjacent formation by a length equivalent to the DNAPL-water capillary fringe height where the top of the pool is under drainage conditions (WCGR, 1991; Adams and Hampton, 1990) as shown in Figure 9-15.

- DNAPL will not flow into a well where it is present at or below residual saturation or at negative pressure.

As demonstrated above, the relationship of the well screen to mobile DNAPL and capillary barriers can govern the thickness of DNAPL in a well and the potential for vertical DNAPL migration in the well environment. A well that is completed to the top of a capillary barrier and screened from the capillary barrier surface to above the DNAPL-water interface is most likely to provide DNAPL thickness and elevation data that are representative of formation conditions. Consideration must be given to the risks and the risk-minimization strategies listed in Chapter 9.2 during well design and construction activities.

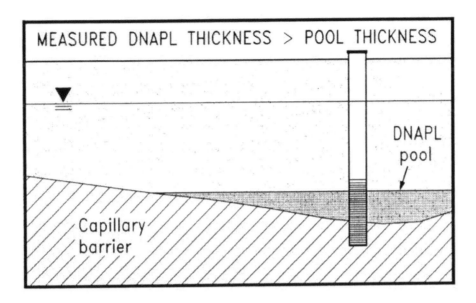

Figure 9-8.   The measured thickness of DNAPL in a well may exceed the DNAPL pool thickness by the length of the well below the barrier layer surface (after Huling and Weaver, 1991).

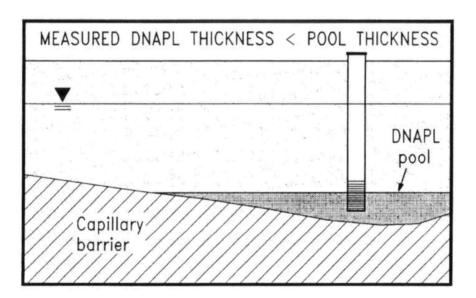

Figure 9-9.   The measured DNAPL thickness in a well may be less than the DNAPL pool thickness by the distance separating the well bottom from the capillary barrier layer upon which DNAPL pools.

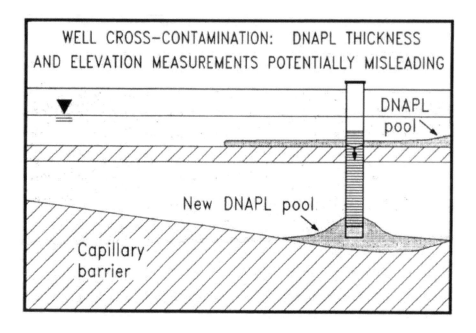

Figure 9-10. DNAPL elevation and thickness measurements in a well are likely to be misleading where the well is screened across a capillary barrier with perched DNAPL.

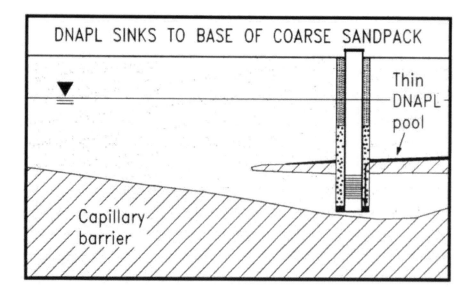

Figure 9-11. DNAPL that enters a coarse sandpack may sink to the bottom of the sandpack rather than flow through the well screen and thereby, possibly escape detection.

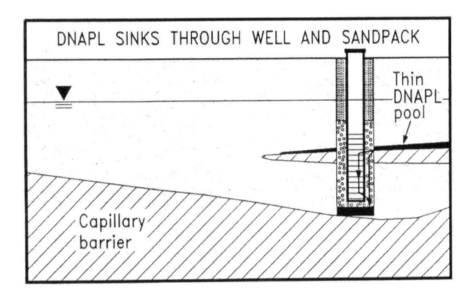

Figure 9-12. DNAPL that enters a well from above may flow out of the base of the well screen and into the underlying sandpack (or formation).

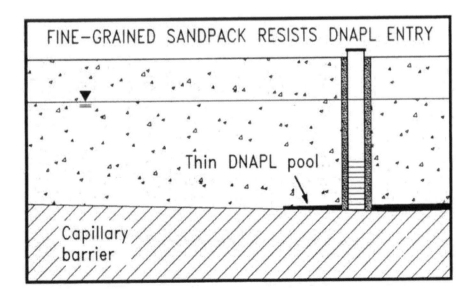

Figure 9-13. Sandpacks should be coarser than the surrounding media to ensure that the sandpack is not a capillary barrier to DNAPL movement.

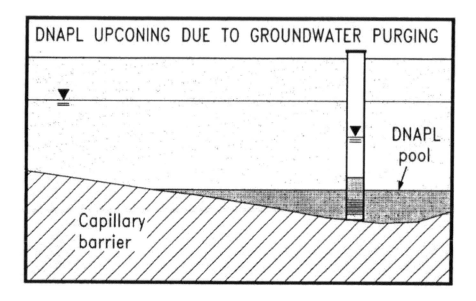

Figure 9-14. Purging groundwater from a well that is screened in a DNAPL pool will result in DNAPL upconing in the well (after Huling and Weaver, 1991).

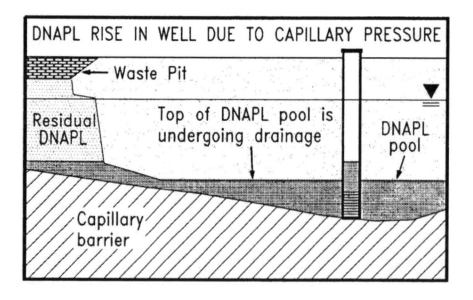

Figure 9-15. The elevation of DNAPL in a well may exceed that in the adjacent formation by a length equivalent to the DNAPL-water capillary fringe height where the top of the DNAPL pool is undergoing drainage (invasion by DNAPL) (after WCGR, 1991).

The compatibility of well materials with NAPLs and highly contaminated groundwater should also be evaluated during well design (Nielsen and Schalla, 1991; Aller et al., 1989; Driscoll, 1986; Barcelona et al., 1983, 1985). Advantages and disadvantages of different casing and screen materials are summarized in Table 9-7. Steel casing is generally recommended where DNAPLs, particularly halogenated solvents, may be present because of its strength and chemical resistance to solvents. Steel, however, is susceptible to corrosion caused by long-term exposure to groundwater with low pH, hydrogen sulfide present, or high concentrations of dissolved oxygen (>2 mg/L), carbon dioxide (>50 mg/L), or chloride (>500 mg/L) (Aller et al., 1989). Stainless steel generally provides better resistance to corrosion than carbon steels. Another potential drawback of steel casing is that hydrophobic organic liquids may preferentially wet steel (Arthur D. Little, Inc., 1981) and result in vertical DNAPL migration along the outside of well casings that are not adequately sealed with grout. Fluoropolymeric (Teflon) and fiberglass materials are also resistant to organic solvents, but lack the strength of steel. The high cost of fluoropolymers also limits their widespread use. Halogenated solvents can degrade PVC and ABS thermoplastics. These materials should not be used for well construction where they will come into contact with halogenated solvents.

Well annular seals are composed of several different stable and low hydraulic conductivity materials. The composition and characteristics of various sealant grouts are given in Table 9-4 and discussed by Edil et al. (1992), Aller et al. (1989), Nielson and Schalla (1991), Driscoll (1986), and Barcelona et al. (1983, 1985). As described in Chapter 9.3.2, hydrophobic organic liquids may cause bentonite to shrink and fracture, and its use, therefore, may be inappropriate at DNAPL sites (except perhaps as a minor component grout additive).

Monitor wells are typically developed after construction to remove fine grained particles and drilling fluid residues from the borehole and near-well environment. Various well development methods, including overpumping, backwashing, mechanical surging, and high velocity jetting, are discussed by Aller et al. (1989), GeoTrans (1989), Kraemer et al. (1991), and Driscoll (1986). Well development should be limited in wells containing DNAPL to gentle pumping and removal of fine particles to minimize DNAPL redistribution. Measurements should be made of immiscible fluid stratification in the well prior to and after development.

9.6 WELL MEASUREMENTS OF FLUID THICKNESS AND ELEVATION

Fluid elevation and thickness measurements are made in wells to assist determination of fluid potentials, flow directions, and immiscible fluid distributions. With knowledge of DNAPL entry areas, the surface slope of capillary barriers, hydraulic data, and other observations of DNAPL presence, well data can be evaluated to infer the directions of DNAPL migration. Interpretation of fluid data from wells containing NAPL may be complicated by several factors related to the measurement method, fluid properties, well design, or well location (Chapter 9.5). DNAPL in wells, therefore, should be evaluated in conjunction with evidence of geologic conditions and DNAPL presence obtained during drilling.

Methods for acquiring groundwater level data are reviewed by Dalton et al. (1991), EPRI (1989), USEPA (1986), and Ritchey (1986). Techniques for measuring immiscible fluid elevations and thicknesses in wells (Mercer and Cohen, 1990; Huling and Weaver, 1991; API, 1989; GRI, 1987; Sanders, 1984) are summarized below.

While conducting immiscible fluid level and thickness measurements, care should be taken to slowly lower and raise the measuring device within the well to avoid disturbing the immiscible fluid equilibrium and creating emulsions. Similarly, measurements should be made prior to purging activities. The cost of purchase and decontamination of the measuring device should be considered when selecting a measurement method, particularly given uncertainties involved in interpreting NAPL thickness and elevation data.

9.6.1 Interface Probes

Most interface probes employ an optical device to distinguish the air-fluid interface and a conductivity sensor to distinguish NAPL-water interfaces (Sanders, 1984). These probes are typically 1 to 1¾ inches in diameter, 9 inches long, and attached to a tape guide that is sequentially marked in 0.01 or 0.05 ft increments. Interface probes should be slowly lowered to the well bottom and then raised to survey the distribution of immiscible fluids within a well. Different audible signals are emitted upon contact with NAPL (e.g., a continuous tone) and water (e.g., an intermittent tone).

Under ideal conditions, interface probes can be used to detect NAPL layers that are as thin as 0.01 ft, and

Table 9-7. Advantages and disadvantages of some common well casing materials (modified from Driscoll, 1986; GeoTrans, 1989; and Nielsen and Schalla, 1991).

TYPE	ADVANTAGES	DISADVANTAGES
FLUOROPOLYMERS such as polytetrafluoro-ethyene (PTFE), tetra-fluoroethylene (TFE), and fluorinated ethylene propylene (FEP)	• Excellent chemical resistance to organic chemicals and corrosive environments; practically insoluble in all organic liquids except a few fluorinated solvents • Lightweight • High impact strength	• Lower tensile strength and wear resistance compared to other plastics, iron, or steel • Expensive relative to steel and other plastics
THERMOPLASTICS: POLYVINYLCHLORIDE (PVC) AND ACRYLONITRILE BUTADIENE STYRENE (ABS)	• Lightweight • Easy workability (with threaded couplings) • Inexpensive compared to fluoropolymers and steel • Resistant to alcohols, aliphatic hydrocarbons, weak and strong alkalies, oils, strong mineral acids, and oxidizing acids • Completely resistant to galvanic and electrochemical corrosion • High strength-to-weight ratios, and resistant to abrasion	• More reactive than PTFE • Poor chemical resistance to aromatic hydrocarbons, esters, ketones, and organic solvents • Much lower tensile, compressive, and collapse strength than steel or iron • May adsorb or elute trace organics • PVC glues, if used, may contribute organic chemicals to well water
STAINLESS STEEL such as Type 304 and Type 316	• Stronger, more rigid, and less temperature-sensitive than plastic materials • Good chemical resistance to organic chemicals • Resistant to corrosion and oxidation • Readily available	• Expensive • May catalyze some organic chemical reactions • May corrode if exposed to long-term corrosive conditions and leach chromium • Heavy
CARBON STEEL	• Stronger, more rigid, and less temperature-sensitive than plastic materials • Less expensive than stainless steel or teflon	• Expensive • Rusts easily, providing high sorptive and reactive capacity for many metals and organic chemicals • Subject to corrosion (under conditions of low pH, high dissolved oxygen, H^2S presence, > 1000 mg/L total dissolved solids, > 50 mg/L CO_2, or > 500 mg/L Cl⁻ • Heavy
GALVANIZED STEEL	• Stronger, more rigid, and less temperature-sensitive than plastic materials	• Expensive • Will rust if galvanized coating is scratched • Resistance to corrosion provided by zinc coating may be short-lived • May be source of zinc • Heavy

measure NAPL layer thickness to within 0.01 to 0.10 ft. A comparison of LNAPL thickness measurements made using an interface probe, hydrocarbon gauging paste, and a bailer in Table 9-8 evidences the utility of an interface probe. Interface probes, however, may produce spurious results in the presence of conductive NAPLs, emulsified NAPL, or viscous NAPL that coats the sensors. To minimize the latter problem, the air-fluid interface is best measured by lowering the probe from air into fluid to avoid the effects of dripping fluid when raising the probe through the air-fluid interface. Similarly the NAPL-water interface is best measured by lowering or raising the probe from water into NAPL to minimize the problem of NAPL coating the conductivity sensor which would increase the measured NAPL thickness (Sanders, 1984). Standard electric line water-level probes can detect interfaces between non-conductive NAPL and water, but may fail to identify interfaces between LNAPL and air.

Interface probes are expensive (typically $900 to $2000) and widely available (e.g., see the *Ground Water Monitoring Review* Annual Buyers Guide Summer issue).

9.6.2 Hydrocarbon-Detection Paste

Hydrocarbon-detection paste changes color within seconds upon contact with liquid hydrocarbons. For DNAPL measurement, a thin film of hydrocarbon-detection paste can be applied to the bottom of a measuring tape, gauge line, or rod that is lowered to the well bottom. Upon retrieval, the DNAPL column depth and thickness will be revealed by the zone of paste color change. If there is LNAPL floating on the water column, all of the paste that is lowered into the fluid column will change color, thereby preventing measurement of DNAPL at the well bottom.

Water-detection paste that changes color upon contact with water can be used to measure the thickness of NAPL floating on the water column in a well. A thin film of the water-detection paste is applied to the bottom of a tape, gauge line, or rod that is lowered to below the expected base of the floating product layer, retrieved, and then inspected to determine the depth to and thickness of the floating NAPL.

Detection paste methods may provide NAPL thickness measurements that are accurate to within 0.01 ft under favorable conditions (API, 1989; Testa and Paczkowski, 1989). Tests conducted by Sanders (1984) suggest that

this method tends to overestimate NAPL thickness (Table 9-8). Measurements made using these pastes are slow relative to use of an interface probe. If the paste is applied to a disposable weighted string or wire, however, the time and cost associated with equipment decontamination may be eliminated.

Hydrocarbon- and water-detection pastes cost approximately $1 to $2 per ounce. Numerous measurements can be made per ounce. According to a detection paste manufacturing representative (Kolor Kut Products Company, Houston, Texas), detection paste compositions are proprietary and include no hazardous ingredients.

9.6.3 Transparent Bailers

Transparent bottom-loading bailers can be used to measure NAPL thickness in a well. For DNAPL thickness measurement, the bailer is gently lowered to the well bottom and then raised back to the surface. The thickness of DNAPL in the well can be estimated by measuring the DNAPL column height in the bailer. If the thickness exceeds the bailer length, then additional samples must be taken using a bailer or other method to determine the top of the DNAPL column. The additional sampling, however, should not be made until the DNAPL returns to equilibrium in the well. NAPL rise due to displacement by the bailer may result in slight overestimation of its thickness in a well. Alternatively, leakage from the bottom of the bailer while it is being raised may result in an underestimation of DNAPL thickness. For floating NAPL, the bailer should be long enough so that its top is in air when its check valve is in water. Care should be taken to ensure that an equilibrium height of LNAPL has entered the bailer before lowering it into the water layer.

Transparent bottom-loading PVC bailers (1.66 inch diameter) cost approximately $40, $50, and $70 apiece for 2, 3, and 5-ft lengths, respectively. Disposable bailers can be obtained for less.

9.6.4 Other Methods

Various other methods are available to estimate DNAPL thickness and elevations in wells. For example, measurements can be made by taking small fluid samples from specific depths using either an inertial lift pump (e.g., Waterra Pumps Ltd, Toronto, Ontario), a peristaltic

Table 9-8.   Comparison of measured LNAPL thicknesses using water-detection paste, a clear bottom-loading bailer, and an interface probe (from Sanders, 1984).

Measurement Method	3 Inches Gasoline		1 Inch Gasoline		3 Inches Kerosene		1 Inch Kerosene	
	Ave. (inches)	Standard Deviation	Ave. (inches)	Standard Deviation	Ave. (inches)	Standard Deviation	Ave. (inches)	Standard Deviation
Water Detection paste on a stick*	3.60	0.21	1.38	0.13	3.36	0.27	1.12	0.75
Clear bottom-loading bailer**	2.58	0.16	0.82	0.13	2.46	0.18	0.80	0.12
Interface probe***	3.18	0.11	1.14	0.11	3.12	0.08	1.10	0.12

Notes: Ave. means average. Five tests were conducted with each method and fluid thickness.
*Gauging stick -- 8-ft Bagby Stick Co.; McCabe, Inc. water-detection paste.
**Surface sampler -- 1¾-inch OD, 12-inches long, bottom-loading bailer.
***ORS interface probe (manufacture date circa 1984).

pump, or a discrete-depth mechanical (Kemmerer type) sampler; or by measuring the length of DNAPL (if it contrasts visually with water) that coats a weighted string that has been retrieved from the bottom of a well.

9.7 WELL FLUID SAMPLING

Fluid sampling surveys should be conducted at potential DNAPL sites to examine wells for the presence of LNAPLs and DNAPLs. NAPL samples can be tested for physical properties and chemical composition (Chapter 10). Well sampling methods are described in numerous references (i.e., Herzog et al., 1991; Barcelona et al., 1985, 1987; USEPA, 1986, 1987, 1991a; GRI, 1987; Rehm et al., 1985).

Prior to sampling, immiscible fluid stratification should be measured in wells as described in Chapter 9.6. Various sampling devices can be employed to acquire fluid samples from the top and bottom of the well fluid column. Villaume (1985) recommends use of a bottom-loading bailer or mechanical discrete-depth sampler for collecting DNAPL samples. Huling and Weaver (1991) suggest that the best DNAPL sampler is a double check valve bailer which should be slowly lowered to the well bottom and then slowly raised to provide the most reliable results. DNAPL can be sampled from wells with a shallow water table (<25 ft deep) with a peristaltic pump and from depths to approximately 300 ft using a simple inertial pump (e.g., Waterra Pumps Ltd, Toronto, Ontario; Rannie and Nadon, 1988). An advantage of the peristaltic and inertial pumps is that fluid contact is confined to inexpensive tubing (and a foot-valve with the inertial pump). The cost to decontaminate or replace DNAPL-contaminated equipment is usually a major factor in selecting a sampling method (Mercer and Cohen, 1990).

9.8 ASSESSING DNAPL MOBILITY

Goals of DNAPL site characterization include determining the extent of mobile DNAPL and the feasibility of DNAPL containment and recovery options. Monitoring DNAPL levels in wells over time may provide evidence of DNAPL mobility (Huling and Weaver, 1991). DNAPL mobility can also be assessed based on: (1) observations and measurements of DNAPL saturation in subsurface media samples and (2) DNAPL pumping experiments. DNAPL containment and recovery pumping methods and considerations are described in Table 6-1.

At many sites, DNAPL migration from a release location may cease at residual saturation or be immobilized in a stratigraphic trap within days to weeks. Factors which may cause DNAPL to flow slowly and therefore remain mobile for longer periods include relatively low DNAPL viscosity and media permeability. Long-term mobility is also promoted where large quantities of DNAPL have been released to the subsurface and where there is an ongoing or episodic release (e.g., a leaking tank or drums that rupture periodically in a landfill). Finally, increased hydraulic gradients, a fluctuating water table, and increased DNAPL wetting of the porous medium with time possibly due to mineral surface chemistry modifications, may cause remobilization of stagnant DNAPL. As such, slow long-term migration of DNAPL is possible at some sites.

9.9 BOREHOLE GEOPHYSICAL METHODS

The application of borehole geophysical methods to groundwater investigations is described by Keys and MacCary (1976), Keys (1988), Benson (1991), Driscoll (1986), GRI (1987), Guyod (1972), Rehm et al. (1985), Cambell and Lehr (1984), Michalski (1989), and USEPA (1987). Borehole geophysical surveys involves lowering a logging tool, also known as a sonde, down a well or boring to make physical measurements as a function of depth (Figure 9-16). A sensing element within the sonde measures the property of interest and converts it into electrical signals. These signals are transmitted to the surface through a cable and recorded digitally or by using an analog strip chart. Several types of geophysical log may be made, sometimes simultaneously, in the same borehole.

Borehole geophysical surveys are conducted to characterize lithologies, correlate stratigraphy between borings, identify fracture zones, estimate formation properties (e.g., porosity and density), identify intervals containing conductive dissolved contaminants, etc. The utility and limitations of various techniques are summarized in Table 9-9 and example logs are shown in Figure 9-17. These methods provide continuous high-resolution measurements of subsurface conditions. As such, they can be used to distinguish thin layers and subtle stratigraphic features which may influence contaminant movement.

The use of geophysical surveys at DNAPL contamination sites is discussed by WCGR (1991) and Annan et al. (1991). Sequential borehole surveys were made before,

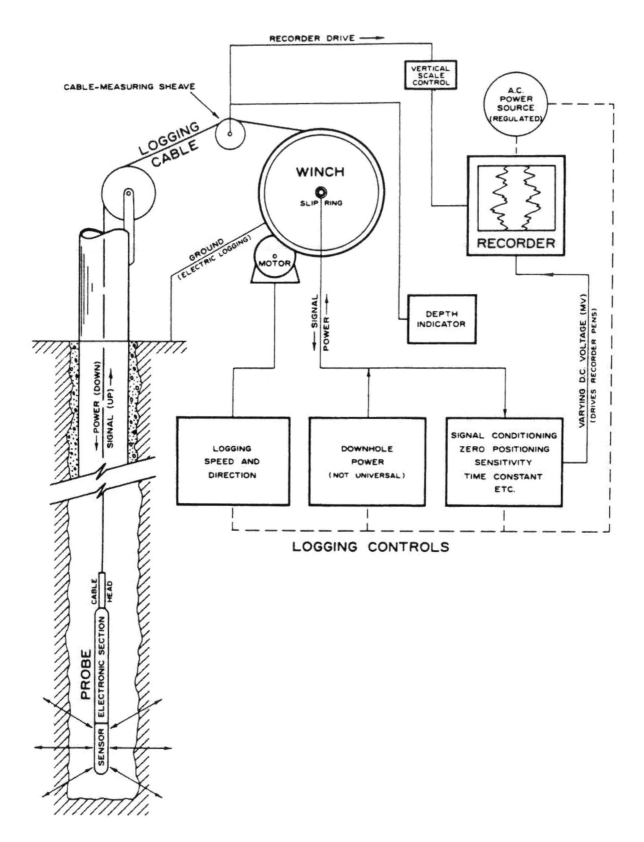

Figure 9-16. Schematic diagram of borehole geophysical well logging equipment (from Keys and MacCary, 1976)

Table 9-9. Utility and limitations of borehole geophysical methods for site characterization (modified from Benson, 1991; Rehm et al, 1985; Keys and MacCary, 1971).

| METHOD | UTILITY | Can method be used in following conditions? | | Radius of Measurement | Affect of Hole Diameter and Mud |
		Casing Uncased/PVC/Steel	Saturated Unsaturated		
Electrical Resistivity	Electrical resistivity logs record apparent electrical resistivity as a function of depth within the saturated zone. The voltage drop due to the electrical resistance of the formation fluids and media is measured using two current and two potential electrodes. Several tools with different electrode spacings are commonly utilized to evaluate stratigraphy (i.e., clay content and fracture density) and fluid conductivity (i.e., high ionic strength contaminated groundwater). Resistivity logs are used by the petroleum industry to evaluate water and oil saturations. High resistivity in porous sands may reflect NAPL presence.	Yes/No/No	Yes/No	12-60"	Significant to minimal depending on probe used
Spontaneous Potential (SP)	Spontaneous potential logs measure natural potential differences between the borehole fluid and adjacent media. SP logs are used to help characterize and correlate stratigraphy (layer type, thickness, etc.).	Yes/No/No	Yes/No	Near borehole surface	Significant and highly variable
Natural Gammma	Natural gamma logs measure the amount of natural gamma radiation emitted along the length of the borehole. The natural gamma-emitting radioisotopes (K-40 and daughter products of thorium and uranium decay) are preferentially adsorbed in clay and shale layers. The natural gamma log, therefore, is primarily used to reveal the presence of clay and shale layers (i.e., DNAPL capillary barriers).	Yes/Yes/Yes	Yes/Yes	6-12"	Moderate
Gamma Gamma (Density)	Gamma gamma logs measure the response of media adjacent to the boring to gamma radiation that is emitted from a radiation source in the logging tool. The amount of radiation detected is inversely related to formation density. Formation porosity can be calculated using the gamma gamma log if the formation bulk density is known.	Yes/Yes/Yes	Yes/Yes	6"	Significant
Neutron-Neutron (Porosity)	The neutron logging tool contains a radiation source and detector. The detector output is proportional to the water content (hydrogen content) of the borehole environment. Neutron logs can provide estimates of moisture content in the vadose zone, total porosity in the saturated zone, the water table elevation, and the rate of fluid infiltration.	Yes/Yes/Yes	Yes/Yes	6-12"	Moderate
EM Induction	EM induction logs record the bulk electrical conductivity of the near borehole environment are used, in conjunction with other data, to identify lithology, correlate stratigraphy, and infer fluid conductance. Highly conductive (high ionic strength) zones of contaminated groundwater can be identified using this method. Additionally, thick intervals with high organic NAPL saturation may be revealed as low conductivity anomalies.	Yes/Yes/No	Yes/Yes	30"	Negligible

Table 9-9. Utility and limitations of borehole geophysical methods for site characterization (modified from Benson, 1991; Rehm et al, 1985; Keys and MacCary, 1971).

| METHOD | UTILITY | Can method be used in following conditions? | | Radius of Measurement | Affect of Hole Diameter and Mud |
		Casing Uncased/PVC/Steel	Saturated Unsaturated		
Temperature	Temperature logs provide a continuous record of fluid temperature with depth. Preferential inflow zones may be indicated by an increase or decrease in groundwater temperature.	Yes/No/No	Yes/No	Within borehole	NA
Fluid Conductivity	A specific conductance probe is used to record fluid conductivity with depth. This data may be used to assess groundwater conductivity, inflow zones, contamination zones, etc.	Yes/No/No	Yes/No	Within borehole	NA
Flow	Fluid movement logging utilizes impeller flowmeters, thermal flowmeters, and various tracer detection systems to measure the groundwater inflow rate as a function of boring depth. Variation in the groundwater inflow rate may derive from well construction details, hydraulic head differences, and/or variable hydraulic conductivity. Most frequently, an impeller-type flowmeter is used to identify zones of preferential inflow.	Yes/No/No	Yes/No	Within borehole	NA
Caliper	Spring-loaded feelers extend from the caliper logging tool, follow the borehole wall, and continuously measure the hole diameter. Caliper logs record the outer diameter of a well or open boring and can be used to locate fractures and cavities in an open borehole. Caliper logs are also used to establish correction factors for other measurements influenced by hole size.	Yes/Yes/Yes	Yes/Yes	To limit of sensor, typically 2-3'	NA
Video	Small-diameter video cameras can be lowered down a borehole to inspect for casing corrosion, well condition, leaks, fractures zones, etc.	Yes/Yes/Yes	Yes/Yes	Within borehole	NA

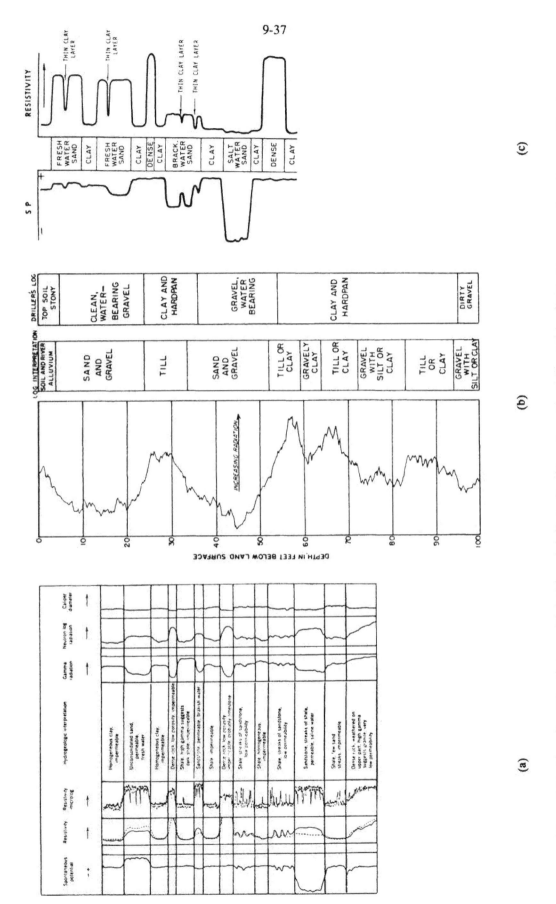

Figure 9-17. Examples of borehole geophysical logs: (a) six idealized logs (from Cambell and Lehr, 1984); (b) a gamma log of unconsolidated sediments near Dayton, Ohio (from Norris, 1972); and (c) an idealized electrical log (from Guyod, 1972).

during, and after a controlled release of tetrachloroethene (PCE) at the Borden DNAPL research site (WCGR, 1991). The infiltration of poorly-conductive PCE into the Borden sand was tracked with some success by EM induction, resistivity, and neutron logging. Generally, however, pre-contamination baseline surveys will be unavailable and subsurface conditions will be more complex than at the Borden site. High resistivity or low EM induction values in coarse soil may reflect NAPL presence. Overall, the potential value of using borehole geophysical surveys to delineate NAPL presence and saturation is poorly defined.

9.10 IDENTIFICATION OF DNAPL IN SOIL AND WATER SAMPLES

Significant cost savings can be realized during a site investigation if DNAPL presence can be determined directly by visual examination of soil and groundwater samples rather than indirectly by more costly chemical analyses (Chapter 10). Direct and indirect methods for detecting and/or suspecting the presence of DNAPL in the subsurface are described briefly in Chapter 7.2 and Table 7-4, and in more detail below.

9.10.1 Visual Detection of NAPL in Soil and Water

Under ideal conditions, NAPL presence can be identified by visual examination of soil or groundwater samples. Direct visual detection may be difficult, however, where the NAPL is clear and colorless, present at low saturation, or distributed heterogeneously. There is little documentation of practical methods to directly identify NAPL in soil or water (Huling and Weaver, 1991).

Although not well-documented, ultraviolet (UV) fluorescence analysis and phase separation techniques such as centrifugation and soil-water shake tests have been used with varied success to identify NAPL presence at some contamination sites. Recently, Cary et al. (1991) demonstrated a method to extract NAPL from soil by shaking a soil-water suspension with a strip of hydrophobic, porous polyethylene in a glass jar for 3 to 4 hours. Estimates of NAPL content were derived by gravimetric analysis of the strips before and after the extraction process. This method is described in Chapter 10.2 and its use at a crude-oil spill site is documented by Hess et al. (1992).

A series of experiments was conducted by Cohen et al. (1992) to test the hypothesis that simple and inexpensive methods can be used to visually identify clear, colorless NAPL in soil and water samples. The procedures and findings reported by Cohen et al. (1992) are condensed below.

Seventy-eight samples were prepared in a random sequence by adding varying amounts of clear, colorless NAPL (kerosene, chlorobenzene, or tetrachloroethene) and water to soil in sealable polyethylene bags. Three soils were utilized in the experiments: a pale-yellowish orange (10YR 8/6, Munsell color notation of hue, value/chroma), well-sorted, subrounded to subangular, medium quartz sand; a moderate brown (5Y 4/4), saprolitic silt loam; and, a black (N1), organic-rich, silt loam top soil. The quantities of soil and fluid used to prepare samples were calculated to provide a range of NAPL saturations from 0.01 to 0.23. Blank samples were prepared by adding water without NAPL to the soil sample. For dissolved contaminant samples, water containing an equilibrium concentration of dissolved NAPL (derived from water in contact with NAPL) was introduced to the soil sample. Of the 78 samples: 56 contained NAPL, 11 contained dissolved contaminant levels, and 11 were blanks.

The soil samples were then examined for NAPL presence by two investigators who were unaware of the sample contents using the following procedures (Figure 9-18):

(1) Three minutes after each sample was prepared, the accumulated organic vapor concentration was measured using a Flame Ionization Detector (Foxboro Century OVA Model 128 calibrated with methane) by inserting the FID probe into an opened corner of the sample bag.

(2) An unaided visual examination was then made through the sample bags to inspect for NAPL presence. Examination results were categorized as: "A" if NAPL was identified as present based on visual evidence; "B" if NAPL presence was suspected based on visual evidence; or "C" if there was no visual evidence of NAPL presence.

(3) An examination was then made of the UV fluorescence of the samples using an inexpensive, portable, battery-powered UV light (Raytech Industries' Versalume™ model; cost ≈ $50). The 4-watt broad spectrum lamp emits both shortwave UV (2536 A) and longwave UV (3000 to 4000 A)

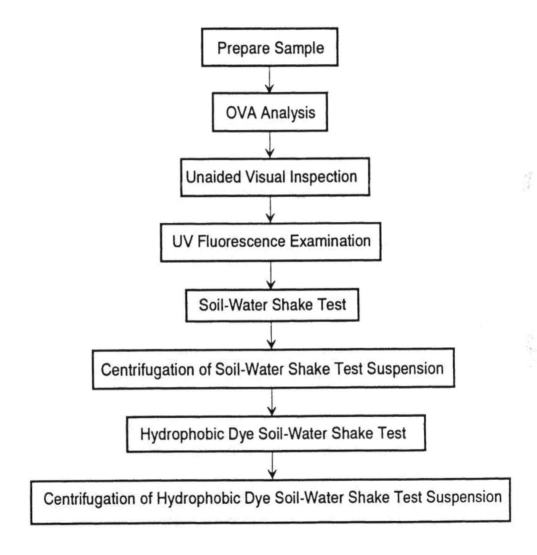

Figure 9-18. Sequence of NAPL detection procedures utilized by Cohen et al. (1992).

simultaneously. The examination was made in a dark room by scanning the sample bag with the UV light. The bags were manipulated during the examination to squeeze fluid against the bag beneath the lamp, and the samples were categorized using the A, B, or C classification. Examination of samples without NAPL provided a check on the presence of mineral or shell fluorescence in the samples.

(4) Following the fluorescence examination, approximately 20 cm^3 of sample was transferred using a spoon into a 50-mL, polypropylene centrifuge tube. Twenty mL of water was added to the subsample and the tube was shaken by hand for approximately 10 seconds to create a soil-water suspension. An unaided visual inspection was then made for NAPL presence (A, B, or C rating) by peering through the tube walls and at the fluid surface.

(5) The subsample was then centrifuged at approximately 1250 rpm for one minute. The subsamples were inspected as described above and rated for NAPL presence.

(6) After the centrifuge test, approximately 2 mg (an amount that would rest on the edge of a toothpick) of Sudan IV, a nonvolatile hydrophobic dye, was placed in the centrifuge tube. Sudan IV is a relatively inexpensive (100 g cost $19 from Aldrich Chemical Co.), reddish-brown powder that dyes organic fluids red upon contact, but is practically insoluble in water at ambient temperatures. Like many other solvent dyes, Sudan IV is an irritant and possible mutagen with which skin or eye contact should be avoided. Although widely used to colorize NAPL flow experiments (Schwille, 1988, for example), minimal use has been made of solvent dyes such as Sudan IV and Oil Red O to detect NAPL in soil and water at contamination sites. The contents of the tube were then mixed by shaking manually for approximately 10 to 30 seconds and examined for NAPL presence. NAPL presence was rated A, B, or C, and, a notation of the relative NAPL density and quantity was made when apparent.

(7) The final step in the soil examination procedure was to centrifuge the dyed subsample at approximately 1250 rpm for one minute, and then peer through the tube walls and at the fluid surface to assess NAPL presence and relative density and quantity.

Examination results are summarized in Table 9-10 and Figure 9-19 and conclusions regarding the utility of each examination method are highlighted below.

• *OVA Measurement* -- Analysis of organic vapors in soil sample headspace is an effective screening procedure which may be used, in some cases, to infer NAPL presence. As shown in Figure 9-20, OVA concentrations for samples containing NAPL ranged from 60 to >1000 ppm compared to maximum concentrations of 30 ppm and <20 ppm in dissolved contamination and blank samples, respectively. The poor correlation between NAPL saturation and OVA concentration indicates that OVA measurements cannot interpreted to estimate NAPL saturation. Measured organic vapor concentrations are sensitive to the effective contaminant volatility, sample temperature, and sample handling.

• *Unaided Visual Examination* -- Identification of clear, colorless NAPL in soil by unaided visual examination is very difficult. Using this method, the sample examiners were unable to determine NAPL presence in any of the 56 NAPL-contaminated soil samples.

• *UV Fluorescence Analysis* -- Fluorescence refers to the spontaneous emission of visible light resulting from a concomitant movement of electrons to higher and lower energy states when excited by UV radiation. Many DNAPLs fluoresce (Konstantinova-Shlezinger, 1961) including: (1) nearly all aromatic or polyaromatic hydrocarbons (having one or more benzene ring) such as coal tar, creosote, and PCBs; (2) nearly all DNAPL mixtures that contain petroleum products; (3) many unsaturated aliphatic hydrocarbons (such as trichloroethene and tetrachloroethene); and, (4) all unsaturated hydrocarbons with conjugated double bonds. Note that: unsaturated refers to hydrocarbons having carbon atoms that are bonded together by one or more double or triple bonds; aliphatic hydrocarbons contain carbon chains and no carbon rings; and conjugated double bonds refer to at least two double bonds that are separated by only one single bond. Saturated aliphatic hydrocarbons, such as carbon tetrachloride and dichloromethane, generally do not fluoresce unless mixed with fluorescent impurities. This can result from industrial processes and waste disposal practices. UV fluorescence has been utilized for decades by the oil industry to identify petroleum presence in drill mud, cuttings, and cores. Several standard examination methods using extractants to

Table 9-10. Summary of Test Results (Note: A = NAPL presence apparent based on visual examination; B = NAPL presence suspected based on visual examination; and C = no visual evidence of NAPL presence).

Method	Sample categories are based upon estimated NAPL saturations as a percent. The volume of NAPL mixed with 172 g of soil and sufficient mL of water to constitute a total fluid content of 35 mL is also given.							Notes and Conclusions
	Blank Samples (No NAPL)	Dissolved Samples (No NAPL)	1% (0.35 mL)	2.86% (1 mL)	5.71% (2 mL)	11.43% (4 mL)	22.86% (8 mL)	
OVA Headspace Analysis using an FID	1.4 - 4.8 ppm (see notes)	1.4-30 ppm	120->1000 ppm	50->1000 ppm	60->1000 ppm	100->1000 ppm	65->1000 ppm	1. An effective screening method which may be used, in some cases, to infer NAPL presence. 2. Organic vapor concentration depends on contaminant volatility: measured concentrations were much higher in chlorobenzene and PCE samples than kerosene samples. 3. Two blank samples had OVA concentrations of <10 and <20 ppm due to residual vapors from prior samples.
Unaided Visual Exam	0 A 0 B 11 C	0 A 0 B 11 C	0 A 0 B 11 C	0 A 1 B 11 C	0 A 3 B 8 C	0 A 6 B 5 C	0 A 7 B 4 C	1. Unable to identify presence of colorless NAPL. 2. NAPL presence was suspected in some samples with higher NAPL saturation based on fluid sudsiness.
UV Fluorescence Exam	1 A 1 B 9 C	0 A 2 B 9 C	4 A 1 B 6 C	9 A 1 B 2 C	9 A 1 B 1 C	11 A 0 B 0 C	11 A 0 B 0 C	1. Very effective simple test for fluorescent NAPLs. 2. One false positive in 22 blank or dissolved samples. 3. Only 3 false negatives in 45 samples with estimated NAPL saturations between 1% and 23%. 4. Sensitivity depends on fluorescent intensity of NAPL: at low NAPL saturations, kerosene and chlorobenzene were easier to detect than tetrachloroethene. 5. Greater visual contrast evident between milky white fluorescence and darker soils. 6. Adding more water to the contaminated soil sample improved the detectability of NAPL in some cases by bringing more fluorescent fluid to the polybag wall.
Soil-Water Shake Test Exam	0 A 1 B 10 C	0 A 2 B 9 C	0 A 4 B 7 C	2 A 6 B 4 C	0 A 9 B 2 C	2 A 8 B 1 C	3 A 7 B 1 C	1. Difficult to positively identify clear, colorless NAPL. 2. At relatively high saturations (between 1% and 23%), NAPL presence was usually suspected based on fluid characteristics at the fluid-air interface. 3. As a result, colorless LNAPL (kerosene) was easier to detect than colorless DNAPL (chlorobenzene and tetrachloroethene) using the shake test.
Centrifugation Exam	0 A 1 B 10 C	0 A 1 B 10 C	0 A 4 B 7 C	2 A 5 B 5 C	4 A 3 B 4 C	6 A 1 B 4 C	3 A 4 B 4 C	1. Fairly effective for identification of LNAPL (kerosene), but not DNAPLs, based on fluid characteristics at the fluid-air interface. 2. Seventeen false negatives in 45 samples with estimated NAPL saturations between 1% and 23%; only 15 positive NAPL identifications in these 45 samples.
Hydrophobic Dye Shake Test Exam	0 A 0 B 11 C	0 A 0 B 11 C	4 A 2 B 5 C	8 A 1 B 3 C	10 A 0 B 1 C	11 A 0 B 0 C	11 A 0 B 0 C	1. Very effective simple test. 2. No false positives in 22 blank or dissolved samples. 3. Identified NAPL presence in 40 of 45 samples with estimated NAPL saturations >1%. False negatives recorded in only 4 of these 45 samples. 4. Dye coloration obvious even in black topsoil samples. 5. NAPL density relative to water was correctly determined in 21 samples and misjudged in 1 sample. 6. Can be used to estimate quantity of NAPL in sample.
Centrifugation of Hydrophobic Dye Shake Test Sample	0 A 1 B 10 C	0 A 0 B 11 C	5 A 1 B 5 C	9 A 2 B 1 C	11 A 0 B 0 C	11 A 0 B 0 C	11 A 0 B 0 C	1. Slight enhancement of hydrophobic dye shake test. 2. No false positives in 22 blank or dissolved samples. 3. Identified NAPL presence in 42 of 45 samples with estimated NAPL saturations >1%. False negative recorded in only 1 of these 45 samples. 4. NAPL density relative to water was correctly determined in 43 samples and misjudged in 3 samples. 5. Can be used to estimate quantity of NAPL in sample.

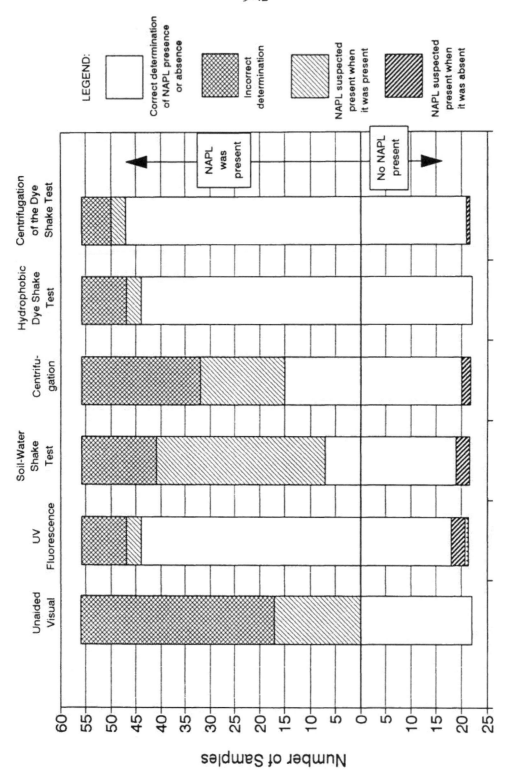

Figure 9-19. Summary of NAPL detection method results (from Cohen et al., 1992).

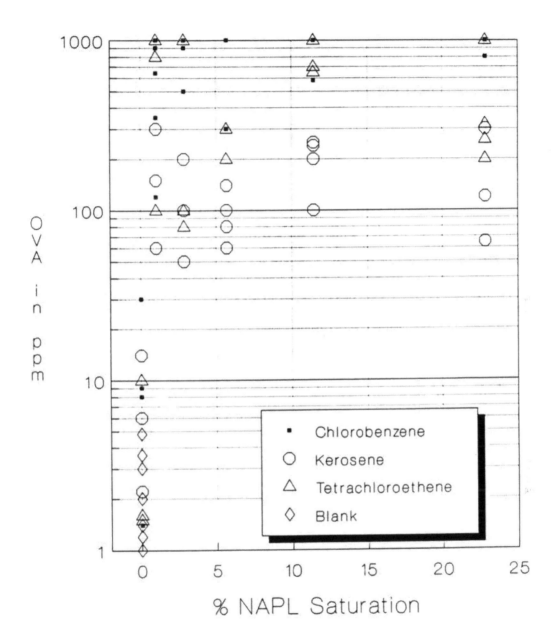

Figure 9-20. OVA concentrations plotted as a function of NAPL type and saturation (from Cohen et al., 1992). OVA measurements shown as 1000 ppm are actually > 1000 ppm. Dissolved contaminant samples are treated as 0% NAPL saturation samples.

leach cuttings within a UV lamp metal viewing box (a fluoroscope) are described by Swanson (1981) and Halliburton Co. (1981).

UV fluorescence examination proved to be a simple and very effective method for identifying the presence of fluorescent NAPLs in dark soils. The fluorescence of each NAPL type was milky white. Due to the greater visual contrast of milky fluorescence against darker soils, this method was much more effective with the moderate brown silt loam and black silt loam samples than with the pale-yellowish orange medium sand samples. Overall, UV fluorescence was somewhat less effective than the hydrophobic dye methods (Figure 9-19 and Table 9-10). The addition of water to some soil samples enhanced the ability to detect NAPL fluorescence.

- *Soil-Water Shake Test* -- Shaking soil with water in a jar can be used to detect NAPL where sufficient visual contrast exists between the NAPL, soil, and water (as was demonstrated by the hydrophobic dye shake test). However, the examiners were only able to positively identify NAPL in 7 of the 56 soil samples contaminated with clear, colorless NAPLs. In 34 of the 56 NAPL-contaminated samples, NAPL presence was suspected at the fluid-air interface based on distinctive fluid characteristics of floating LNAPL (kerosene) or DNAPL held by surface tension. This was most apparent in samples containing relatively large LNAPL (kerosene) saturations.

- *Centrifugation* -- Centrifugation of soil samples mixed with water was an effective method for identifying clear, colorless LNAPL (kerosene) based on fluid characteristics at the fluid-air interface, but was ineffective for identifying DNAPL (chlorobenzene and tetrachloroethene). Its utility probably could have been enhanced by using a syringe needle to extract fluid from the top and bottom of the fluid column in the centrifuge tube for further inspection.

- *Hydrophobic Dye Soil-Water Shake Test* -- Mixing a tiny amount of hydrophobic dye (Sudan IV) with NAPL-contaminated soil and water proved to be a simple and effective means for identifying the presence of NAPLs in soil, particularly in the medium sand samples. This method allowed for determination of NAPL density relative to water and estimation of NAPL volume in many samples. Comparison of visual estimates of NAPL volume (or saturation) with actual

measurements (such as described by Cary et al., 1991) may provide a basis for "calibrating" visual estimates.

- *Centrifugation of Hydrophobic Dye Shake Test Suspension* -- Centrifugation of the hydrophobic dye shake test samples provided some enhancement of NAPL detectability (Figure 9-19), and, like the hydrophobic dye shake test, was particularly effective with the medium sand samples. Overall, this combined procedure provided the most accurate results of the visual methods tested to determine NAPL presence in soil.

The ease of determining NAPL presence in soil is directly related to the magnitude of visual contrast between NAPL, water, and soil. Cohen et al. (1992) recommend that dye shake tests be conducted in plastic containers (e.g., polypropylene tubes) because hydrophobic NAPLs generally wet plastic better than glass, thereby enhancing NAPL detection on the container wall. The visual contrast afforded by using hydrophobic dye to tint NAPL red is generally greater than that provided by UV fluorescence and much greater than that associated with the interfacial characteristics of colorless immiscible fluids. Of the methods tested, therefore, the hydrophobic dye techniques, followed by UV fluorescence, most facilitated the determination of NAPL presence in the soil samples.

After completing the soil testing program, 0.05 mL of kerosene, chlorobenzene, and tetrachloroethene were added individually to 40 mL of water in three 50 mL centrifuge tubes. A blank sample was prepared using 40 mL of water and no NAPL. The samples were visually examined after: (a) the initial mixing; (b) centrifugation at 1250 rpm for one minute; (c) adding approximately 2 mg of Sudan IV dye; and, (d) centrifugation of the dye-fluid mix at 1250 rpm for one minute.

During the unaided visual examination of the water samples, the presence of NAPL was suspected in each of the three samples with NAPL, but not in the blank, due to the contrast of fluid characteristics at the fluid-air interface. Phase separation caused by centrifugation eliminated this contrast and the visual evidence of NAPL presence. Mixing hydrophobic dye with the samples instantaneously revealed the presence and density relative to water of the 0.05 mL of kerosene, chlorobenzene, and tetrachloroethene present in the three contaminated 40 mL samples (volumetric NAPL content = 0.125%). Subsequent centrifugation separated the dyed NAPL from the clear water, thereby facilitating determination of the volumetric NAPL to water ratio in a graduated centrifuge tube.

Of the methods tested, the hydrophobic dye methods, followed by UV fluorescence (for fluorescent NAPLs), offer the most simple, practical, and effective means for direct visual identification of clear, colorless NAPL in contaminated soil samples. These methods can be utilized in the field or in a lab with minimal time and material expense. Known background and NAPL-contaminated samples should be examined in addition to unknown samples, where possible, to check for interferences and site-specific NAPL responses. For volatile NAPLs, analysis of organic vapors in soil sample headspace can be used to screen samples for further examination and, possibly, to infer NAPL presence. The NAPL in water experiments demonstrate that the presence of very small quantities of clear, colorless NAPL in water can be quickly identified by mixing in a tiny amount of hydrophobic dye. At some sites, however, careful unaided visual examination of soil and water samples, or use of an undyed shake test, may be sufficient to detect NAPL presence.

9.10.2 Indirect Detection of NAPL Presence

As noted in Table 7-4, NAPL presence may be inferred where:

(1) groundwater concentrations exceed 1% of the pure phase or effective aqueous solubility of a NAPL chemical;

(2) NAPL chemical concentrations in soil exceed 10,000 mg/kg (>1% of soil mass);

(3) NAPL chemical concentrations in groundwater calculated from soil-water partitioning relationships and soil sample analyses exceed their effective solubility;

(4) organic vapor concentrations detected in soil gas (or sample head space) exceed 100 to 1000 ppm; or,

(5) observed chemical distribution patterns suggest NAPL presence (see Table 7.4).

Examples are provided below to further explain the concept of effective solubility and how soil-water partitioning relationships can be used in conjunction with soil analyses to assess DNAPL presence.

9.10.2.1 Effective Solubility

Effective solubility based on Raoult's Law is discussed in Chapter 4.7, Chapter 5.3.17, and Worksheet 7-1, and by Feenstra et al. (1991), Feenstra (1990), Mackay et al. (1991), Sitar et al. (1990), Banerjee (1984), Leinonen and Mackay (1973) and Shiu et al. (1988). For mixtures of liquid chemicals, the dissolved-phase concentrations in equilibrium with the NAPL mixture can be estimated by

$$S^e_i = X_i S_i \qquad (9-3)$$

where S^e_i is the effective aqueous solubility of liquid constituent i (mg/L), X_i is the mole fraction of liquid constituent i in the NAPL mixture, and, S_i is the pure-phase solubility of liquid constituent i (mg/L). Aqueous solubility data for numerous DNAPL chemicals are provided in Appendix A. Example calculations of effective solubilities for multi-liquid DNAPLs are given in Table 9-11.

Laboratory analyses suggest that Equation 9-3 is a reasonable approximation for mixtures of sparingly soluble hydrophobic organic liquids that are structurally similar (Banerjee, 1984) and that effective solubilities calculated for complex mixtures (e.g., petroleum products) are unlikely to be in error by more than a factor of two (Leinonen and Mackay, 1973). Greater errors are expected where high concentrations of cosolvents, such as alcohols, may significantly increase the solubility of DNAPL components (Rao et al., 1991).

Many DNAPL mixtures are comprised of liquid and solid chemicals (e.g., coal tar and creosote). For organic chemicals that are solids at the temperature of interest but in liquid solution within a NAPL, the correct value of S_i for calculating S^e_i is that of the subcooled liquid chemical (Mackay et al., 1991). Solubilities reported in chemical data handbooks (including Appendix A) are usually for the pure solid compound in contact with water (Feenstra et al., 1991). The S_i of a subcooled liquid chemical exceeds the solid solubility at temperatures below the compound's melting point. It can be estimated by

$$S_{i\text{-subcooled liquid}} = S_{i\text{-solid}} / [\exp[(6.79(1-T_m/T)]] \qquad (9-4)$$

where $S_{l-solid}$ is the solid solubility at temperatures below melting point, T_m is the compound melting point (°K), and T is the system temperature (°K) (Mackay et al., 1991). For example, naphthalene with a solid solubility of 33 mg/L and a melting point of 80.2 °C (353.35 °K) has a subcooled liquid solubility at 25 °C (298.15 °K) of 118 mg/L (33 mg/L divided by 0.28). Values of subcooled liquid solubility are also provided by Miller et al. (1985) and Eastcott et al. (1988). Example calculations of effective solubilities for DNAPL components which include organic compounds that are solids at the temperature of interest are given in Table 9-11.

At sites with complex and varied DNAPL mixtures, or where DNAPL samples cannot be obtained for analysis, reliable estimation of the effective solubilities of DNAPL components will be confounded by deficient knowledge of DNAPL composition (constituent mole fractions). The significance of this uncertainty regarding DNAPL composition can be assessed by sensitivity analysis (i.e., making bounding assumptions regarding composition to determine the range of possible effective solubilities). Given site-specific samples of DNAPL and groundwater, effective solubilities can also be determined by conducting equilibrium dissolution experiments (Chapter 10.11).

9.10.2.2 Assessing NAPL Presence in Soil Based on Partitioning Theory

Chemical analyses of soil typically indicate the total quantity of each analyte determined as a chemical mass per unit dry weight of soil sample. The analytical determination includes, and does not distinguish between, mass sorbed to soil solids, dissolved in soil water, volatilized in soil gas, and, contained in NAPL, if present. Feenstra et al. (1991) describe a method for evaluating the possible presence of NAPL in soil samples based on equilibrium partitioning theory which is summarized in Worksheet 7-2 and below. NAPL presence can be inferred where chemical concentrations determined in a soil sample exceed the theoretical maximum chemical mass that can be adsorbed to soil solids, dissolved in soil water, and volatilized in soil gas.

The method requires measurements or estimates of: total chemical concentrations in the soil sample, soil moisture content (n_w, volume fraction), soil dry bulk density (ρ_b, g/cm^3), soil porosity (n, volume fraction), soil organic carbon content (f_{oc}, mass fraction), and, the organic carbon to water partition coefficient (K_{oc}, cm^3/g), dimensionless Henry's Constant ($K_{H'}$) partitioning

coefficient, and effective solubilities for the compounds of interest. Using these data, the apparent soil water concentration, C_{asw}, of a particular compound can be calculated from the total soil concentration by assuming equilibrium partitioning (Feenstra et al., 1991):

$$C_{asw} = (C_t \, \rho_b) \, / \, (K_d \, \rho_b + n_w + K_{H'} \, n_a) \qquad (9\text{-}5)$$

where C_t is the total soil concentration of a particular analyte, K_d (cm^3/g) is the soil-water partition coefficient ($K_d = K_{oc} \times f_{oc}$), and n_a (volume fraction) is the air-filled porosity ($n_a = n - n_w$). For saturated soil samples (Feenstra et al., 1991),

$$C_{asw} = (C_t \, \rho_b) \, / \, (K_d \, \rho_b + n_w) \qquad (9\text{-}6)$$

NAPL presence can be inferred if the calculated apparent soil water concentration, C_{asw}, exceeds the effective solubility, S^e_i, of compound of interest. The reliability of conclusions regarding NAPL presence depends on the validity of partitioning coefficients, effective solubilities, chemical analyses, and other data utilized in this method. Example applications of this method are illustrated in Figure 9-21. As shown, log-log graphs of C_t versus C_{asw} for different values of f_{oc} (or K_d) can be used to facilitate rapid evaluation of soil analyses and sensitivity analysis.

9.11 INTEGRATED DATA ANALYSIS

There is no practical cookbook approach to site investigation or data analysis. In addition to the noninvasive, invasive, and laboratory procedures described in Chapters 8, 9, and 10, respectively, many additional methods can be used to enhance contamination site evaluation. In particular, consideration should be given to the use of tracers (Davis et al., 1985; Hendry, 1988; Uhlmann, 1992), interpreting chemical distributions and ratios (e.g., Hinchee and Reisinger, 1987), and conducting hydraulic tests.

Each site presents variations of contaminant transport conditions and issues. Site characterization, data analysis, and conceptual model refinement are iterative activities which should satisfy the characterization objectives outlined in Figure 6-1 as needed to converge to a final remedy.

During the process, acquired data should be utilized to guide ongoing investigations. For example, careful examination of soil, rock, and fluid samples obtained as

Table 9-11. Example effective solubility calculations (using Equations 9-3 and 9-4) for a mixture of liquids with an unidentified fraction (DNAPL A) and a mixture of liquid and solid chemicals (DNAPL B).

Compound	Molecular Weight (g)	Melting Point °C	% Mass	Moles per Kg of DNAPL	Mole Fraction	Aqueous Solubility (mg/L)	Subcooled Liquid Solubility (mg/L)	Effective Solubility (mg/L)
DNAPL A								
Trichloroethene	131.39	-73	0.23	1.751	0.2296[1] 0.2882[2] 0.3150[3]	1100	NA	252.6[1] 317.0[2] 346.5[3]
Tetrachloroethene	165.83	-19	0.46	2.774	0.3638[1] 0.4567[2] 0.4991[3]	150	NA	54.6[1] 68.5[2] 74.9[3]
Unidentified Compounds (assume different molecular weights)	100[1] 200[2] 300[3]		0.31	3.100[1] 1.550[2] 1.033[3]	0.4066[1] 0.2552[2] 0.1859[3]			
DNAPL B								
Chlorobenzene	112.56	-46	0.43	3.820	0.599	500	NA	299.
1,2,4-Trichlorobenzene	181.45	17	0.15	0.827	0.130	19	NA	2.46
1,2,3,5-Tetrachlorobenzene	215.89	54.5	0.21	0.973	0.152	2.89	5.66	0.863
Pentachlorobenzene	250.34	86	0.05	0.200	0.031	0.83	3.33	0.104
Hexachlorobenzene	284.78	230	0.16	0.562	0.088	0.005	0.534	0.047

Notes: NA = not applicable; for DNAPL A, superscripts [1], [2], and [3] correspond to assumptions that the molecular mass of the unidentified DNAPL fraction is 100, 200, and 300 g., respectively.

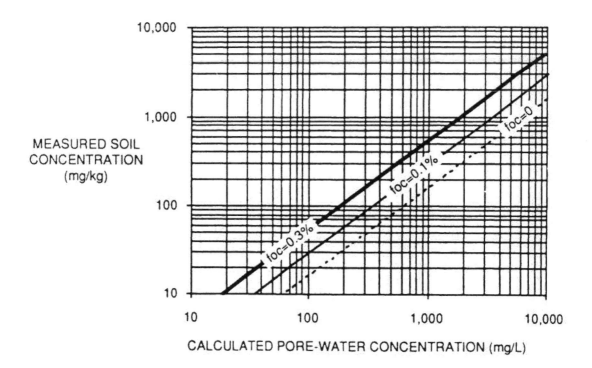

Figure 9-21. Relationship between measured concentration of TCE in soil (C_t) and the calculated apparent equilibrium concentration of TCE in pore water (C_{apw}) based on: K_{oc} = 126, bulk density (ρ_b) = 1.86 g/cm³; water-filled porosity (n_w) = 0.30; air-filled porosity (n_a) = 0; and three different values of organic carbon content (f_{oc}) (from Feenstra et al., 1991). NAPL presence can be inferred if C_{apw} exceeds the TCE effective solubility.

drilling progresses should be made to identify DNAPL presence and potential barrier layers and thereby guide decisions regarding continued drilling, well construction, and/or borehole abandonment. Geologic, fluid elevation, and chemical distribution data should be organized (preferably using database, CAD, and/or GIS programs) and displayed on maps that are updated periodically to help determine the worth of additional data collection activities. With continued refinement of the site conceptual model, the benefit, cost, and risk of additional work can and should be evaluated with improved accuracy. This is the advantage of a flexible, phased approach to site characterization.

10 LABORATORY MEASUREMENTS: METHODS AND COSTS

Chemical and physical properties of contaminants and media are measured to evaluate chemical migration and clean-up alternatives at DNAPL-contaminated sites. These chemical and physical properties are defined and their significance with regard to DNAPL contamination is discussed in Chapters 4 and 5. Methods used to determine DNAPL composition and DNAPL-media properties are described briefly in this chapter. Estimated costs in 1992 dollars are provided for some of the test apparatus and method determinations based on quotations from equipment catalogs and laboratories specializing in analytical chemistry or petroleum reservoir analysis. Actual costs will vary and may be increased due to special hazardous material handling requirements. Additional information regarding the advantages, limitations, and costs of using different analytical methods can be obtained from analytical laboratories and instrument vendors. Physical properties of pure DNAPLs are given in Appendix A.

10.1 DNAPL COMPOSITION

DNAPL chemicals may be non-hazardous and require no special treatment, or, they may be found on USEPA's Priority Pollutant List (PPL) or the Target Compound List (TCL). The chemical makeup of the DNAPL sample will affect the Health and Safety Plan, the Risk Analysis, the Remedial Action Plan (RAP), and the Feasibility Study (FS).

The organic content of soils or groundwater at contamination sites is often a complex mixture of chemicals. Immiscible fluid samples can be fractionated, or split into several portions, using a separatory funnel (Figure 10-1) or a centrifuge. A methodological flow chart for analysis of complex hydrocarbon mixtures is given in Figure 10-2. This flow chart was developed for a manufactured gas site with low and high molecular weight organic compounds and can be adapted to site-specific conditions. All analytical results are considered to evaluate the total fluid chemistry of complex mixtures (labeled the "Puzzle Fitting and Structural Parameter Correlation" in Figure 10-2).

For NAPL organic samples, the customary methods of analysis are infrared (IR) spectrometry, high resolution nuclear magnetic resonance (NMR) spectrometry, gas chromatography (GC), high performance liquid chromatography (HPLC), and mass spectrometry (MS).

These analyses require expensive, high sensitivity instrumentation maintained in controlled environments and are best performed by trained technicians in fixed laboratories. Laboratories may request several liters of a sample but can analyze much smaller specimens if large volume samples cannot be obtained. Inorganic contaminants, not considered in this discussion, can be determined by atomic absorption (AA), differential thermal analysis (DTA), ion chromatography (IC), and X-ray diffraction (XRD).

Composition analyses may be either qualitative or quantitative, or both. A qualitative analysis will only identify a compound's presence and will not determine the purity or percentage composition of a mixture. Preliminary site investigations may focus on a qualitative analysis to indicate the constituents of concern. Quantitative analyses identify a compound's presence and determine the purity or the percentage composition of each component of a mixture. Quantitative analyses are usually required to provide data for site characterization, feasibility, risk assessment, and treatability studies. A brief description of a qualitative method, infrared (IR) spectrometry, and two quantitative methods, gas chromatography (GC) and high performance liquid chromatography (HPLC) coupled with mass spectrometry (MS), are presented below. The applicability and cost of various analytical methods for characterizing different types of DNAPLs are summarized in Table 10-1.

10.1.1 Infrared (IR) Spectrometry

Infrared spectrometry is primarily used as a non-destructive method to determine the structure and identity of compounds; it may also be used to determine compound concentrations. The total infrared spectrum extends from 0.75 to 400 micrometers (μm) but the area of most usefulness lies between 2.5 to 16 μm. Infrared means "inferior to red" because the radiation is adjacent to but slightly lower in energy than red light in the visible spectrum. Identification and quantitation of compounds results from comparing spectral data of sample unknowns with known compounds. Most molecules are in modes of rotation, stretching, and bending at frequencies found in the infrared spectral region and each molecule has its own distinctive infrared spectrum or "fingerprint". Infrared spectrophotometers emit an infrared beam that is split, half passing through the sample and half by-passing the sample. The difference in energy of the two halves is compared and reduced to a plot of percent absorbance versus wavenumber or wavelength. Two types of IR spectrophotometers are used: dispersive and

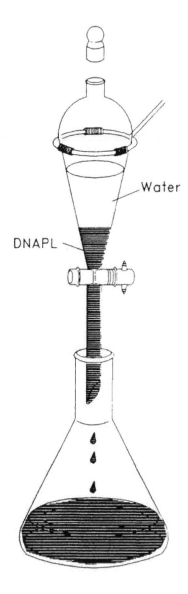

Figure 10-1. Use of a separatory funnel to separate immiscible liquids (redrawn from Shugar, G.J., and J.T. Ballinger, 1990. Chemical Technician's Ready Reference Handbook. Reprinted with permission from McGraw-Hill Book Co.).

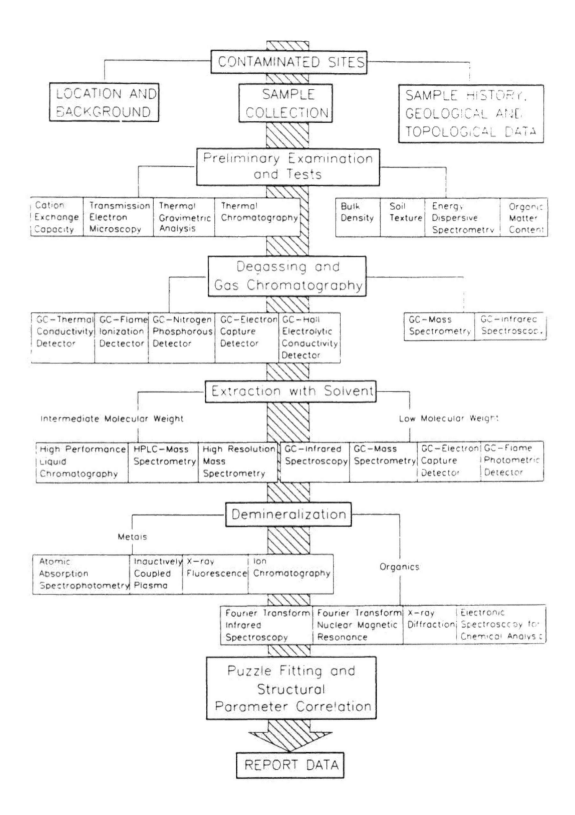

Figure 10-2. Flow chart for analysis of complex mixtures (from Devinny et al., 1990).

Table 10-1. Analytical methods and estimated costs (in 1992 dollars) to determine chemical composition.

Analysis	Matrix	Method	Sample Size and Container	Preservation	Holding Time	Est. Cost
Acid Extractables	w/ww	EPA 625	3 X 1000 mL, G(am)	Cool, 4_oC; $Na_2S_2O_3$	7/40 d	$350
	w/ww	SW-846 8270	3 X 1000 mL, G(am)	Cool, 4_oC; $Na_2S_2O_3$	7/40 d	$400
	sw	SW-846 8270	100 g, G	Cool, 4_oC	14/40 d	$470
Appendix IX Volatiles Organic Compounds (VOCs)	w/ww	SW-846 8240	2 X 40 mL, G	Cool, 4_oC, HCl to pH<2; (No Headspace)	14 d	$415
	sw	SW-846 8240	100 g, G	Cool, 4_oC	14 d	$450
Appendix IX Semivolatiles (SVOCs)	w/ww	SW-846 8270	3 X 1000 mL, G(am)	Cool, 4_oC; $Na_2S_2O_3$	7/40 d	$690
	sw	SW-846 8270	100 g, G	Cool, 4_oC	14/40 d	$760
Appendix IX Organochlorine Pesticides	w/ww	SW-846 8080	3 X 1000 mL, G(am)	Cool, 4_oC; $Na_2S_2O_3$	7/40 d	$270
	sw	SW-846 8080	100 g, G	Cool, 4_oC	14/40 d	$315
Appendix IX Herbicides	w/ww	SW-846 8150	3 X 1000 mL, G(am)	Cool, 4_oC; $Na_2S_2O_3$	7/40 d	$225
	sw	SW-846 8150	100 g, G	Cool, 4_oC	14/40 d	$270
Appendix IX Organo-phosphate Pesticides	w/ww	SW-846 8140	3 X 1000 mL, G(am)	Cool, 4_oC; $Na_2S_2O_3$	7/40 d	$200
	sw	SW-846 8140	100 g, G	Cool, 4_oC	14/40 d	$245
Appendix IX Metals	w/ww	SW-846	1000 mL, P	Cool, 4_oC; HNO_3 to pH<2	6 mo.	$294
	sw	SW-846	100 g, G	Cool, 4_oC	6 mo.	$374
Appendix IX Cyanide	w/ww	SW-846 9012	500 mL, P	Cool, 4_oC; NaOH to pH>12	14 d	$55
	sw	SW-846 9012	100 g, G	Cool, 4_oC	14 d	$75
Appendix IX Sulfide	w/ww	EPA 376.1	500 mL, G	Cool, 4_oC; NaOH, ZnAc	7 d	$16
	sw	SW-846 9030	100 g, G	Cool, 4_oC	7 d	$60
Biochemical Oxygen Demand	w/ww	EPA 405.1	100 mL, P/G	Cool, 4_oC	2 d	$33
BTEX (Benzene, Toluene, Ethylbenezene, Xylenes)	w/ww	EPA 602 or SW-846 8020	2 X 40 mL, G	Cool, 4_oC; HCl to pH<2 (No Headspace)	14 d	$84
	sw	SW-846 5030/8020m	100 g, G	Cool, 4_oC	14 d	$115
BTEX and Naphthalene	w/ww	EPA 602 or SW-846 8020	2 X 40 mL, G	Cool, 4_oC; HCl to pH,2 (No Headspace)	14 d	$100
	sw	SW-846 5030/8020m	100 g, G	Cool, 4_oC	14 d	$135
BTEX and MTBE	w/ww	EPA 602 or SW-846 5030/8020	2 X 40 mL, G	Cool, 4_oC; HCl to pH<2 (No Headspace)	14 d	$90
	sw	SW-846 5030/8020m	100 g, G	Cool, 4_oC	14 d	$125
Carbamate Pesticides	w/ww	EPA 531.1	2 X 250 mL, G(am)	MCA to pH of 3	28 d	$200
	sw	California SOP 734	100 g, G	Cool, 4_oC	7/40 d	$245
Carbon (Total Organic)	w/ww	EPA 415.1	125 mL, G	Cool, 4_oC; H_2SO_4 to pH<2	28 d	$25
	sw	EPA 415.1m	20 g, G	Cool, 4_oC	28 d	$95
Chemical Oxygen Demand	w/ww	EPA 410.4	100 mL, P/G	Cool, 4_oC; H_2SO_4 to pH<2	28 d	$29
	sw	EPA 410.1	100 g, G	Cool, 4_oC	28 d	$29
Chlorinated Pesticides/PCBs plus Organophosphates	w/ww	SW-846 8080m, 8140	2 X 1000 mL, G(am)	Cool, 4_oC; $Na_2S_2O_3$	7/40 d	$225
	sw	SW-846 8080m, 8140	100 g, G	Cool, 4_oC	14/40 d	$275

Table 10-1. Analytical methods and estimated costs (in 1992 dollars) to determine chemical composition.

Analysis	Matrix	Method	Sample Size and Container	Preservation	Holding Time	Est. Cost
Cyanide (Total)	w/ww	SW-846 9012	500 mL, P	Cool, 4_oC; NaOH to pH>12	14 d	$55
	sw	SW-846 9012	100 g, G	Cool, 4_oC	14 d	$75
Dioxin (Screen)	w/ww	EPA 625	2 X 1000 mL, G	Cool, 4_oC; $Na_2S_2O_3$	7/40 d	$350
	sw	EPA 625 m	100 g, G	Cool, 4_oC	14/40 d	$420
Flashpoint	w/ww	ASTM D-93	200 mL, G	NA	30 d	$38
GC Fingerprint (Qualitative)	oil	SW-846 8015m	20 mL, G	NA	NA	$70
GC Fingerprint (Quantitative)	w/ww	SW-846 8015m	2 X 1000 mL, G	Cool, 4_oC (No Headspace)	14 d	$120
	sw	SW-846 8015m	200 g, G	Cool, 4_oC	14 d	$140
	oil	SW-846 8015m	20 mL, G	NA	NA	$120
GC VOCs Purgeable Aromatics-Halocarbons	w/ww	SW-846 5030/8010/8020	2 X 40 mL, G	Cool, 4_oC; HCl to pH<2 (No Headspace)	14 d	$125
GC VOCs Purgeable Aromatics	w/ww	SW-846 5030/8020	2 X 40 mL, G	Cool, 4_oC; HCl to pH<2 (No Headspace)	14 d	$90
	sw	SW-846 5030/8020m	100 g, G	Cool, 4_oC	14 d	$110
GC VOCs Purgeable Halocarbons	w/ww	SW-846 5030/8010	2 X 40 mL, G	Cool, 4_oC; HCl to pH<2 (No Headspace)	14 d	$110
	sw	SW-846 5030/8010m	100 g, G	Cool, 4_oC	14 d	$145
GC VOCs Library Search	w/ww/sw	NA	NA	NA	NA	$55
HPLC (High Performance Liquid Chromatography) Polynuclear Aromatic Hydrocarbons (PAHs)	w/ww	SW-846 8310	2 X 1000 mL, G(am)	Cool, 4_oC	28 d	$200
	sw	SW-846 8310	100 g, G	Cool, 4_oC	7/40 d	$245
Metals by ICP (Inductively Coupled Plasma)	w/ww	SW-846 6010	500 ml, P/G	Cool, 4_oC; HNO_3 to pH<2	6 mo	$325
	sw	SW-846 6010	100 g, G	Cool, 4_oC	6 mo	$345
Naphthalene by GC	w/ww	SW-846 8020	2 X 40 mL, G	Cool, 4_oC; HCl to pH<2 (No Headspace)	14 d	$100
	sw	SW-846 5030/8020m	100 g, G	Cool, 4_oC	14 d	$135
Naphthalene by GC/MS	w/ww	SW-846 8270	3 X 1000 mL, G	Cool, 4_oC; $Na_2S_2O_3$	7/40 d	$400
	sw	SW-846 8270	100 g, G	Cool, 4_oC	14/40 d	$470
Oil & Grease (Gravimetric)	w/ww	EPA 413.1	2 X 1000 mL, G	Cool, 4_oC	28 d	$55
	sw	SW-846 9071	50 g, G	Cool, 4_oC	28 d	$85
Oil & Grease/Total Petroleum Hydrocarbons	w/ww	EPA 413.2 EPA 418.1	2 X 1000 mL, G	Cool, 4_oC	28 d	$100
PCBs	w/ww	SW-846 8080	2 X 1000 mL, G(am)	Cool, 4_oC; $Na_2S_2O_3$	7/40 d	$115
	sw	SW-846 8080	100 g, G	Cool, 4_oC	14/40 d	$145
	oil	EPA 600/4-81-045	20 mL, G	NA	NA	$80
Petroleum Materials Contamination -- GC Fingerprint (Qualitative)	oil	SW-846 8015m	20 mL, G	NA	NA	$70

Table 10-1. Analytical methods and estimated costs (in 1992 dollars) to determine chemical composition.

Analysis	Matrix	Method	Sample Size and Container	Preservation	Holding Time	Est. Cost
Petroleum Materials Contamination -- GC Fingerprint (Quantitative)	w/ww	SW-846 8015m	2 X 1000 ml, G	Cool, 4_oC (No Headspace)	14 d	$120
	sw	SW-846 8015m	200 g, G	Cool, 4_oC	14 d	$140
	oil	SW-846 8015m	20 ml, G	NA	NA	$100
Total Petroleum Hydrocarbons (TPH)	w/ww	EPA 418.1	2 X 1000 mL, G	Cool, 4_oC	28 d	$70
	sw	EPA 418.1m	100 g, G	Cool, 4_oC	28 d	$105
Phenolics	w/ww	SW-846 9066	500 mL, G	Cool, 4_oC; H_2SO_4 to pH<2	28 d	$55
	sw	SW-846 9066	100 g, G	Cool, 4_oC	28 d	$75
Priority Pollutants -- VOCs by GC/MS	w/ww	SW-846 8240	2 X 40 mL, G	Cool, 4_oC; HCl to pH<2 (No Headspace)	14 d	$270
	sw	SW-846 8240	100 g, G	Cool, 4_oC	14 d	$305
Priority Pollutants -- Acid/Base Neutral Extractables by GC/MS	w/ww	SW-846 8270	3 X 1000 mL, G(am)	Cool, 4_oC; $Na_2S_2O_3$	7/40 d	$540
	sw	SW-846 8270	100 g, G	Cool, 4_oC	14/40 d	$610
Priority Pollutants -- Metals	w/ww	SW-846	1000 mL, P	Cool, 4_oC; HNO_3 to pH<2	6 mo	$210
	sw	SW-846	100 g, G	Cool, 4_oC	6 mo	$270
Target Analyte List (TAL) Metals	w/ww	EPA CLP	1000 mL, P	Cool, 4_oC; HNO_3 to pH<2	6 mo	$376
	sw	EPA CLP	100 g, G	Cool, 4_oC	6 mo	$456
Target Compound List (TCL) -- VOCs	w/ww	EPA CLP 3/90 SOW	2 X 40 mL, G	Cool, 4_oC; HCl to pH<2 (No Headspace)	10 d	$270
	sw	EPA CLP 3/90 SOW	100 g, G	Cool, 4_oC	10 d	$305
Target Compound List (TCL) -- SVOCs	w/ww	EPA CLP 3/90 SOW	3 X 1000 mL, G (am)	Cool, 4_oC; $Na_2S_2O_3$	ex. 5 d	$540
	sw	EPA CLP 3/90 SOW	100 g, G	Cool, 4_oC	ex. 10 d	$610
Target Compound List (TCL) -- Pesticides/PCBs	w/ww	EPA CLP 3/90 SOW	2 X 1000 mL, G(am)	Cool, 4_oC; $Na_2S_2O_3$	ex. 5d	$290
	sw	EPA CLP 3/90 SOW	100 g, G	Cool, 4_oC	ex. 10 d	$330
Total Organic Halogen (TOX)	w/ww	SW-846 9020	4 X 250mL, G	Cool, 4_oC; H_2SO_4 to pH<2 (No Headspace)	28 d	$70
	sw	SW-846 9020m	50 g, G	Cool, 4_oC	NA	$130
	oil, solvent	SW-846 9020m	20 g, G	Cool, 4_oC	NA	$70
Total Toxicity Characteristic Leaching Procedure (TCLP) Analysis	w/ww	SW-846 8080/8150/8240/8270	3 liters			$1800
	sw	Federal Register 6/29/90	2 liters			$2000

NOTES: w/ww indicates water or wastewater; sw indicates soil or solid waste; G indicates glass; P indicates plastic; (am) indicates amber; NA indicates not applicable; the cost to analyze a NAPL sample will generally approximate that shown for sw media.

fourier transform. Results are available on-line when using a dedicated microprocessor.

Sample volumes of 500 to 1000 mL are recommended for laboratory analytical work. Simple FT/IR analyses are priced from $50 to $75 each unless the unknowns are not available in the laboratory's library. FT/IR offers a rapid preliminary screening of organic compounds. FT/IR equipment can be used onsite by a trained technician or scientist in a climate-controlled trailer.

10.1.2 Chromatography

"Chromatography" was derived from the words "chromatus" and "graphien" meaning "color" and "to write" in Tswett's technique for separating plant pigments. Tswett's experiment discovered the varying color affinities for a column packing when the color mixture was washed through the column. All chromatographic processes have a fixed (stationary) phase and a mobile (fluid) phase. Several analytical methods and separation processes have been developed from this phenomenon.

USEPA methods are primarily based on gas or liquid chromatography (GC or HPLC), with enhancements such as mass spectrometry to determine organic chemicals. Peaks observed by these GC and HPLC instruments are not readily identifiable in most cases unless another instrument is coupled to the chromatograph. Of several possible couplings, mass spectrometry (MS) is the most generally useful. With MS, these analytical methods are referred to as GC/MS and HPLC/MS. GC is typically used for compounds that readily vaporize (either fully or partially), are not thermally labile, and may be carried through the instrument in the gas phase. HPLC is used for compounds that are nonvolatile, thermally labile, and solubilized so that a solution of the sample and a solvent carry the unknown through the instrument.

10.1.2.1 Gas Chromatography/Mass Spectrometry (GC-MS)

One of the most rapid and useful separation techniques is GC. Samples are injected into a heated block, vaporized, and transported to a separation column by a carrier gas (usually helium if mass spectrometry is to follow). The sample is fractionated in the separation column. Column effluent proceeds through a detection device and exits the chromatograph (Figure 10-3).

Two types of GC units are used: gas-liquid and gas-solid chromatography. The most frequently employed method, gas-liquid chromatography, uses a liquid deposited on a solid support and is also called partition chromatography. The various components of the sample are separated by the liquid as they pass through the column. In gas-solid (also known as absorption) chromatography, the various sample components are absorbed on the surface of the solid as they pass through the column.

In both types of GC, various compounds are selectively absorbed by the stationary phase and desorbed by fresh carrier gas as the sample passes through the column. This process occurs repeatedly as the compounds move through the column. Compounds having a greater affinity for the stationary phase reside longer in the column than those that have lesser affinity for the stationary phase. As the individual compounds exit the column, they pass through a detector that emits a signal to a recording device.

Thermal conductivity and flame ionization detection devices are most commonly utilized with GC units. Thermal conductivity is used to quantify major components and flame ionization provides high sensitivity to trace amounts of compounds. Other detection devices are available that are sensitive to sulfur, nitrogen, phosphorous, and chlorine.

When GC/MS is used, the sample that exits the GC is forwarded to the mass spectrometer. The MS produces gas phase ions, separates the ions according to their mass-to-charge ratio, and emits a flux detected as an ion current. A mass spectrum is generated by measuring ion currents relating to the values of the mass-to-charge ratio over a particular mass range. Fast repetitive scanning and computerized data acquisition allow up to 1600 MS scans in an half-hour GC run.

Generally, compounds may be classified as inorganic or organic. Organic compounds are generally subclassified as volatiles, semi-volatiles, and pesticides in the regulatory field for contaminated sites. DNAPLs are organic and may be found in all three subclassifications.

Most laboratories use USEPA's Statement of Work 846 (SW-846) as the source of their standard analytical routines. A typical analytical methodology for liquid or solid waste samples includes determination of volatiles, semi-volatiles, and pesticides that are found on the Target Compound List (TCL) or Priority Pollutant List (PPL). Sample sizes range from 100 g of soil or solid materials

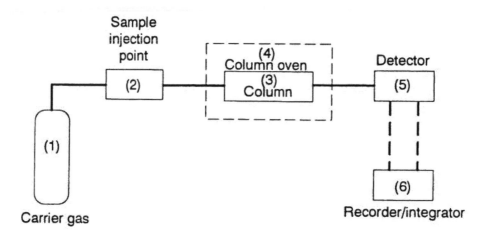

Figure 10-3. Schematic of a gas chromatograph (redrawn from Shugar, G.J., and J.T. Ballinger, 1990. Chemical Technician's Ready Reference Handbook. Reprinted with permission from McGraw-Hill Book Co.).

to between 3 and 5 L for liquid samples. The laboratory cost for determination of aqueous TCL compounds (including volatiles, semi-volatiles, pesticides/PCBs, and inorganics) typically ranges from $1500 to $2000. Soil/solid samples cost between $1700 and $2200 for similar analysis of volatiles, semi-volatiles, pesticides/PCBs, and inorganics. Of these costs, between $500 and $600 is for inorganic (metals) analyses. Analysis of volatile organic compounds using Method 8240 costs between $150 and $300 per sample, depending the requested number of analytes. Similarly, the cost for analysis of semi-volatile organic compounds using Method 8260 ranges between $350 and $600. If the site has been targeted by USEPA and requires a contract lab program (CLP), the estimated cost to analyze 20 samples for TCL parameters, with site-specific quality control and additional required QC samples, is $40,000 to $45,000.

10.1.2.2 High Performance Liquid Chromatography with Mass Spectrometry (HPLC/MS)

Liquid chromatography (LC) is used to analyze compounds within a sample that are not volatile or thermally labile, but are soluble in a solvent suitable for the chromatograph. Whereas GC can be used to analyze only about 20% of all organic compounds, 80% of all organic compounds can be determined using LC. High Performance (also known as High Pressure) Liquid Chromatography (HPLC) is used to determine formaldehyde and polynuclear aromatic hydrocarbons following USEPA methods 8315 and 8310, respectively. The HPLC column is packed with small beads, either coated or a homogeneous resin, that require high pressure to force the liquid to flow through the column. Most analyses are run at pressures in the range of 4,000 psi using pumps that produce an almost pulse-free flow rate of several mL per minute. Ultraviolet (UV) spectrophotometer or differential refractometer detectors provide signals to a computerized recording and analyzing system. When coupled with a mass spectrometer, the column effluent passes through a thermospray device to vaporize the sample and then the vaporized sample enters the mass spectrometer for continued analysis.

Laboratories typically request 100 g of solid or 2 L of liquid for analysis. Analytical costs range from $150 to determine the concentration of a single compound to $700 to determine the concentrations of a mixture of compounds such as polynuclear aromatic hydrocarbons (PAHs).

10.2 SATURATION

Several methods have been described in literature for determining DNAPL saturation in porous media (Amxy et al., 1960). The most common method is to extract the organic DNAPLs from the soil with a suitable organic solvent (i.e., ethanol) and make a gravimetric or HPLC determination of the DNAPL in the solvent. Analytical laboratories charge between $300 to $800 and require a minimum of 1 kg of solid sample for this analysis. Solvent extraction of DNAPL and standard gravimetric analysis to provide saturation estimates may be made in an onsite trailer if personnel have a balance (accurate to 0.0001 g), constant weight crucibles, separatory funnels, and ovens or evaporation chambers.

A standard laboratory method for determining immiscible fluid saturation that has been utilized by the petroleum industry employs a modified ASTM method (Dean-Stark) as shown in Figure 10-4. The test employed is essentially a distillation where an LNAPL solvent (toluene, naphtha, or gasoline) in a heated reservoir is vaporized and then passes through a core (or plug) of porous media. As the solvent vapor passes through the core, water and NAPL are evaporated from the core and carried off by the solvent vapor. The solvent vapor and any vaporized NAPL and water are condensed and drains into a graduated receiving tube where the water settles to the bottom of the tube and the solvent is refluxed back into the reservoir over the core sample. The core can be completely cleaned in this apparatus (removing all water and NAPL to produce a clean, dry sample for porosity and other tests).

The water saturation can be determined directly by measuring the volume of condensed water and dividing by sample porosity. NAPL saturation, s_n, is determined indirectly by weighing the core sample (1) prior to extraction and (2) after extraction, cleaning, and drying of the core sample, by

$$s_n = (m_T - m_d - m_w) / (v_n \, \rho_n) \qquad (10\text{-}1)$$

where m_T is the wet sample mass, m_d is the dry sample mass, m_w is the extracted water mass, v_n is the sample pore volume, and ρ_n is the NAPL density.

If this analysis is completed by chemical analysis rather than gravimetrically, then it will cost approximately $25 per sample by spectrophotometric determination. However, if chlorinated hydrocarbons are present, the

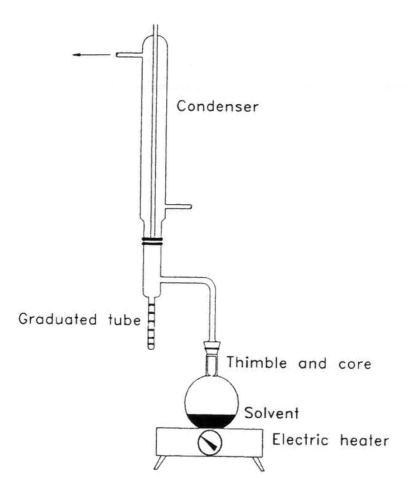

Figure 10-4. Modified Dean-Stark Apparatus for extracting NAPL from soil or rock sample (redrawn from Amyx, J.W., D.M. Bass, Jr., and R.L. Whiting, 1960. Petroleum Reservoir Engineering. Reprinted with permission from McGraw-Hill Book Co.).

analysis is completed with GC/MS at a cost of $500 per sample.

Saturation can also be determined using a centrifuge and solvent to extract and trap water and NAPL from a core sample. The fluid volumes are measured, and used in conjunction with mass and porosity measurements to determine NAPL and water saturation.

Several experimental methods to determine NAPL saturation are presented in the literature. Two such methods described below are not common analytical laboratory determinations and may be costly to replicate.

Cary et al. (1991) developed a method to determine NAPL saturation that relies on water to displace organic compounds from hydrophilic soils and porous polyethylene to absorb the displaced organic liquid. The method is reported to be adaptable to field conditions and should be applicable to volatile, semivolatile, and non-volatile organic compounds. The method procedure involves the following steps.

(1) Cut several strips of porous polyethylene from a 3.2 mm thick sheet having a pore size range between 10 μm and 20 μm. Efficient extraction is reported by using at least 1 g of dry porous polyethylene for each 0.5 mL or NAPL in the sample. Porous polyethylene sheets are available from Porex Technologies (Fairburn, Georgia) and other vendors, and cost approximately $120 for a 44" X 44" X 3.2 mm thick sheet.

(2) Oven dry the strips, record their weights, and pretreat the strips by wrapping in oil-wet (such as Soltrol 220) tissue paper. Allow the strips to equilibrate overnight and then reweigh them. The oil treatment is done to ensure maximum hydrophobicity.

(3) Weigh, to the nearest 0.001 g, 20 g of the soil sample and place it and a polyethylene strip in a 50 mL glass vial fitted with a Teflon cap.

(4) Add 20 mL of water to the vial and stopper the vial. Place the vial on a mechanical shaker and gently rock it for 3-4 hours.

(5) Remove the porous polyethylene strip from the vial, wash any soil particles from the strip using a stream of water from a wash bottle, and then brush off any water droplets with a tissue paper. Weigh the strip and determine the mass that has been adsorbed.

(6) Oven dry the soil at 105°C for a minimum of 12 hours to determine the initial water content of the sample.

(7) Determine the percent saturation of water and DNAPL by using correction factors to account for errors induced by the presence of hydrophobic soil sectors (see Cary et al., 1991).

A second experimental method is adapted from a procedure described by Cary et al. (1989b) that is applicable to DNAPLs with low (semivolatile) to negligible (non-volatile) vapor pressures. A simple description of the method is as follows.

(1) Weigh 100 g sample of soil to the nearest 0.001 g and record the weight.

(2) Place the sample in petri dish and the petri dish in a silica gel desiccator under vacuum. Allow the sample to remain in the desiccator for five days during which time the water will vaporized from the soil and be deposited on the silica gel.

(3) After five days, reweigh the soil sample to determine the water content of the soil.

(4) Place the soil sample in a centrifuge tube, add a solvent such as ethanol, and agitate the sample for several minutes. Centrifuge the tube and then decant the ethanol. Place the solvent in a weighed petri dish, allow the ethanol to evaporate at room temperature, and then reweigh the petri dish to determine the amount of organic material extracted.

(5) The extraction technique should be repeated from four to seven times until there is no evidence of further DNAPL extraction.

10.3 DENSITY (SPECIFIC GRAVITY)

Several methods for determining the density (and/or specific gravity) of liquids and solids are described below.

10.3.1 Displacement Method for Solids

In the displacement method, a solid sample is weighed to determine its mass and then immersed in a graduated cylinder containing a known volume of water. Density is calculated by dividing the mass of the object by the change in volume in the graduated cylinder. A balance suitable for density analysis costs between $150 and $200 and a graduated cylinder costs between $10 and $20.

10.3.2 Density of Liquids by Westphal Balance Method

The density of a liquid can be determined by employing Archimedes' principle that the mass of a floating object is equal to the mass of the liquid it displaces. Based on this principle, Westphal and chain balances are used to determine the volume displacement of liquids by an object of known, constant mass. If one of the liquids is water, the density and/or specific gravity may be determined. A Westphal balance is illustrated in Figure 10-5.

This type of balance should be calibrated with water prior to each determination and the index end of the balance should be positioned for equilibrium. Once the index is set, the container of water (SG=1.000) is removed, the plummet dried, and the plummet is immersed in the liquid sample. The moveable weight on the beam is then adjusted until the balance is once again in equilibrium and the specific gravity is read directly from the beam scale. A chain balance uses a moveable chain to establish balance equilibrium.

Westphal balances cost approximately $250 complete with carrying case. Chain balances cost approximately $800 without a carrying case.

10.3.3 Density of Liquids by Densitometers

Hand-held portable densitometers are available that give direct LCD readouts of specific gravity and the temperature. The liquid is aspirated into the measurement tube and the determination is performed automatically. The units cost approximately $2000.

10.3.4 Specific Gravity Using a Hydrometer

Perhaps the simplest and least expensive for field use, this method involves use of a hydrometer (calibrated, weighted, glass float), thermometer, and cylinder. After the cylinder is filled with sufficient NAPL to allow the hydrometer to float, it is placed in the NAPL and allowed to come to rest without touching the walls of the cylinder (Figure 10-6). The liquid specific gravity is read from the graduated scale on the hydrometer stem at the liquid-air interface, and the temperature is recorded. Wide and narrow range hydrometers are available. Individual hydrometers cost approximately $20 and a set of eight hydrometers (covering the specific gravity range from 0.695 to 2.000) with a thermometer included costs approximately $160.

10.3.5 Density of Liquids by Mass Determination

The density of a liquid at a measured temperature can be determined by weighing the mass of a known volume of liquid in a container of known mass. The mass of the empty container is subtracted from the mass of the full container and the resultant liquid mass is divided by its volume. The cost of a balance suitable for density analysis can be obtained for less than $200.

10.3.6 Certified Laboratory Determinations

If certifiable and high precision measurements are required, 100 g of solid samples or 1 L of liquid samples are typically required by laboratories. Certified results using American Society of Testing Materials (ASTM) methods usually cost between $10 and $24 per sample.

10.4 VISCOSITY

Methods of viscosity measurement include the following.

10.4.1 Falling Ball Method

A ball falling through a viscous liquid will accelerate until it attains a constant velocity that is inversely proportional to fluid viscosity. The amount of time required to fall a known distance can be measured using a stopwatch. A falling ball viscometer is shown in Figure 10-7. After inverting the tube to capture the ball in the cap, it is returned to the upright position and filled with the liquid sample. The top cap is then twisted to release the ball and the time required to fall between the two sets of parallel lines is recorded. A set of three tubes can be obtained having viscosity ranges of 0.2 to 2.0 cp, 2.0 to 20 cp, and 20 to 1000 cp, at a cost of approximately $120 each. These tubes are easy to use and convenient for field measurements.

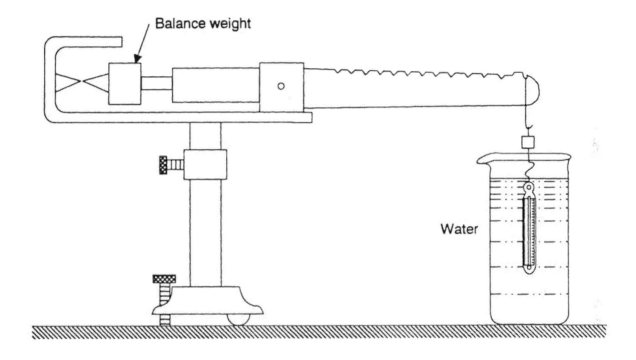

Figure 10-5. Schematic of a Westphal balance (redrawn from Shugar, G.J., and J.T. Ballinger, 1990. Chemical Technician's Ready Reference Handbook. Reprinted with permission from McGraw-Hill Book Co.).

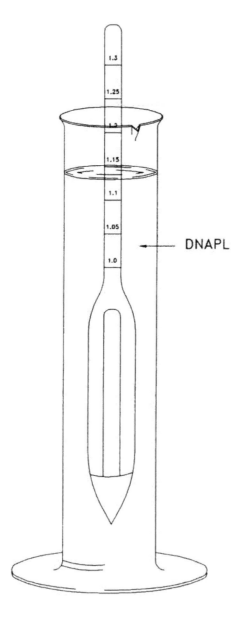

Figure 10-6. Use of a glass hydrometer for specific gravity determination (of a DNAPL with a specific gravity of approximately 1.13 at the sample temperature).

Figure 10-7. Schematic of a falling ball viscometer. Viscosity is measured by determining how long it takes a glass or stainless steel ball to descend between the reference lines through a liquid sample.

10.4.2 Falling Needle Method

An adaptation of the falling ball method uses glass needles of varying density measuring 4 inches long and 0.065 inches in diameter. This method was designed to reduce wall interferences and eddy currents possible in falling ball units. The falling needle units cost approximately $3800 if the time is measured manually with a stopwatch, or $7900 with automatic timing and viscosity readout. The units can measure viscosities ranging from 10 to 2,400,000 cp.

10.4.3 Rotating Disc Viscometer

Viscosities from 10 to 8,000,000 cp can be measured with a rotating disc viscometer. A spindle is selected and mounted on the unit, a rotating speed is selected, and then the spindle is immersed in the liquid. The amount of time required for the spindle to return to constant speed is measured and converted to a direct digital readout in centipoises. Units cost approximately $1750.

10.4.4 Viscosity Cups

Kinematic viscosities from 15 to 1627 cSt (centistokes) can be measured using viscosity cups (Figure 10-8). Direction for using 44 mL capacity Zahn (or equivalent) viscosity cups are as follows.

(1) The viscosity cup is completely immersed vertically in the test liquid.

(2) The cup is then quickly raised above the liquid surface and a timer is started simultaneously.

(3) The time for the liquid to drain through a hole at the cup bottom is measured until observation of the first distinct break in the efflux stream.

(4) The drainage duration is then used to calculate the kinematic viscosity using either a table or equation.

Five Zahn cups are available for approximately $80 each to measure various liquid viscosity ranges: 15 to 78 cSt (Cup #1); 40 to 380 cSt (Cup #2); 90 to 604 cSt (Cup #3); 136 to 899 cSt (Cup #4); and 251 to 1627 cSt (Cup #5).

10.4.5 Certified Laboratory Analyses for Viscosity

Independent testing laboratories use various methods, typically require a 500 mL sample, and will charge approximately $30 to $40 per viscosity determination.

10.5 INTERFACIAL TENSION

Methods to determine surface tension and interfacial liquid tension are described below.

10.5.1 Surface Tension Determination by Capillary Rise

The wetting force of a liquid is equal to the gravitational force on a liquid that has risen in a capillary tube. After measuring the capillary rise of a fluid sample in a tube, surface tension can be calculated by

$$\sigma = (r\ h\ \rho\ g)\ /\ (2\cos\phi) \qquad (10\text{-}2)$$

where σ is the surface tension in dynes/cm, ρ is the density of the liquid in g/mL, h is the height of the capillary rise in cm, r is the internal radius of the tube in cm, ϕ is the contact angle, and, g is the acceleration due to gravity ($980\ cm/s^2$). The apparatus for determining surface tension by capillary rise is shown in Figure 10-9.

To measure capillary rise, gentle suction is applied to raise the fluid sample to the top of the capillary tube, and then released to allow the fluid to decline to an equilibrium position. This measurement should be repeated several times to provide reliable data. The surface tension of the liquid is then calculated using Equation 10-2. The presence of air bubbles within the liquid column will result in inaccurate data. Between samples, the capillary tube should be cleaned and dried, and accuracy checks should be made periodically using a liquid of known surface tension (such as water). The temperature of samples should be maintained and documented because surface tension decreases with rising temperature. Capillary rise measurement apparatus are available from laboratory supply vendors for about $65.

10.5.2 du Nouy Ring Tensiometer Method

Surface or liquid interfacial tension can be determined directly using a tensiometer employing the du Nouy ring method in accordance with ASTM D971 and D1331. The force necessary to separate a platinum-iridium ring from

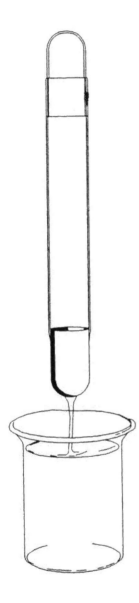

Figure 10-8. Use of a viscosity cup to determine kinematic viscosity. Liquid viscosity is determined by measuring how long it takes for liquid to drain from a small hole at the bottom of a viscosity cup.

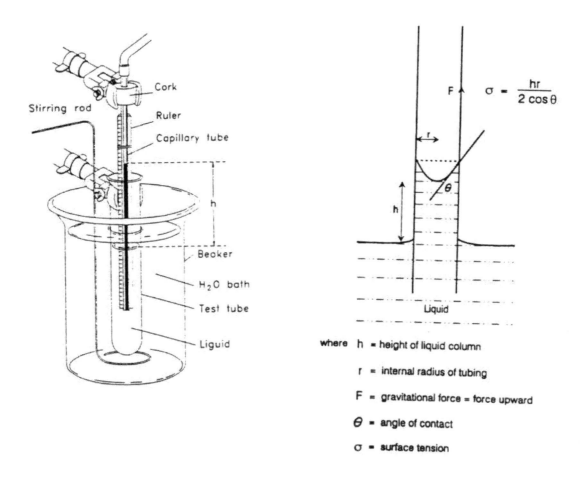

Figure 10-9. Determination of surface tension by measuring capillary rise and contact angle (redrawn from Shugar, G.J., and J.T. Ballinger, 1990. Chemical Technician's Ready Reference Handbook. Reprinted with permission from McGraw-Hill Book Co.).

the liquid's surface (either at a liquid-air or liquid-liquid interface) is measured to the nearest +/- 0.25 dynes/cm. Manual and semi-automatic units are available and range in price from $2200 to $2800. Interfacial tension determinations made by a testing laboratory cost approximately $40 per sample.

10.6 WETTABILITY

Numerous quantitative and qualitative methods have been proposed for measuring wettability (Anderson, 1986b; Adamson, 1982). Five quantitative methods that are generally used are described below.

10.6.1 Contact Angle Method

Using the sessile drop contact angle method, a drop of DNAPL is formed at the end of a fine capillary tube and brought in contact with the smooth surface of a porous medium under water within a contact angle cell (Figure 10-10). The drop of DNAPL is allowed to age on the medium surface and the contact angle can be measured and documented by taking photographs, preferably with enlargement using special photomacrographic apparatus.

Generally, a single, flat, polished mineral crystal and oil brine water is used in petroleum engineering studies; but for applications to DNAPL-groundwater-media systems, the smooth surface may be a rock thin section, clay smeared on a glass slide, the top of a cohesive soil sample that has been sliced with a knife, or a relatively flat surface of silt or sand. The water should be actual or simulated groundwater. Water advancing and water receding contact angles typically vary due to hysteresis and can be measured by using the capillary tube to expand and contract the volume of the DNAPL drop. Alternatively, a modified sessile drop method can be utilized whereby the drop of DNAPL is positioned between two flat, substrates that are mounted parallel to each other on adjustable posts in the contact angle cell.

As noted in Chapter 4.3, NAPL wetting has been shown to increase with aging (contact time) during contact angle studies. Thus, an assessment of the significance of aging should be considered, and the contact duration associated with each measurement should be noted.

The representativeness of contact angle measurements is uncertain. Although Melrose and Brandner (1974) contend that contact angles provide the only unambiguous measure of wettability, Anderson (1986b)

notes that contact angle measurements cannot take into account the effects of media heterogeneity, roughness, and complex pore geometry.

Contact angle measurements range in price from $1200 to $1500 when accomplished by petroleum laboratories.

10.6.2 Amott Method

The Amott method measures the wettability of a soil or rock core based on the immiscible fluid displacement properties of the NAPL-water-media system (Amott, 1959; Anderson, 1986b). Four fluid displacements are performed combining imbibition and forced drainage to measure the average wettability of the sample. The method relies on the spontaneous imbibition of the wetting fluid into the sample and, for consolidated media, involves the following steps (Anderson, 1986b).

(1) The core sample is prepared by centrifugation under water until residual NAPL saturation, s_m is attained.

(2) The core sample is then immersed in NAPL and the volume of water that is spontaneously displaced by the NAPL is measured after 20 hours.

(3) The core sample is then centrifuged under NAPL until the irreducible water saturation, s_{rw} is reached, at which time, the total volume of water displaced is measured, including that spontaneously displaced by immersion in the NAPL.

(4) The core sample is then immersed in water and the volume of NAPL spontaneously displaced by the water is measured after 20 hours.

(5) Finally, the core sample is centrifuged under water until s_m is reached and the total volume of NAPL displaced is measured.

For unconsolidated sediments which cannot be centrifuged, the displacements are achieved by forced flow through the sample core.

Although Amott (1959) selected an arbitrary time period of 20 hours for the spontaneous NAPL and water imbibition steps, Anderson (1986b) recommends that the core samples be allowed to imbibe to completion (which may take from several hours to several months) or for a period of 1 to 2 weeks.

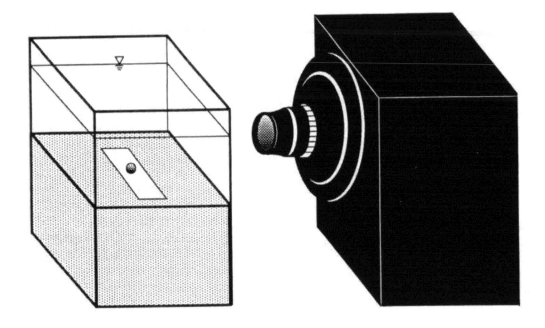

Figure 10-10. Use of a contact angle cell and photographic equipment for determination of wettability. A small DNAPL drop is aged on a flat porous medium surface under water.

Results of these displacements are expressed by the "displacement by NAPL" ratio (δ_n) and the "displacement by water" ratio (δ_w). The displacement by NAPL ratio is the water volume displaced by spontaneous imbibition alone (V_{wsp}) to the total volume displaced by imbibition and forced displacement (V_{wt}):

$$\delta_n = V_{wsp} / V_{wt} \qquad (10\text{-}3)$$

The displacement by water ratio (δ_w) is the ratio of the NAPL volume displaced by spontaneous imbibition alone (V_{nsp}) to the total NAPL volume displaced by imbibition and forced displacement (δ_{nt})

$$\delta_w = V_{nsp} / V_{nt} \qquad (10\text{-}4)$$

For water wet media, δ_w is positive and approaches unity with increasing water wetness, and δ_n is zero. Similarly, for NAPL wet media, δ_n is positive and approaches unity with increasing NAPL wetness, and δ_w is zero. The main limitation of the Amott wettability test is its insensitivity for near neutral wettability (i.e., contact angles between 60° and 120°). The approximate relationship between wettability, contact angle, and the Amott and USBM wettability indexes is shown in Table 10-2.

Petroleum laboratories charge approximately $1500 for wettability determinations by the Amott Method.

10.6.3 Amott-Harvey Relative Displacement Index

A relative wettability index known as the Amott-Harvey modification can be calculated by first driving the cores to s_{rw} prior to initiating the test. A relative wettability index is then calculated where:

$$I = \delta_w - \delta_n = V_{nsp}/V_{nt} - V_{wsp}/V_{wt} \qquad (10\text{-}5)$$

A system is defined as water wet when $0.3 \leq I \leq 1$, intermediate wet when $-0.3 < I < 0.3$, and DNAPL wet when $-1 \leq I \leq -0.3$ (Table 10-2).

10.6.4 United States Bureau of Mines (USBM) Wettability Index

The USBM method (Donaldson et al., 1969; Anderson, 1986b) is applicable only to porous media that can be cut into plugs for centrifugation (rock and clay), but it is more sensitive to the intermediate (or neutral) wetting area than the Amott methods. Like the Amott methods,

the USBM test measures the average wettability of the core sample. Capillary forces are varied over a 20 psi range by incrementally raising the centrifuge rotation velocity. Displaced fluid volumes are measured at each incremental capillary pressure point.

The plugs are prepared by placing them under NAPL and centrifuging at high speed to reach s_{rw}. The plugs are then centrifuged under water at incrementally increasing speeds until reaching a capillary pressure of -10 psi. The plug saturation is determined at each incremental capillary pressure by measuring the volume of expelled NAPL. The plugs are then centrifuged under NAPL at incrementally increasing capillary pressure until a maximum of 10 psi is attained. After constructing a graph of capillary pressure from -10 to 10 psi versus average water saturation (Figure 10-11), the positive and negative areas of the curve are calculated to determine the USBM Wettability index (W)

$$W = \log(A_1/A_2) \qquad (10\text{-}6)$$

where A_1 is the area of the NAPL drive curve and A_2 is the area under the water drive curve. Values of W near 1, 0, and -1 indicate water-wet, neutral, and NAPL-wet conditions, respectively.

10.6.5 Modified USBM Wettability Index

A modification of the USBM method allows the calculation of both the USBM and Amott indexes (Sharma and Wunderlich, 1985; Anderson, 1986b). The procedure involves five steps: initial NAPL drive, spontaneous imbibition of water, water drive, spontaneous imbibition of NAPL, and NAPL drive. The Amott and USBM indexes are then calculated as described above.

10.6.6 Other Methods

Wettability is assessed as a byproduct of determining relative permeability by some laboratories. A variety of additional qualitative methods are described by Anderson (1986b).

10.7 CAPILLARY PRESSURE VERSUS SATURATION

Two main types of methods are available to determine capillary pressure-saturation $P_c(s_w)$ relations in porous media (Bear, 1972; Corey, 1986; Amxy et al., 1960): (1)

Table 10-2. Approximate relationship between wettability, contact angle, and the USBM and Amott wettability indexes (from Anderson, 1986b).

Method	Water-Wet	Neutrally Wet	NAPL-Wet
Contact Angle: Minimum Maximum	0° 60 to 75°	60 to 75° 105 to 120°	105 to 120° 180°
USBN wettability index	W near 1	W near 0	W near -1
Amott wettability index Displacement-by-water ratio Displacement by NAPL ratio	Positive Zero	Zero Zero	Zero Postive
Amott-Harvey wettability index	0.3 ≤ I ≤ 1.0	-0.3 < I < 0.3	-1.0 ≤ I ≤ -0.3

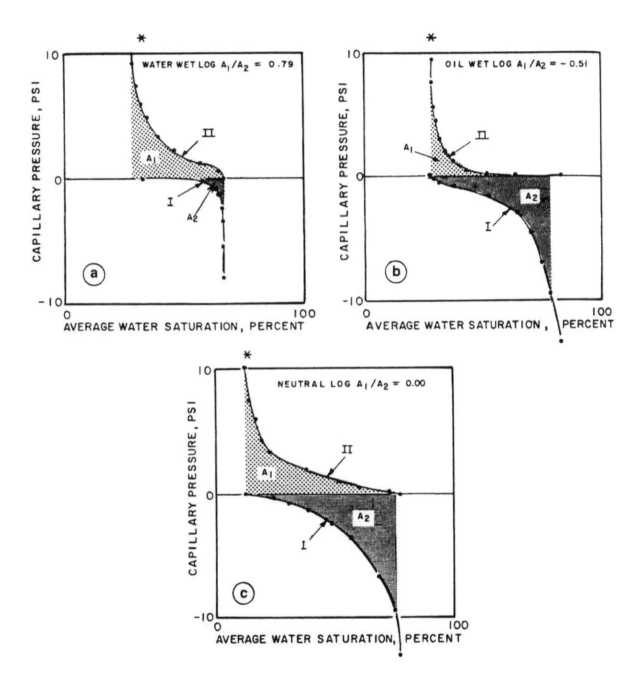

Figure 10-11. USBM wettability measurement showing (a) water-wet, (b) NAPL-wet, and (c) neutral conditions (reprinted with permission from Society of Petroleum Engineers, 1969).

displacement methods based on the establishing successive states of hydraulic equilibrium, and (2) dynamic methods based on establishing successive states of steady flow of wetting and nonwetting fluids. Displacement methods are utilized more commonly than dynamic methods.

Several methods to determine $P_c(s_w)$ relations are described briefly below. Some of these methods can utilize the actual immiscible fluids of interest (i.e., the cylinder and diaphragm methods) to determine $P_c(s_w)$ curves; others utilize alternative nonwetting fluids (mercury) or air to displace water from an initially water-saturated sample. As discussed in Chapter 4.4, the capillary pressure measured for a particular nonwetting fluid saturation can be scaled using interfacial tension measurements to estimate the capillary pressure for a different NAPL for the same saturation and soil. Amxy et al. (1960) provide a comparison of $P_c(s_w)$ curves determined for the same media using several different methods.

10.7.1 Cylinder Methods for Unconsolidated Media

Methods to determine drainage and imbibition $P_c(s_w)$ curves for DNAPL-water-unconsolidated media samples were recently described by Kueper and Frind (1991), Wilson et al. (1990), and Guarnaccia et al. (1992). The procedure utilized by Kueper and Frind (1991) is described below.

A schematic of a $P_c(s_w)$ test cell is provided in Figure 10-12. As shown, the porous media sample is placed in the stainless steel cylinder and the water burette and NAPL reservoir are connected to the cell. The porous ceramic disk has a DNAPL breakthrough pressure exceeding the range of test pressures thereby preventing DNAPL passage.

After the sample is completely saturated with water, the initial water and DNAPL pressures are recorded. The water pressure is equal to the water level in the outflow burette. The DNAPL pressure can be calculated based on the DNAPL density and the levels of the DNAPL-water interface and the overlying water surface. The value of P_c is then calculated as the DNAPL pressure minus the water pressure.

The P_c is then increased incrementally by raising the height of the DNAPL reservoir and the system is allowed to come to equilibrium. As the DNAPL reservoir is raised, the threshold entry pressure, P_d, is eventually exceeded and the DNAPL begins to displace water from the porous media sample. The volume of water displaced as a result of each incremental increase in P_c is measured to determine the relationship between P_c and saturation. Saturation can also be determined by gravimetric analysis of the core (Wilson et al., 1990) or by gamma radiation techniques (Guarnaccia et al., 1992). Eventually, upon reaching the irreducible water saturation (s_{rw}), no additional water is displaced by further raising the DNAPL pressure.

To determine the imbibition curve, the DNAPL/water reservoir is then lowered in increments until the residual DNAPL saturation (s_m) is reached and lowering of the DNAPL/water reservoir does not allow further imbibition of water into the sample. The raw data are plotted as shown in Figure 4-9 producing the main drainage and imbibition $P_c(s_w)$ curves from which the values of s_{rw}, s_m, P_d, can be determined.

10.7.2 Porous Diaphragm Method (Welge Restored State Method)

Very similar to the cylinder method described above, the Welge static method is also based upon the drainage of a sample initially saturated with water (Welge and Bruce, 1947; Welge, 1949). A schematic diagram of the test apparatus is given in Figure 10-13. The methodology involves the same general procedure as the cylinder method.

This procedure may take 10 to 40 days to yield a $P_c(s_w)$ drainage curve due to the slow equilibration process that follows incremental pressure adjustment (which takes longer and longer as s_{rw} is approached). Suitable modifications to the test apparatus and procedure can be made to derive imbibition curves.

Determination of $P_c(s_w)$ drainage curves by the Welge method costs approximately $350 for a 6-point air-water or air-NAPL curve, $770 for a 6-point water-NAPL curve, and $1525 for drainage and imbibition water-NAPL curves.

10.7.3 Mercury Injection Method

Another displacement method, the mercury injection method (Purcell, 1949), can be used to reduce the time needed to determine a $P_c(s_w)$ curve. Mercury is a very

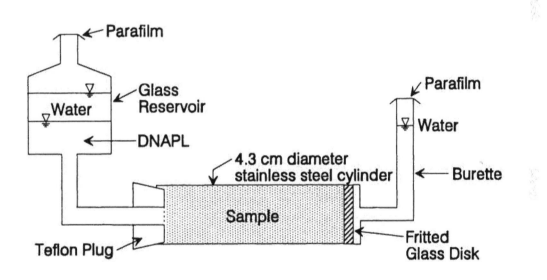

Figure 10-12. Schematic of a $P_c(s_w)$ test cell (modified from Kueper et al., 1989).

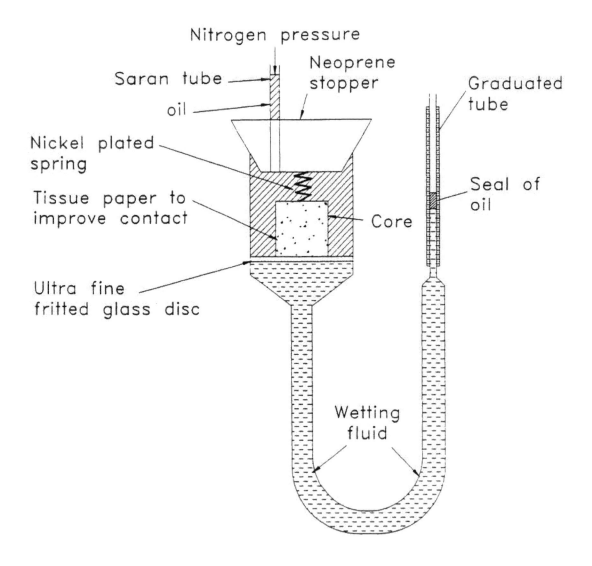

Figure 10-13.   Schematic of the Welge porous diaphragm $P_c(S_w)$ device (modified from Bear, 1972; and, Welge and Bruce, 1947).

dense, nonwetting fluid even in mercury-air-media systems. The test procedure is as follows.

(1) The core or porous medium sample is placed in a sample chamber and the chamber is evacuated to a very low pressure using a vacuum pump.

(2) Mercury is then forced into the sample under incremental pressure steps and the nonwetting fluid saturation is determined by measuring the volume of mercury injected during each step.

(3) Mercury injection is continued in small incremental pressure steps until the sample is saturated with mercury or a predetermined maximum pressure is attained.

Using this method, equilibrium is attained during each pressure step in a matter of minutes rather than hours or days and much higher capillary pressures can be tested than when using the porous diaphragm.

Capillary pressure curve determinations by mercury injection cost approximately $375 for a drainage curve and $800 for both drainage and imbibition curves.

10.7.4 Centrifuge Method

The centrifuge method is another static test used to reduce the equilibration time associated with determining the $P_c(s_w)$ curve (Amyx et al., 1960; Slobod and Chambers, 1951). Using this method, a complete $P_c(s_w)$ can be derived within a few hours. The procedure is outlined below.

(1) A sample is saturated with water (to determine the drainage curve) and placed in a centrifuge tube with a small, graduated burette on the end as shown in Figure 10-14.

(2) The centrifuge is started and maintained at a predetermined rotation velocity. Fluid displaced by the increased gravitational force collects in the burette. When no more fluid drains from the sample, the drained wetting fluid volume is measured to determine its saturation at the corresponding rotation velocity. The centrifugal force is increased incrementally by raising the rotation velocity until there is no additional fluid displacement.

Centrifuge rotation velocity is related to P_c by (Bear, 1972):

$$P_c(r_1) = (r_2^2 - r_1^2)\omega^2 \Delta p/2$$

where ω is the angular velocity of the centrifuge and r_1 and r_2 are the radii of rotation to the inner and outer faces of the sample.

Laboratory determination by the centrifuge method costs approximately $500 for determination of air-water, air-NAPL, or water-NAPL curves, and $1000 for water-NAPL drainage and imbibition curves.

10.7.5 Dynamic Method Using Hassler's Principle

Brown (1951) reported a dynamic method based on Hassler's principle that is used for both capillary pressure and relative permeability determinations. This method controls the capillary pressure at both ends of a test sample. A schematic of Hassler's apparatus is presented in Figure 10-15. The test procedure is as follows (Bear, 1972; Osoba et al., 1951).

(1) A sample is placed between two porous membranes or plates permeable only to the wetting fluid (Figure 10-15). These membranes allow the wetting phase (water), but not the nonwetting phase (gas) to pass, and facilitate uniform saturation throughout the sample even at low test flow rates.

(2) The sample is first saturated by the wetting fluid prior to placement in the test apparatus. Each membrane is divided into an inner disc (B) and an outer ring (A).

(3) To initiate the test, the wetting fluid and nonwetting fluid are introduced through ring A and via the radial grooves on the inner face of this ring, respectively. Wetting phase pressure is measured through disc B and the capillary pressure equals to difference in pressure between the wetting and nonwetting phases at the inflow face.

(4) When equilibrium of wetting and nonwetting phase flow rates is attained, the sample is removed and saturation is determined gravimetrically.

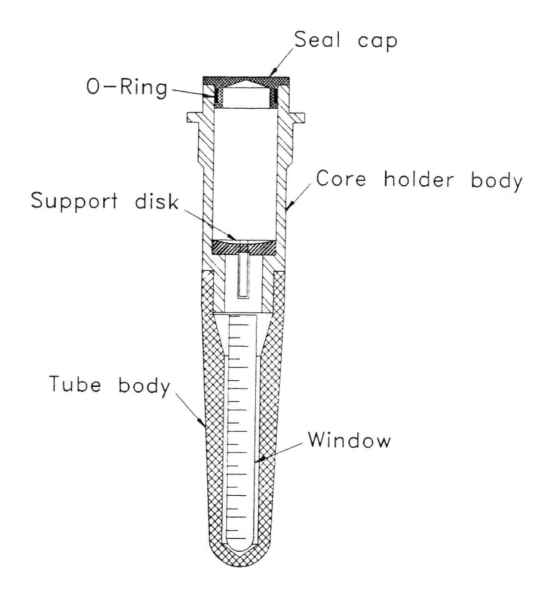

Figure 10-14. Schematic of a centrifuge tube with graduated burette for determining $P_c(s_w)$ relations (redrawn from Amyx, J.W., D.M. Bass, Jr., and R.L. Whiting, 1960. Petroleum Reservoir Engineering. Reprinted with permission from McGraw-Hill Book Co.).

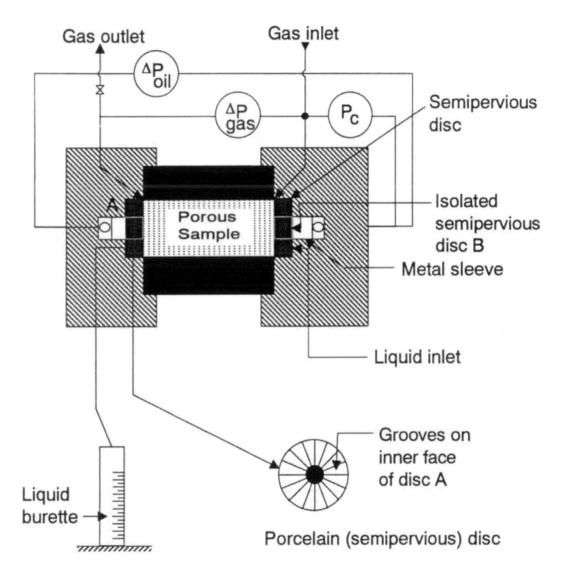

Figure 10-15. A schematic of Hassler's apparatus used for $P_c(s_w)$ and relative permeability measurements (modified from Bear, 1972; and Osoba et al., 1951).

10.8 RELATIVE PERMEABILITY VERSUS SATURATION

There are four primary means to acquire relative permeability data (Amyx et al., 1960; Honapour et al., 1986): (1) laboratory measurements using steady-state fluid flow, (2) laboratory measurements using transient fluid flow, (3) calculations based on $P_c(s_w)$ data, and (4) calculations from field data. The utilization of $P_c(s_w)$ data to estimate relative permeability relationships is discussed in Chapter 4.6. The calculation of relative permeability based on field data is described by Honarpour et al. (1986). Most commonly, relative permeability is measured using laboratory methods, particularly steady-state tests, which are described below. For a detailed discussion of many methods to determine two- and three-phase relatively permeability and related factors, refer to Honarpour et al. (1986).

10.8.1 Steady State Relative Permeability Methods

The many steady state relative permeability methods utilize the same general procedure (Bear, 1972). A soil or rock sample is mounted in a test cell as shown in Figure 10-15. Two fluids (NAPL and water) are injected into and through the sample at a steady rate via different piping systems. After 2 to 40 hours depending on the media permeability, steady flow rates are obtained such that the ratio of NAPL:water is the same in the inflow and outflow. At this equilibrium, the flow rates are measured and saturations are determined by fluid balance calculations, gravimetrically, or other methods. The injection ratio is then modified incrementally to remove more of the wetting fluid. Flow, pressure, and saturation measurements are made for each fluid upon equilibration following each injection rate change. Desaturation tests involve draining water from an initially water-saturated sample. Resaturation refers to tests that begin with an initially NAPL-saturated sample.

Permeability values associated with each saturation are calculated by

$$k_n(s_w) = q_n \, \mu_n \, / \, i_n \qquad (10\text{-}7)$$

and

$$k_w(s_w) = q_w \, \mu_w \, / \, i_w \qquad (10\text{-}8)$$

where: $k_n(s_w)$ and $k_w(s_w)$ are the permeabilities of NAPL and water at the given water saturation (s_w), respectively;

q_n and q_w are the measured NAPL and water flow rates, respectively; μ_n and μ_w are the absolute viscosities of the NAPL and water, respectively; and i_n and i_w are the pressure gradients imposed on the NAPL and water, respectively. Relative permeabilities of NAPL, k_{rn}, and water, k_{rw}, are equal to these permeability values divided by the respective intrinsic fluid permeabilities at complete saturation. Two-phase relative permeability curves, like that shown in Figure 4-11a, are derived by measuring relative permeability at many s_w values.

Steady-state relative permeability tests costs approximately $3000 each for the desaturation and resaturation curves.

10.8.2 Unsteady Relative Permeability Methods

Unsteady flow relative permeability methods are reviewed by Honarpour et al. (1986). More recently, Wilson et al. (1990) provide a detailed description of apparatus and a procedure developed to perform unsteady relative permeability measurements in soil-water-NAPL systems.

The cost to determine NAPL-water relative permeability curves using unsteady flow methods is approximately $2000 each for desaturation and resaturation curves.

10.9 THRESHOLD ENTRY PRESSURE

The threshold entry capillary pressure, P_d, that must be exceeded to drive nonwetting DNAPL into a water-saturated medium is determined during the $P_c(s_w)$ tests described in Chapter 10.7. If only the value of P_d is of interest, the test can be concluded after the DNAPL pressure has been raised incrementally to a sufficient level to initiate DNAPL movement into the core sample. The initial fluid movement can be determined by: (1) observing the first drainage of water from the core sample; and/or, (2) the initial movement of an air bubble placed using a syringe needle within a capillary tube upstream from the core sample plug.

10.10 RESIDUAL SATURATION

Residual saturation is normally determined in conjunction with relative permeability or $P_c(s_w)$ testing at no additional cost. If the entire relative permeability or $P_c(s_w)$ curves are not required, residual NAPL saturation can be obtained by (1) raising the NAPL pressure and allowing water drainage to proceed until the maximum NAPL saturation and irreducible water content are obtained, and then, (2) waterflooding the sample at hydraulic gradients representative of field conditions until no additional DNAPL flows from the core sample. The core sample can then be analyzed for DNAPL saturation by one of the methods described in Chapter 10.2.

10.11 DNAPL DISSOLUTION

DNAPLs are soluble to a variable degree in water and will be leached by infiltration and groundwater as discussed in Chapter 4.7. The dissolved chemistry derived from water contact with DNAPL can be assessed directly by analysis of DNAPL-contaminated groundwater samples, and indirectly by equilibrium calculation methods or laboratory dissolution studies. A few dissolution study options are described briefly below.

The equilibrium aqueous concentrations of DNAPL components in groundwater can be assessed by placing DNAPL and (real or simulated) groundwater in a closed jar at the prevailing groundwater temperature. After four hours of contact, samples of water, excluding DNAPL, can be taken for chemical analysis to determine the dissolved phase composition.

Simple leaching experiments can also be conducted in which water is passed through a sample of DNAPL-contaminated porous media to simulate vadose zone or saturated conditions. The experiments can be designed to represent various field conditions (e.g., DNAPL pools or ganglia, variable flow rates, and variable background groundwater chemistry). Details of laboratory studies on NAPL dissolution are described by van der Waarden et al. (1971), Fried et al. (1979), Pfannkuch (1984), Hunt et al. (1988a,b), Schwille (1988), Anderson (1988), Miller et al. (1990), Zalidis et al. (1991), and Mackay et al. (1991).

Leachability is also a regulatory basis for classifying contaminated soil as a hazardous waste. The USEPA Toxic Characteristic Leach Procedure (TCLP) involves leaching a solid sample with an acidic leaching solution to simulate climatic conditions expected to occur at landfills. Remediation of contaminated soils must meet the TCLP regulatory levels for the metals and organic compounds listed in Table 10-3. TCLP analyses are performed by independent laboratories and require elaborate sample extraction, quality control, and chain-of-custody records.

Table 10-3.   **TCLP regulatory levels for metals and organic compounds (Federal Register -- March 29, 1990).**

Constituent	mg/L	Constituent	mg/L
Arsenic	5	2,4-Dinitrotoluene	0.13
Barium	100	Endrin	0.02
Cadmium	1	Heptachlor (and its epoxide)	0.008
Chromium	5	Hexachlorobenzene	0.13
Lead	5	Hexachlorobutadiene	0.5
Mercury	0.2	Hexachloroethane	3.
Selenium	1	Lindane	0.4
Silver	5	Methoxychlor	10.
Benzene	0.5	Methyl ethyl ketone	200.
Carbon tetrachloride	0.5	Nitrobenzene	2.
Chlordane	0.03	Pentachlorophenol	100.
Chlorobenzene	100.	Pyridine	5.
Chloroform	6.	Tetrachloroethene	0.7
o-Cresol	200.	Toxaphene	0.5
m-Cresol	200.	2,4,5-TP (Silvex)	1.
p-Cresol	200.	Trichloroethene	0.5
2,4-D	10.	2,4,5-Trichlorophenol	400.
1,4-Dichlorobenzene	7.5	2,4,6-Trichlorophenol	2.
1,2-Dichloroethane	0.5	Vinyl Chloride	0.2
1,1,-Dichloroethene	0.7		

11 CASE STUDIES

Case histories can illustrate special problems associated with DNAPL sites. Lindorff and Cartwright (1977) discuss 116 case histories of groundwater contamination and remediation. USEPA (1984a and b) presents 23 case histories of groundwater contamination and remediation. Groundwater extraction was evaluated at 19 sites by USEPA (1989c). These 19 sites were reexamined and expanded to 24 sites in Phase II of the USEPA study (Sutter et al., 1991). At 20 of the 24 sites, chemical data collected during remedial operations exhibit trends consistent with the presence of DNAPL (USEPA, 1991b). However, except at the sites contaminated with creosote, DNAPL was generally not discovered directly during site sampling activities. Chemical waste disposal sites where DNAPL was observed are discussed by Cohen et al. (1987). In the discussion that follows, several sites described in these references are used to illustrate characterization aspects of DNAPL sites.

11.1 IBM DAYTON SITE, SOUTH BRUNSWICK, N.J.

The IBM Dayton site was featured in an article in the *Wall Street Journal* (Stipp, 1991) describing the limitations of aquifer restoration caused by the presence of DNAPL. This case history was selected to illustrate how undetected and unsuspected DNAPL can influence remediation and monitoring.

11.1.1 Brief History

The site is located in the Township of South Brunswick, New Jersey (see Figure 11-1). In December 1977, organic contaminants, primarily 1,1,1-trichloroethane (TCA) and tetrachloroethene (PCE), were discovered in public supply well SB11. Subsequent to this discovery, several investigations were conducted to locate source(s) of contamination and remediate groundwater contamination. These investigations identified three industries that were contamination sources. Figure 11-2 shows these facilities and their associated TCA plumes. Plumes of other chemicals showed similar distributions; TCA is used because it was the most widely distributed and almost always occurred with the highest concentration (Roux and Althoff, 1980). Note that Plant A in this figure is the IBM facility that was determined to be the major contributor to the contamination at SB11.

In January 1978, SB11 was shut down and IBM began a site assessment (USEPA, 1989c). During 1978, more

than 60 monitor wells and 10 onsite recovery wells were installed; the first groundwater extraction began in March 1978. To limit the spread of contamination, pumping resumed from SB11 in June 1978 with discharge going to the sanitary sewer system. In addition, buried chemical storage tanks that were the suspected source of IBM's groundwater contamination were removed during the summer of 1978 (USEPA, 1989c).

An additional four onsite extraction wells and ten monitor wells were installed in 1979. Seven more extraction wells were installed offsite in 1981 to intercept the plume movement toward SB11. At this time, nine injection wells also were installed along the northeast boundary of the IBM property (USEPA, 1989c). The pump-and-treat system continued until September 9, 1984, at which time all parties agreed that further reductions in contaminant concentrations could not be achieved by continued operation of the six onsite extraction wells and the seven offsite injection wells. The parties also agreed to reactivate the extraction system if offsite TCA concentration increased to above 100 µg/l. Production well SB11 continued to produce, but with a well-head treatment system installed and an increased pumping rate. Figure 11-3 shows the TCA plume as of January 1985. In this figure, north has been rotated and the buried chemical storage tanks that were removed were located near wells GW32 and GW04.

Groundwater monitoring continued after the pump-and-treat system was terminated. Results showed a gradual increase in concentrations and a reemergence of the contaminant plume, presumably due to the presence of DNAPL in the aquifer. Figure 11-4 shows the TCA plume as of June 1989. As a result, a new pump-and-treat system began operation in the Fall of 1990 (Robertson, 1992). The new system is designed to remediate the aquifer and contain the source (DNAPL). As the aquifer is cleaned up, this portion of the remediation system will be terminated; low-yield, source-control pumping near GW32 and GW04 is expected to continue indefinitely.

11.1.2 Site Characterization

According to Roux and Althoff (1980), the investigations were designed to (1) define the subsurface lithology, (2) determine directions and rates of groundwater movement, and (3) define the extent of contaminant plumes. General hydrogeology of the area was known based on data from existing wells. The site is underlain by a water-table aquifer known as the Old Bridge aquifer; this is underlain by a confining bed known as the Woodbridge

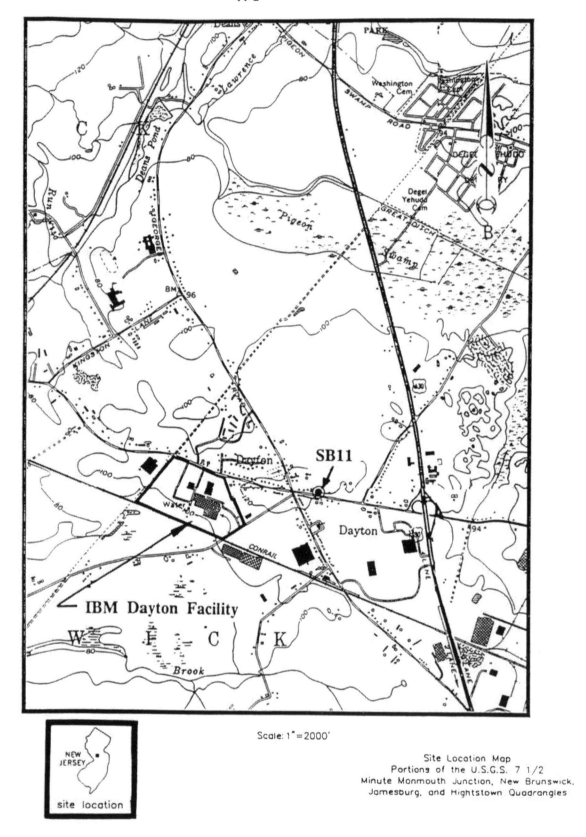

Scale: 1" = 2000'

Site Location Map
Portions of the U.S.G.S. 7 1/2
Minute Monmouth Junction, New Brunswick,
Jamesburg, and Hightstown Quadrangles

Figure 11-1.   Dayton facility location map showing public water supply well SB11 (from Robertson, 1992).

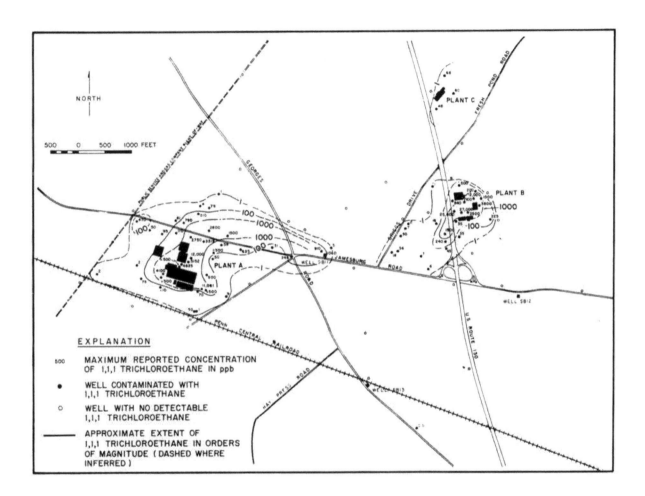

Figure 11-2. TCA distribution in the Old Bridge aquifer in January 1978-March 1979 associated with three facilities near SB11 (from Roux and Althoff, 1980).

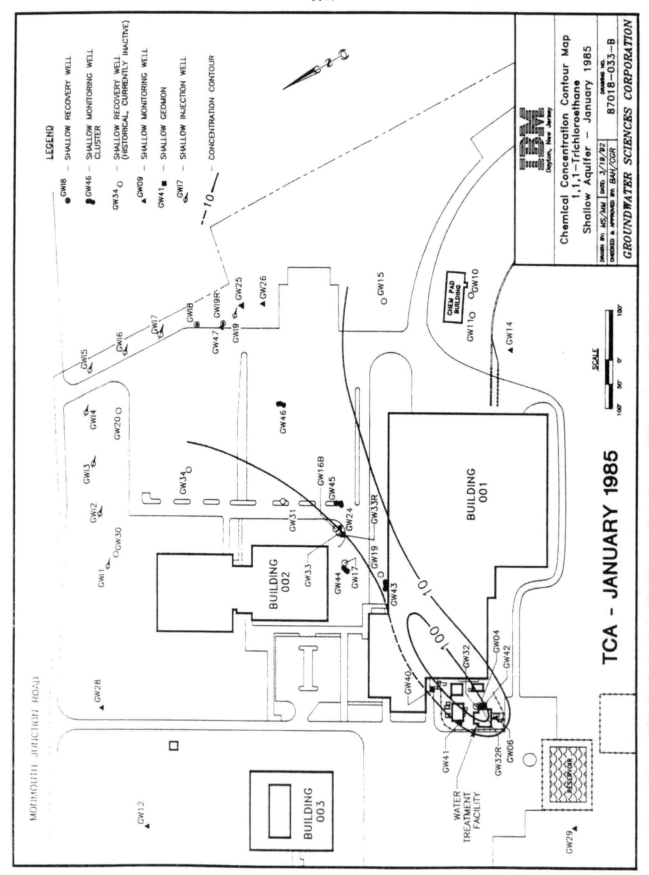

Figure 11-3. TCA distribution in the Old Bridge aquifer in January 1985 associated with IBM facilities (Plant A in Figure 9.2) (from Robertson, 1992).

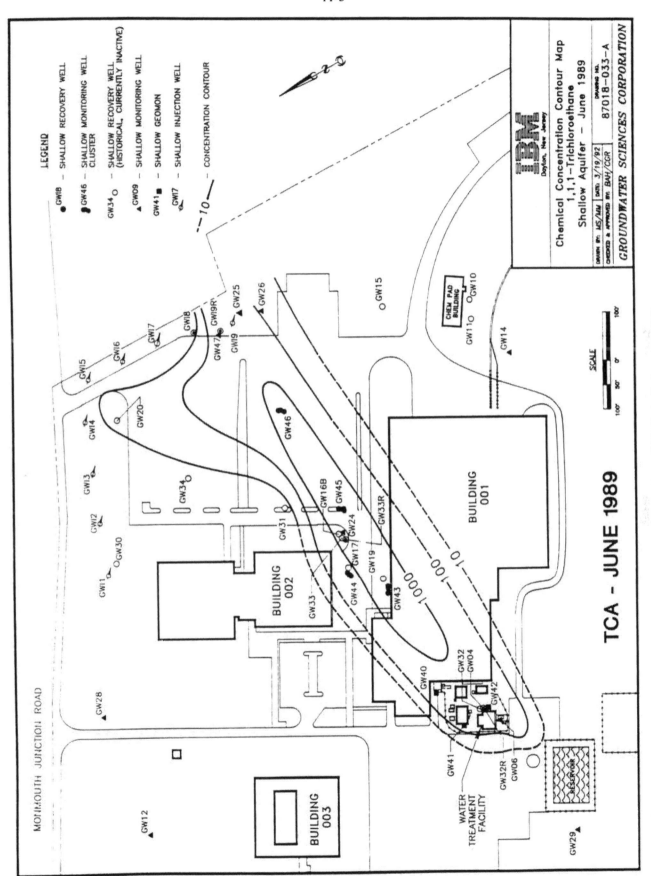

Figure 11-4.   TCA distribution in the Old Bridge aquifer in June 1989 (from Robertson, 1992).

clay; and this is underlain by a confined aquifer known as the Farrington aquifer. Highly productive, the Farrington aquifer is tapped by many large-capacity wells, such as SB11.

Eventually, 104 monitor wells were installed in the Old Bridge aquifer, 44 monitor wells were installed in the Farrington aquifer, and more than 25,000 groundwater samples were analyzed (Robertson, 1992). Using lithologic logs, it was found that the Woodbridge clay was discontinuous and was absent in the area of SB11. This early finding helped explain why SB11 was contaminated and is demonstrated in Figure 11-5. This absence of clay provides direct access for large volumes of recharge from the Old Bridge aquifer to the lower Farrington aquifer. Thus, contaminants in the shallow aquifer can migrate directly into the deeper aquifer.

As indicated, considerable water quality data were collected, some of which is presented in the figures for this case history. Although DNAPL was never directly observed, TCA concentrations in the Old Bridge aquifer were as high as 12,000 µg/l (Roux and Althoff, 1980) and PCE concentrations were as high as 8,050 µg/l (Robertson, 1992). These concentrations are 0.8% and 5.37% of the aqueous solubilities for TCA and PCE, respectively. Contamination in the Farrington aquifer generally involves lower concentrations of these same contaminants.

The suspected source of the contamination, buried chemical storage tanks near monitor well GW32, was removed in 1978, but no records have been obtained indicating that soil samples were collected at that time (USEPA, 1989c). In 1985 and 1986, soil samples taken from boreholes contained a maximum soil concentration of 13,255 µg/kg of total VOCs at a depth of 22.5 ft (USEPA, 1989c). More details on soil sampling were not available.

11.1.3 Effects of DNAPL Presence

Figures 11-6 through 11-8 show time versus six-month average TCA and PCE concentrations for wells GW32, GW16B, and GW25. The locations of these wells are given in Figure 11-3. As shown, GW32 is an extraction well near the suspected source area, where the highest contaminant concentrations were detected. Well GW16B is an onsite extraction well located downgradient of the source. Monitor well GW25 is located further downgradient near the property boundary.

Figure 11-6 shows a fairly steady decline in concentrations from 1978 to 1984. After the pump-and-treat system was terminated in 1984, concentrations generally increased with time. The PCE concentrations in 1988 are higher than those determined before the onset of remediation. The concentrations of TCA, however, have risen only slightly and are well below those determined before remedial activity. Site investigators speculate that this indicates TCA depletion in the source area. TCA would be expected to leach out more rapidly than PCE because it is more soluble in water and less strongly sorbed to soil than PCE.

The trends in Figure 11-7 (well GW16B) are similar to the trends observed in GW32, except for the decline in concentration at the end of 1988. The concentrations in Figure 11-8 are lower than those for wells closer to the suspected source. The increases in concentrations after 1984 are more pronounced in GW25 than in GW16B and GW32. In this case, concentrations after cessation of extraction are higher than they were at the initiation of remediation for both TCA and PCE. This might be due to changes in flow conditions due to the increased pumping at SB11, which is indicated on the figure.

The reappearance of elevated contaminant concentrations after the onsite groundwater extraction system was shut off led site investigators to suspect the presence of DNAPL. In retrospect, and given present-day knowledge of DNAPLs, the types of chemicals (TCA and PCE), the operational history (underground storage tanks), and the relatively high concentrations (for PCE, 5.37% of its aqueous solubility) should have indicated the potential for DNAPL in the subsurface. This, in turn, should have triggered more thorough source investigation and remediation than what apparently was performed. From a remediation viewpoint, this case history demonstrates the need for source control. Ironically, this case history also demonstrates that pump-and-treat technology did work for cleaning up dissolved contamination.

11.2 UP&L SITE, IDAHO FALLS, IDAHO

This site was included in the 19-site study by USEPA (1989c). It illustrates some of the problems associated with creosote (dissolved and DNAPL) in a fractured media. The material presented follows the discussion in USEPA (1989c).

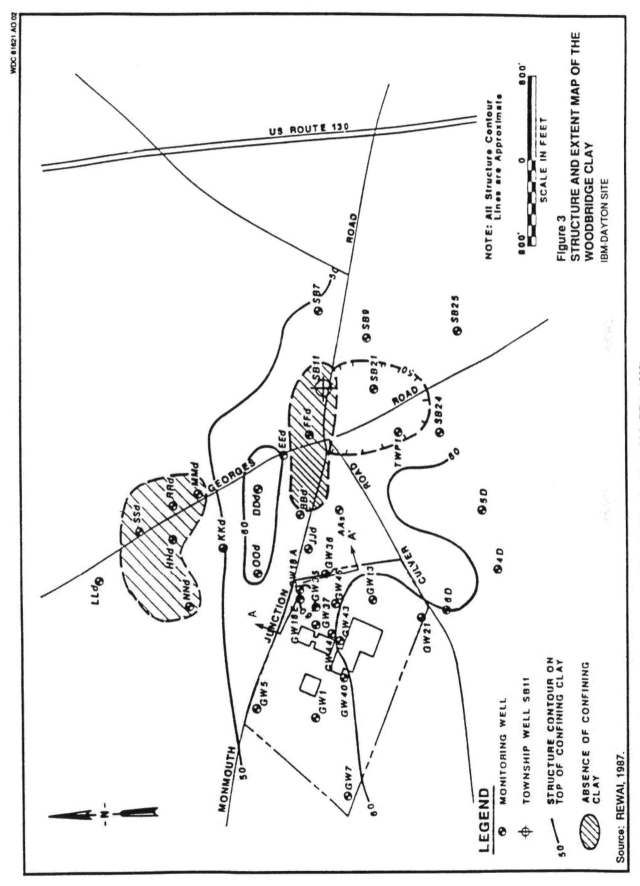

Figure 11-5.   Structure and extent map of Woodbridge clay (from U.S. EPA, 1989).

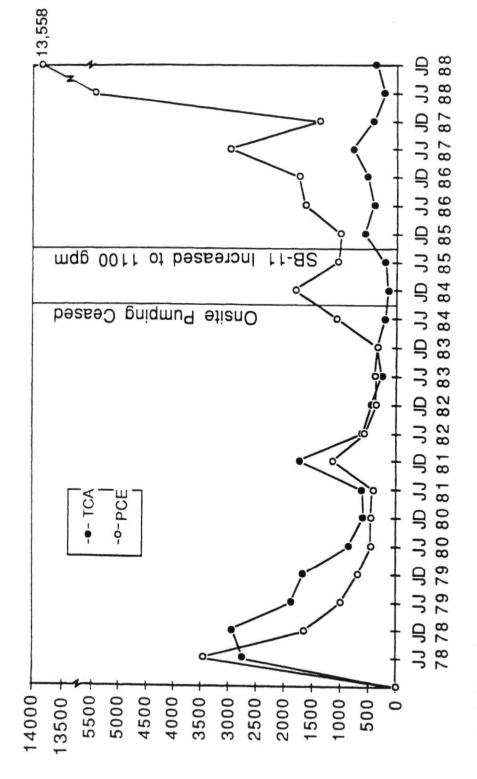

Figure 11-6.   History of TCA ad PCE variations in extraction well GW32, six-month average concentrations in ppb (from U.S. EPA, 1989).

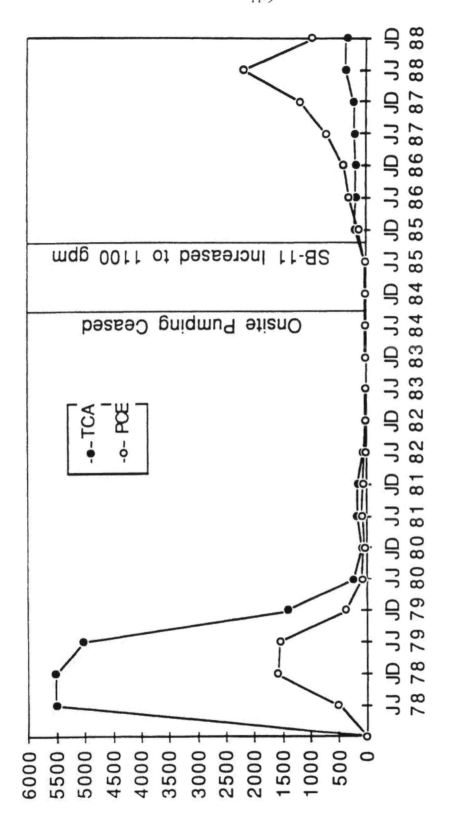

Compiled from various sources.

Figure 11-7.　History of TCA and PCE variations in extraction well GW168B, six-month average concentrations in ppb (from U.S. EPA, 1989).

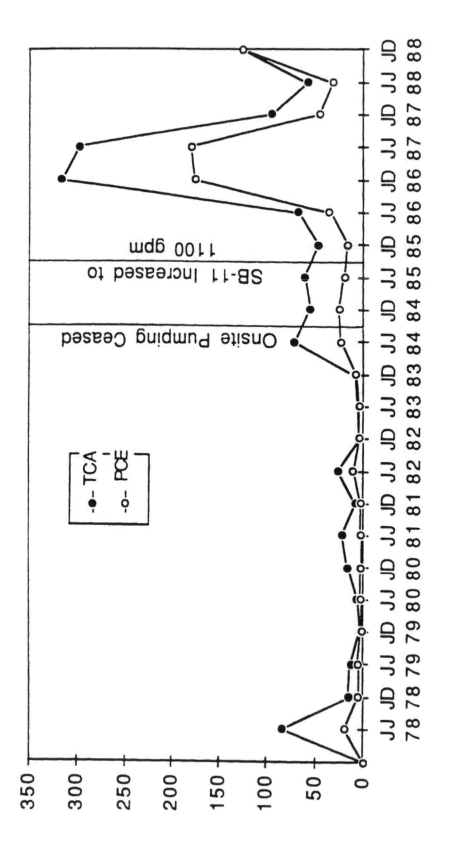

Compiled from various sources.

Figure 11-8.   History of TCA and PCE variations in extraction well GW25, six-month average concentrations in ppb (from U.S. EPA, 1989).

11.2.1 Brief History

As shown in Figure 11-9, the Utah Power & Light Company (UP&L) pole treatment yard is located in the southern part of Idaho Falls, Idaho, near the east bank of the Snake River. The site was used for treating electrical power poles by soaking them in a vat of heated creosote and then allowing the excess creosote to drip off into a receiving tank, before stockpiling the poles onsite. In July 1983, creosote was found to be leaking from underground piping connecting the treatment vat to a storage tank. Prior to this date, the facility had been operating for approximately 60 years. In response to the discovery of the leak, a corrective action was initiated that involved removal of the pole treatment equipment, excavation of contaminated soil, and installation of a bedrock (fractured basalt) pump-and-treat system.

11.2.2 Site Characterization

Subsequent to discovering the leak, a major effort was initiated to remove creosote-contaminated soil and rock. Approximately 37,000 tons of soil were excavated between July and September 1983 creating a pit 25 ft deep. Soil samples were taken from 15 locations in the bottom and walls of the pit to monitor the adequacy of the contaminated soil removal. In addition, 15 borings were drilled into the soil surrounding the pit and the bedrock beneath it. A total of 21 soil and rock samples were taken and analyzed for creosote compounds. Borings into bedrock beneath the pit, but above the water table, revealed the presence of creosote odors and DNAPL creosote coating the drill rods. Further excavation into the bedrock was thought to be impractical and the bottom of the pit was lined with a 12-foot layer of compacted clay in February 1984. The pit was subsequently backfilled with clean gravel and capped in 1985.

In May 1984, eight additional bedrock borings were drilled in the bottom of the excavation. These borings ranged from 55 to 140 ft deep. Only the deepest boring reached the water table at a depth of 122 ft. Evidence of creosote was found in all eight borings based on odor and creosote coating the drill rods. In addition, creosote accumulations were found in the bottom of one boring, which was sampled and analyzed. As shown in Table 11-1, the highest organic concentrations are polycyclic aromatic hydrocarbons (PAHs). Properties of creosote are discussed in Chapter 3.3.

Fifteen monitor wells were installed and one aquifer test was conducted on the UP&L property. Four sets of groundwater samples were collected in 1984 from the onsite wells. In addition, 21 offsite wells were sampled. Stratigraphic data from these wells and background information on regional and site hydrogeology were used to construct the east-west cross section given in Figure 11-10.

The interlayered basalts form the Snake River Plain Aquifer, a regional source of water supply. The interflow zones between the basalt flows are generally permeable and allow horizontal movement of groundwater. Fractures within the basalt flows tend to be concentrated along the upper and lower surfaces of the flow. Vertical movement of water between the interflow zones is via fractures in the basalt. Excavation of the creosote-contaminated gravel at the site in 1983 exposed the top of a basalt flow, where vertical fractures spaced two to four ft apart were observed. These fractures were filled with sand and silt.

The basalt layers beneath the site were classified into groups, labeled Basalt A through Basalt E, in Figure 11-10. Each group may include several individual basalt flows. The fracture zones and interflow zones in the lower part of Basalt B, below the water table, were designated as Aquifer #1. This zone is heterogeneous, but relatively permeable, and generally occurs between the water table and a depth of about 160 ft. Aquifer #1 is separated from the next lower aquifer by a very dense basalt flow, which generally extends from about 160 to 240 ft below ground surface. Below this is Aquifer #2, corresponding to the interflow zone and weathered basalt between the bottom of Basalt B and the top of Basalt C.

The water table is more than 100 ft below ground surface and fluctuates seasonally with an amplitude of about 25 ft. There is a downward vertical hydraulic gradient. Horizontally, flow is generally toward the southwest, but the magnitude of the gradient is variable, reflecting the heterogeneity of the basalt.

11.2.3 Effects of DNAPL Presence

Recognizing that DNAPL creosote was present above the water table, it was suspected that lateral migration of the creosote might be controlled by the northwest dip of the interflow zones between the basalt flows. However, outside the immediate leak area, creosote was only found above the water table in borings located to the south and

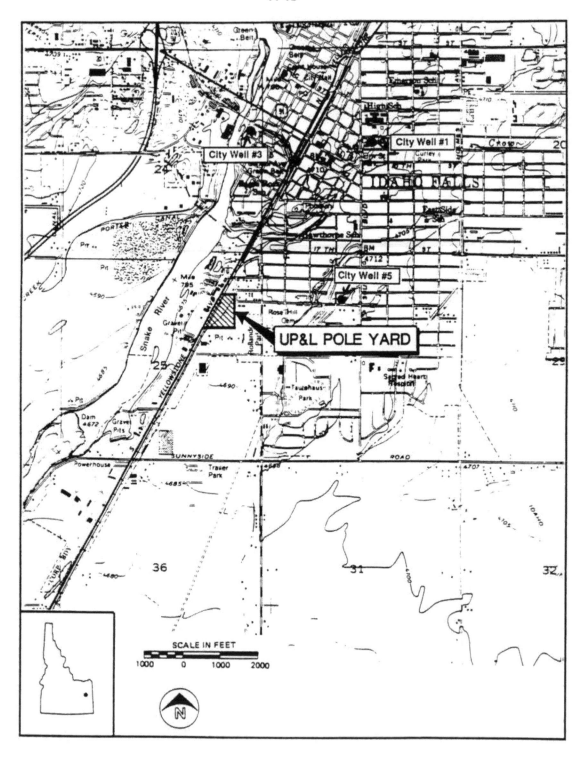

Figure 11-9.   Location of the Utah Power and Light Pole Yard site in Idaho Falls, Idaho (from USEPA, 1989).

Table 11-1. Chemical analysis of a creosote sample taken from a borehole drilled into bedrock at the UP&L Site (from USEPA, 1989). All values in mg/L unless otherwise noted.

INORGANICS		MISCELLANEOUS		Fluoranthene	2140.
Calcium	54.	pH	8.1	Pyrene	1730.
Magnesium	22.	Sp.Cond.(umhos/cm)	805.	Bis(2-ethylhexyl) phthalate	<1.
Sodium	85.	Total Organic Carbon	31000.	Chrysene	300.
Potassium	78.	TOH (mg/L as Cl)	4700.	Benzo(a)anthracene	580.
Chloride	46.	Total Phenols	0.120	Di-n-octyl phthalate	<1.
Fluoride	37.	Coliform (MPN/100mL)	<2.	Chrysene	300.
Sulfate (SO₄)	27.	Total Suspended Solids	380000.	Benzo(k)fluoranthecene	230.
Nitrate (N)	1.7	Settleable Solids	850.	Benzo(a)pyrene	170.
Total Dissolved Solids	1800.	BOD (5-Day)	22000.	Indeno(1,2,3-c,d)pyrene	32.
Alkalinity (CaCO₃)	53000.	COD	77000.	Phenol	85.
Arsenic	1.200	Ammonia (N)	0.50	2,4-Dimethylphenol	21.
Barium	3.200	Sulfides	<2.0	Dibenzofuran	1180.
Cadmium	<0.001	Oil & Grease by IR	67.0	2-Methylnapthalene	1600.
Total Chromium	2.100	Petrol. Hydrocarbons	39.0	2-Methylphenol	25.
Iron	4900.00	Total Cyanide (CN)	<0.110	4-Methylphenol	<1.
Lead	3.000	PURGEABLES - METHOD 624		TENTATIVE IDENTIFICATION	
Manganese	110.00	Methylene Chloride	79.	1-Methyl Napthalene	650.
Mercury	0.0050	Acetone	98.	1,2-Dimethylnapthalene	380.
Selenium	0.007	Benzene	8.	4-Methyl phenanthrene	360.
Silver	0.012	4-Methyl-2-pentanone	8.	1,1'-Biphenyl	350.
PESTICIDES		2-Hexanone	<1.	4-Methyl-dibenzofuran	320.
Endrin	<0.001	Tetrachloroethene	<1.	1,3-Dimethyl napthalene	310.
Lindane	<0.001	Toluene	27.	1-Methylphenanthrene	310.
Methoxychlor	<0.01	Ethylbenzene	22.	1-Phenyl napthalene	300.
Toxaphene	<0.10	o-Xylene	<1.	2-Ethylnapthalene	290.
2,4-D	<1.0	DETECTED BASE/NEUTRALS, ACIDS		2-Methyl,1,1'-biphenyl	190.
2,4,5-TP	<0.10	Napthalene	6400.	1-(2-Propenyl)napthalene	155.
RADIOLOGICAL		Acenapthylene	<1.	Isoquinoline	120.
Gross Alpha	<2.	Acenapthene	2630.	3-Methyl phenol	105.
Gross Beta	8.	Fluorene	<1.	2-Ethenyl-2-methylbenzene	77.
Radium-226	1.4	Phenanthrene	5000.	1H-Indene	17.
Radium-228	<1.	Anthracene	<1.	2,6-Dimethyl phenol	13.

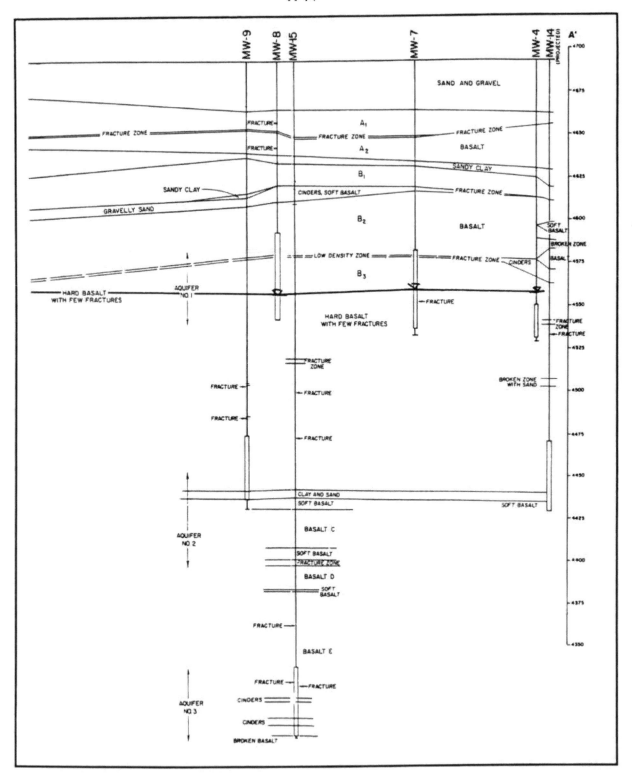

Figure 11-10. East-west geologic cross section across a portion of the Utah Power and Light Pole Yard site (modified from USEPA, 1989).

southwest. It was concluded that the basalt is so densely fractured and has enough vertical permeability that creosote would sink to the water table, rather than migrate laterally (Dames & Moore, 1985). This conclusion has not been verified, however, because no wells have been drilled into the bedrock to the northwest in the direction of the dip of the interflow zones.

Pump-and-treat remediation was selected for this site. Between October 1985 and April 1986, a six-month pilot study of groundwater extraction and treatment was conducted. The pilot study showed that more DNAPL creosote was produced from the wells than had been expected. The production of DNAPL creosote slugs caused various operational problems. Creosote was found to be incompatible with PVC, causing the piping to become brittle and crack. DNAPL creosote in the waste stream also caused some of the treatment processes to clog, requiring more frequent maintenance than expected. In response to these problems, numerous changes to the treatment plant were made during the period between the end of the first pilot phase on April 29, 1986, and the start of the second pilot phase in February 1987. Additional wells were drilled and data collected prior to the second pilot phase.

By the time of the second pilot phase, the treatment plant had been expanded and the treatment processes modified in response to experience gained from the first pilot study. Most of the groundwater in the second phase was produced from wells designed for extraction rather than from monitor wells, as had been the case in the first pilot phase. The plan for this phase was to expand the number of recovery wells incrementally as experience was gained concerning the behavior of the bedrock aquifers and the treatment system. The groundwater extraction and treatment system has continued operating since the beginning of the second pilot study in February 1987. It appears that the recovery system will continue operation into the foreseeable future.

In addition to illustrating certain problems associated with creosote, this example also illustrates how stressing the groundwater system during remediation can produce data that greatly aids understanding the groundwater system. For complex sites, this type of phased approach that combines aspects of site characterization with remediation is recommended.

11.3 HOOKER CHEMICAL SITES, NIAGARA FALLS, NEW YORK

Hooker Chemical and Plastics Corporation (now known as Occidental Chemical Corporation or OCC) buried large quantities of organic chemical manufacturing wastes at the Love Canal, Hyde Park, S-Area, and 102nd Street landfill Superfund sites in Niagara Falls between 1942 and 1975 (Figure 11-11). Based on fragmentary chemical production and process residue data, OCC estimated the types and quantities of wastes buried at each site (Table 11-2). These wastes contain thousands of tons of DNAPLs, including chlorinated benzenes, chlorotoluenes, and chlorinated solvents such as carbon tetrachloride, tetrachloroethene, and trichloroethene. DNAPLs encountered in the subsurface at the Hooker sites are typically brown-black mixtures containing numerous compounds. Representative chemical analyses of DNAPL samples from each site are given in Table 11-3.

Extensive remedial investigations have been conducted at each site, in part, to characterize the nature and extent of subsurface DNAPL contamination. The results of these studies and hydraulic containment remedies developed therefrom are described by USEPA (1982), Faust (1984; 1985a), Faust et al. (1990), Cohen et al. (1987), Pinder et al. (1990), OCC/Olin (1990), Conestoga-Rovers and Associates (1988a,b), and others. Selected findings regarding the presence and distribution of DNAPL at the Hooker sites are described below.

11.3.1 Love Canal Landfill

The stratigraphy of the Love Canal area is illustrated in Figure 11-12. The youngest sediments consist of approximately 5 ft of silt loam and sandy loam. These loams are underlain by older, varved glaciolacustrine silty clay sediments. The upper 6 ft of silty clay (between about 563 and 569 ft above MSL) are stiff and contain interconnected prismatic fractures (i.e., major fracture spacings on the order of 1 to 3 ft) and thin silt laminations. Transitionally below the stiff silty clay are 6 to 14 ft of soft silty clay above 2 to 20 ft of glacial till. The till is underlain by the Lockport Dolomite, a fractured aquifer of regional extent.

Excavation of the Love Canal to enable generation of hydroelectric power was abandoned after having barely begun in the 1890s. Its dimensions prior to waste disposal were approximately 3000 ft long, 40 to 100 ft

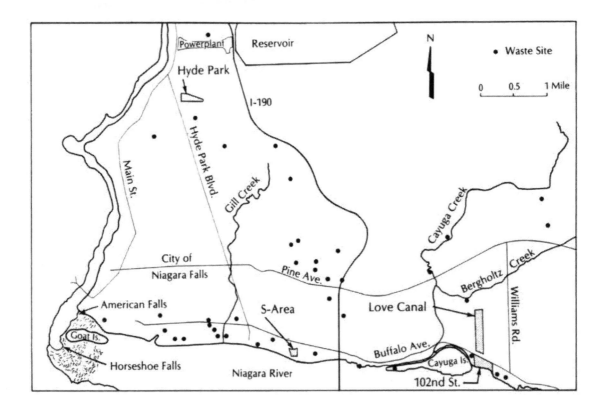

Figure 11-11.   Locations of waste disposal sites in Niagara Falls, including the Love Canal, 102nd Street, Hyde Park, and S-Area landfills (from Cohen et al., 1987).

Table 11-2. Estimated quantities and types of buried wastes at the Love Canal, Hyde Park, S-Area, and 102nd Street Landfills (Interagency Task Force, 1979).

Type of Waste	Love Canal	Hyde Park	S-Area	102nd Street
Chlorobenzenes	2,000	16,500	19,900	--
Benzylchlorides	2,400	3,400	1,600	--
Benzoylchlorides	800	6,200	3,300	--
Thionyl chloride	500	--	4,100	--
Trichlorophenol	200	3,300	200	--
Liquid disulfides, monochlorotoluene, and chlorotoluenes	700	2,600	2,200	--
Miscellaneous chlorinations	1,000	1,600	400	--
Metal chlorides	400	100	900	--
Miscellaneous acid chlorides	400	1,200	400	--
BHC cake including Lindane	6,900	2,000	--	300
Dodecyl mercaptans, chlorides, and miscellaneous organo-sulfur compounds	2,400	4,500	600	--
Sulfides and sulfhydrates	2,000	6,600	4,200	--
C-56 (hexachlorocyclopentadiene) and derivatives	--	5,600	17,400	--
Thiodan (Endosulfan)	--	1,000	700	
HET acid	--	2,100	500	--
Na hypophosphite mud	--	1,000	--	20,000
BTFs and derivatives	--	8,500	--	--
Dechlorane (Mirex)	--	200	--	--
Calcium fluoride	--	400	--	--
Mercury brine sludge	--	100	--	--
Inorganic phosphorus	--	100	--	1,300
Organic phosphorus	--	4,400	200	<100
Phenol tars	--	--	800	--
Brine sludge and gypsum	--	--	--	53,200
Tetrachlorobenzene	--	--	--	2,327
BHC, trichlorophenol, trichlorbobenzene, and benzene	--	--	--	2,000
Na chlorite black cake	--	--	--	18,673
Graphite	--	--	--	742
Lime sludge	--	--	--	22,978
Brine sludge	--	--	--	67,186
Miscellaneous	2,000	7,300	5,700	2,200
ESTIMATED TOTAL	22,000	80,000	63,000	200,000

Table 11-3. Chemical analyses of DNAPL sampled from the 102nd Street, S-Area, Love Canal, and Hyde Park Landfills in Niagara Falls, New York (from OCC/Olin, 1990; Conestoga-Rovers and Associates, 1988a; Herman, 1989; and Shifrin, 1986).

COMPOUND	102ND STREET OW-38	S-AREA OW-213F	LOVE CANAL CW-90	HYDE PARK
Benzene	0.07		0.25	0.09
Chlorobenzene	0.64	0.8	1.1	0.36
Dichlorobenzene	1.9	0.6	0.91	0.19
Trichlorobenzene	7	11.5	20.5	1.50
Tetrachlorobenzene	33.3	32	71.6	2.31
Pentachlorobenzene	11	6.4	18	0.43
Hexachlorobenzene	0.52	1.6	0.59	0.32
Toluene	0.037	0.1	0.96	1.29
Chlorotoluene		0.9	1.17	3.00
Dichlorotoluene		0.9		4.90
Total BHCs		0.4	2.8	0.08
Chloroform	0.057			0.09
Carbon Tetrachloride		1.0		0.06
Trichloroethene		0.4		0.16
Tetrachloroethene		13		2.10
Hexachloroethane		1.1		0.12
Hexachlorobutadiene	0.057	4.2		0.62
Hexachlorocyclopentadiene		12		
Octachlorocyclopentadiene		14		
Hexadecane	0.27		1.7	
Cyclohexadecane	0.46			
Trimethylpentene	0.053			
Chloromethylbenzene	0.046			
Trichloro(methyl, ethyl)benzene	0.056			
Tetrachlorothiophene	0.21			
Trichlorobiphenyl	0.24			
Tetrachlorobiphenyl	0.07			
Pentachlorocyclohexane	0.54			
Endosulfan II		0.3		
Mirex		0.1		
Trichlorophenols				1.20
Phenol				1.20
Dichlorobenzotrifluorides				1.70
Chlorobenzotrifluorides				1.20
Methylbenzoate				1.30
Butylbenzoate				1.09
Trichlorotoluene				1.60
SUBTOTAL	56.5	101.3	119.6	26.91

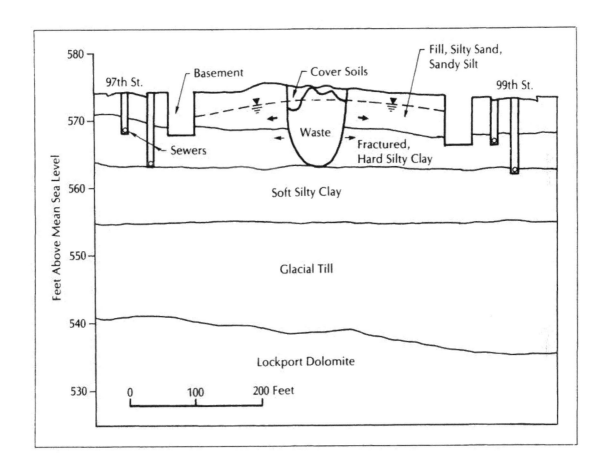

Figure 11-12.   A schematic geologic cross-section through the Love Canal landfill (from Cohen et al., 1987).

wide, and 7 to 16 ft deep. Drummed solid and liquid chemical wastes were rolled from trucks into the abandoned canal excavation and into pits dug adjacent to the canal between 1942 and 1954. Some uncontained solids and liquids were also disposed of in the landfill.

The waste disposal practice created an irregular landfill surface that was elevated a few feet above adjacent residential properties along portions of the site. Chemical wastes, including DNAPL in drums, were buried as high as the original ground surface at some locations. The disposal of liquid chemicals in drums resulted in drum failure due to corrosion and consolidation. This, in turn, caused subsidence of cover soil, creation of 'potholes' and caved areas, exposure of displaced chemicals including NAPL at the surface, emission of chemical vapors, and increased recharge to the landfill. Combined with the shallow water table, humid climate, and limited permeability of the native soils, heavy recharge promoted the development of a water table mound at the site and periodic leachate seeps at the landfill perimeter. Early water-level monitoring at the site in the Spring of 1978 indicated that "the groundwater table [was] located immediately below or at the land surface" of the southern section of the landfill causing overland flow and the development of leachate ponds in low areas along the landfill perimeter (Conestoga-Rovers and Associates, 1978). Basement drains and sewers surrounding the landfill acted both as hydraulic sinks and large monitor wells. DNAPL and chemical odors were observed in various basement sumps and sewer locations.

During the first major offsite drilling program in which soil borings were augered every 10 ft along a proposed barrier drain line in backyards adjacent to the southern sector of the landfill in late 1978, Earth Dimensions (1979) discovered that an oily film appeared to displace water in the fractures in the stiff silty clay layer at most of the augered sites. Pungent odors were reported at nearly all of the drill sites. These early observations of DNAPL presence in the fractured silty clay layer resulted in a decision to lower the leachate collection system drain to the top of the soft silty clay layer. The perimeter drain system was installed between 1978 and 1979 to provide hydraulic containment.

A detailed study of chemical conditions in Love Canal soils was undertaken jointly by Earth Dimensions (1980) and the New York State Department of Health (1980). Continuous split-spoon samples were taken at 64 locations in the immediate vicinity of the landfill. Detailed logs were prepared to document the vertical distribution of soil types, chemical odors, and NAPL; and samples were analyzed for Love Canal chemicals (Figure 11-13). This and other studies documented that:

(1) DNAPL migration is highly correlated with portions of the site where Hooker disposed of chemical wastes;

(2) DNAPL was commonly found in fractures and in thin silty laminations (varves) in the stiff silty glaciolacustrine clay;

(3) DNAPL was rarely found in deep root channels in the upper portion of the soft silty clay;

(4) DNAPL generally did not penetrate the soft silty clay or underlying glacial till layers; and,

(5) The upper surface of DNAPL beneath residential properties adjacent to the landfill (typically 565 to 567.5 ft) was at a lower elevation than the upper surface of DNAPL observed during subsequent drilling in the landfill (typically 568 to 571.5 ft).

The presence of DNAPL in hairline fractures and thin laminations required careful dissection and inspection to ensure its detection in the stiff silty clay soil samples. Relatively low chemical concentrations (low ppm range) were determined in some fractured silty clay samples with DNAPL present. Factors contributing to the low concentration of Love Canal chemicals detected in these samples apparently include: dilution of NAPL concentration during sample homogenization, loss of volatiles during sample handling, and the failure to conduct analyses for all NAPL components.

In 1986, 21 borings were made with continuous split-spoon sampling and 15 monitor wells were completed directly into waste (Figure 11-14) to examine fluid elevations in the landfill, waste materials, and soils beneath the landfill (E.C. Jordan Co., 1987). To avoid creating vertical pathways for chemical migration, each boring was terminated upon retrieval of the first split-spoon sample of undisturbed soil beneath the waste. Elevations of immiscible fluid surfaces, the landfill bottom, and DNAPL encountered in split-spoon samples at each drilling location are given in Table 11-4. DNAPL was observed in the chemical waste disposal areas from the canal bottom to elevations ranging between 567 and 574 ft. Many of the split spoons and spoon samples were coated with dripping DNAPL. DNAPL surface elevations measured in the landfill wells ranged from approximately 566 to 574 ft.

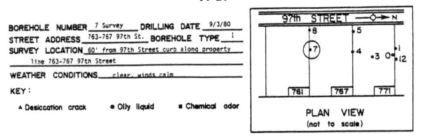

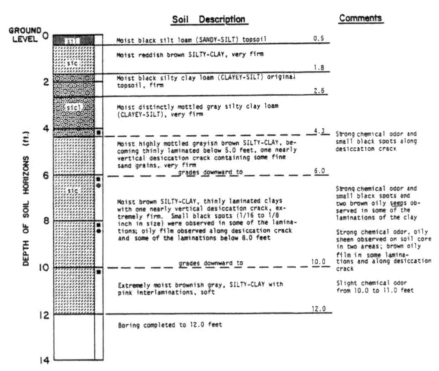

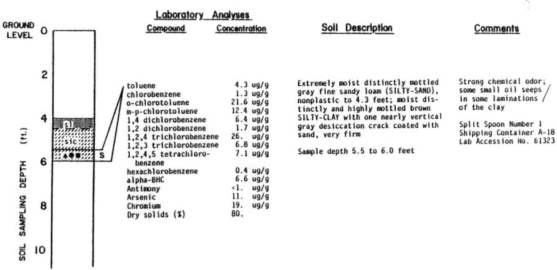

Figure 11-13.   Example boring and laboratory log of soils sampled adjacent to the Love Canal landfill (from New York State Department of Health, 1980).

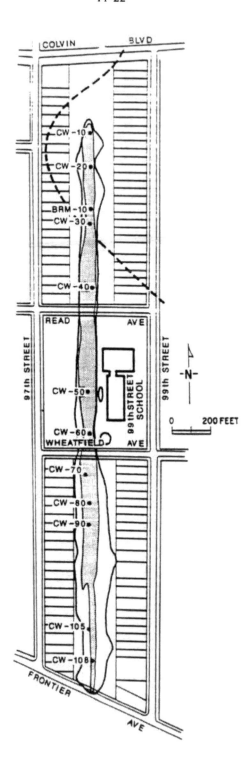

Figure 11-14. Locations of well completed directly into the Love Canal landfill. The original canal excavation is shaded, and is enclosed by an outer line designating the approximate landfill limits.

Table 11-4.   DNAPL distribution and properties in wells completed directly into the Love Canal landfill (modified from Pinder et al., 1990).

WELL	FLUID SURFACE IN WELL	DNAPL SURFACE IN WELL	CANAL BOTTOM	INTERVAL WITH NAPL IN SPLIT-SPOON SAMPLES	MAIN WASTE TYPE	DNAPL DENSITY	DNAPL VISCOSITY (cp)	DNAPL SAMPLE TEMPERATURE (°F.)	FLUID PUMPED FROM WELL
CW10	571.3	571.3	563.5	562.9-568.3	Chemical	1.15-1.18	75-81	60-62	4-5 gallons pumped: all DNAPL
CW20	572.5	572.5	560.9	562.2-567.0	Chemical	1.12	<17	60	4-5 gallons pumped: all DNAPL
BRM10	570.5	570.5	562.9	563.6-571.4	Chemical	1.09-1.12	44-50	60-64	3-4 gallons pumped: all DNAPL
CW30	567.3	None	563.9	None	Municipal				
CW40	--	--	560.5	561.9-562.7	Municipal				
CW50	568.7	None	558.5	None	Soil fill				
CW60	567.6	566.6	557.9	556.9-571.4	Chemical	1.21-1.25	21-36	61-62	6 gallons pumped: all DNAPL
CW70	569.1	566.5	558.8	558.8-571.3	Chemical	>1.50	<23	53-54	Approx. 3 gallons pumped: more water than DNAPL
CW80	570.0	567.3-569.6	560.4	558.5-572.5	Chemical				
CW90	570.2	569.4	560.4	560.1-569.1	Chemical	1.18-1.20	<18	61-62	1.5 gallons pumped: DNAPL and water
CW105	574.1	573.1	566.7	562.7-574.2	Chemical	1.09-1.22	37-270	59-62	Approx. 3 gallons pumped: mainly DNAPL
CW108	571.2	--	565.2	562.7-563.2	Soil fill	1.08	<16	71	4 gallons pumped: approx. 30% DNAPL and 70% water

Pumping experiments were conducted in the Love Canal wells during 1988 and 1990 to examine DNAPL fluids (GeoTrans, 1988; J.R. Kolmer and Associates, 1990). Typically, 3 to 5 gallons of fluid (equivalent to about 10 to 20 well volumes) were extracted from many of the 2-inch diameter wells. DNAPL densities ranged from 1.09 to >1.50, and viscosities ranged from very thick (270 cp) to watery (Table 11-4). At several wells, 3 to 6 gallons of DNAPL were extracted without any water.

The disposal of large quantities of DNAPL created a substantial DNAPL pool within Love Canal. Although wastes were buried between 1942 and 1954, ongoing drum corrosion provides a long-term mechanism for the release of mobile DNAPL. Density-driven DNAPL migration occurs primarily through the fractured silty clay layer at the base of the DNAPL pool. Fractures, and probably the silt laminations to a lesser extent, conducted DNAPL toward basement and sewer drains. The soft silty clay layer is an effective capillary barrier to DNAPL penetration: it forms the bottom of the DNAPL site. DNAPL that migrated to offsite properties (Figure 11-15) was a moving contaminant source from which chemicals dissolved into groundwater and volatilized into soil gas and basement air.

Since the installation of the tile-drain system during the late 1970s, approximately 30,000 gallons of DNAPL have been collected and stored in holding tanks at the onsite leachate treatment plant. The perimeter drain system appears to be an effective hydraulic barrier based on fluid elevation and chemical monitoring. The pumping experiments conducted in landfill wells demonstrate that large quantities of DNAPL can probably be recovered directly from wells in the landfill if incineration or some other treatment technology can be shown to be cost-effective at a site where long-term hydraulic containment will be needed in any event.

11.3.2 102nd Street Landfill

The 102nd Street landfill is adjacent to the upper Niagara River and a short distance south of Love Canal (Figure 11-11). Hooker disposed of approximately 77,00 tons of predominantly inorganic wastes on the western 15.6 acres of the 22-acre site between 1943 and 1971. During a similar time period, Olin Chemical Corporation dumped an estimated 66,000 tons of wastes on the eastern portion. Approximately 10 to 15 ft of wastes were deposited above alluvium at the river edge, raising a swampy area to the grade of Buffalo Avenue to the north. The landfill was closed in 1971 after the U.S. Army Corps of Engineers ordered the companies to construct a bulkhead to prevent erosion of wastes into the river.

The stratigraphy of the 102nd Street landfill site differs from that at Love Canal due to erosion and sedimentation by the Niagara River. As shown in Figure 11-16, the Lockport Dolomite is overlain by glacial till, soft silty clay, river alluvium, and fill materials. The river eroded the soft silty clay in the southern portion of the site and replaced it with silt, sand, and gravel alluvium that coarsens with depth and proximity to the river. The alluvium pinches out near the northern site boundary. In some areas along the southern site boundary, the silty clay was completed eroded and the alluvium is underlain directly by glacial till.

OCC/Olin (1990) undertook a comprehensive remedial investigation at the site between 1985 and 1988 to augment prior studies and facilitate evaluation of remedial alternatives. Based on a review of historic company documents, interviews with former personnel, and interpretation of historic aerial photographs, the companies identified suspected NAPL disposal areas at the site as shown in Figure 11-17. All samples collected during the remedial investigation, regardless of matrix (i.e., soil, rock, groundwater, river sediment, waste, etc.) were examined visually for the presence of NAPL. Static groundwater was extracted from the top and bottom of the water column in numerous overburden wells to examine for NAPL presence. Where adequate volumes could be obtained, DNAPL samples were taken for laboratory examination of chemical and physical properties.

DNAPL samples were analyzed for the USEPA Contract Laboratory Protocol Target Compound List parameters and an effort was made to identify the remaining chromatogram peaks. Water content and major element composition were also determined for some samples. Resulting analytical mass balances ranged from 58 to 137 percent. OCC/Olin (1990) concluded that aliphatic compounds, and high-molecular weight polymeric compounds to a lesser extent, probably constitute the majority of compounds not quantified by GC/MS analysis. Water contents of the DNAPL samples ranged from 0.1 to 64%. Overall, DNAPL at the 102nd Street site was determined to be composed primarily of chlorinated benzenes. DNAPL densities and absolute viscosities generally ranged between 1.3 to 1.6 g/cm^3, and between 2 and 8 cp (at 25 to 40° C), respectively.

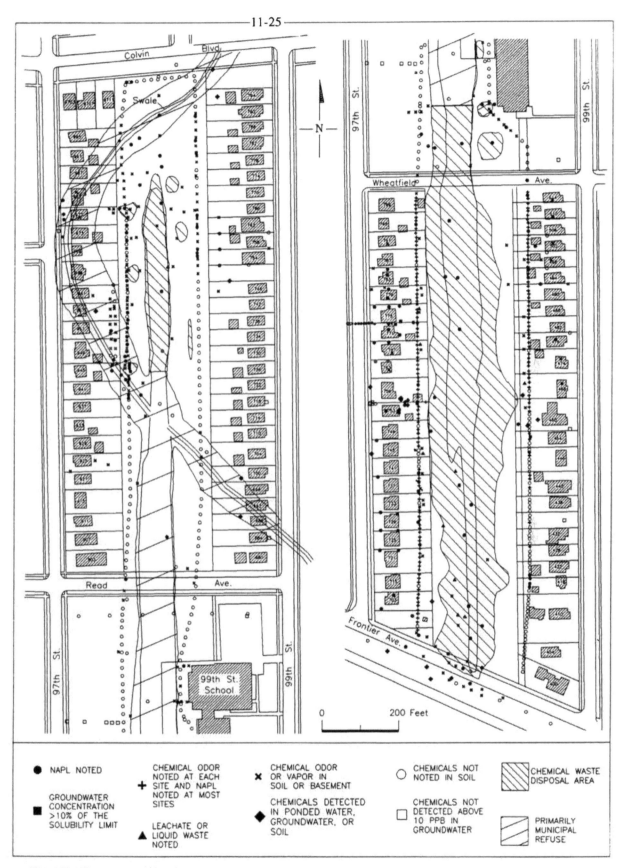

NAPL NOTED

GROUNDWATER
CONCENTRATION
>10% OF THE
SOLUBILITY LIMIT

CHEMICAL ODOR
NOTED AT EACH
SITE AND NAPL
NOTED AT MOST
SITES

LEACHATE OR
LIQUID WASTE
NOTED

CHEMICAL ODOR
OR VAPOR IN
SOIL OR BASEMENT

CHEMICALS DETECTED
IN PONDED WATER,
GROUNDWATER, OR
SOIL

CHEMICALS NOT
NOTED IN SOIL

CHEMICALS NOT
DETECTED ABOVE
10 PPB IN
GROUNDWATER

CHEMICAL WASTE
DISPOSAL AREA

PRIMARILY
MUNICIPAL
REFUSE

Figure 11-15.   Areal distribution of DNAPL and chemical observations at Love Canal (north half
to the left; south half to the right).

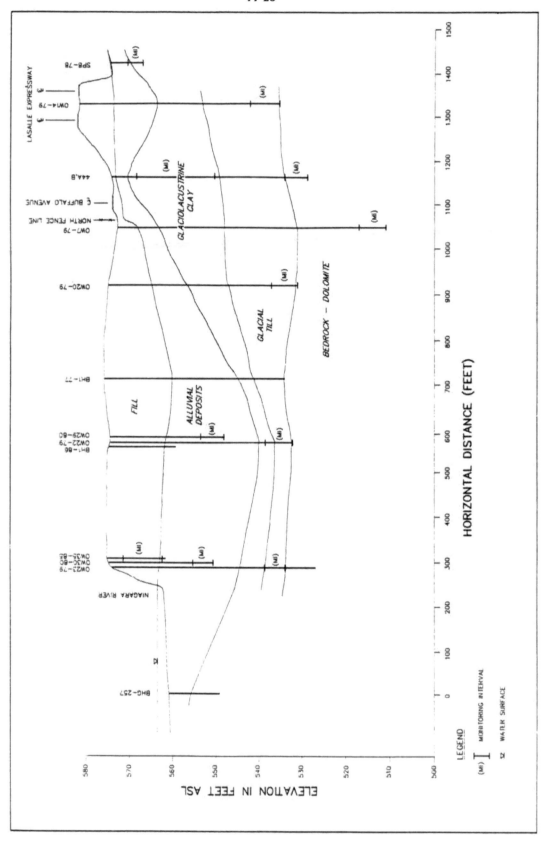

Figure 11-16. South-North geologic cross section through the 102nd Street Landfill (from OCC/Olin, 1980).

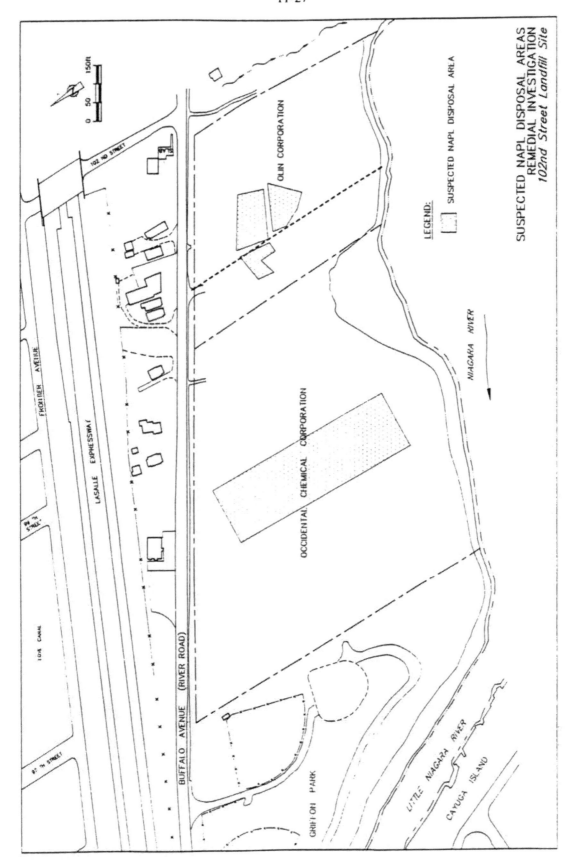

Figure 11-17. Suspected NAPL disposal areas at the 102nd Street Landfill (from OCC/Olin, 1990).

Five separate areas of DNAPL presence were interpreted at the site (Figure 11-18) based on differences in DNAPL chemistry, knowledge of waste disposal history, and the many data points (OCC/Olin, 1990). Within each area, the DNAPL distribution is expected to be highly complex. Simplified conceptual cross sections were developed as illustrated in Figure 11-19. As shown, DNAPL has migrated downward by gravity and appears to have spread along the stratigraphic interfaces at the top and bottom of the alluvium. In the largest and most complex DNAPL area (Area 5 on Figure 11-18), DNAPL appears to be accumulating in a stratigraphic trap formed by the surface of the clay-till capillary barrier (Figures 11-19 and 11-20). DNAPL was not observed within the silty clay, glacial till, or bedrock at the site.

Based on borings in the Niagara River, and the apparent stratigraphic trap present beneath the southern edge of the site, it appears that the potential for offsite DNAPL migration to beneath the Niagara River is very limited. This potential will be further examined when additional borings are installed offshore adjacent to the site to facilitate detailed design of the site containment system (OCC/Olin, 1990).

11.3.3 Hyde Park Landfill

The Hyde Park landfill is 2000 ft east of the deep gorge formed by the Niagara River downstream from the Falls (Figure 11-11). Between 1953 and about 1975, Hooker disposed of an estimated 80,000 tons of liquid and solid chemical wastes in trenches and pits at the this 15-acre site (Table 11-2).

Approximately 15 to 30 ft of chemical waste at the site are underlain by 0 to 10 ft of silty clay sediments which overlie the Lockport Dolomite. The Lockport Dolomite ranges from 60 to 140 ft thick in the vicinity of the landfill. Groundwater flow in the bedrock can be idealized as a sequence of flat-lying, water-bearing zones (with significant bedding plane fractures) that are sandwiched between water-saturated layers of reduced permeability (Faust et al., 1990). Vertical fractures provide an avenue for fluid transmission through the less permeable layers.

Unlike the Love Canal and 102nd Street landfills where fine-grained capillary barrier layers limit downward DNAPL migration, there is no bottom to the DNAPL site in the overburden at Hyde Park. The Hyde Park landfill was excavated to bedrock in some areas, providing

a direct route for density-driven DNAPL migration into the fractured Lockport Dolomite aquifer. As a result, extensive and extremely complex distributions of DNAPL and dissolved chemicals derived from the landfill are present in the dolomite in the vicinity of the site.

The extent and degree of chemical contamination emanating from the Hyde Park site was evaluated during studies conducted by OCC in 1982 and 1983. A major component of these studies involved drilling to determine the extent of chemical contamination in the overburden and bedrock. Borings were cored and tested in 15-ft sections to the top of the Rochester Shale (bottom of the Lockport Dolomite) in a manner similar to that described in Chapter 9.4 along ten vectors radiating out from the landfill perimeter (using the "inside-out" approach discussed in Chapter 9.2). As noted in Chapter 9.4, the degree of DNAPL migration induced by the drilling program is unknown. Brown-black DNAPL was obvious on many of the contaminated core sections. Along each vector, survey wells were installed at 800 ft intervals until a well showed no chemical parameters above the survey levels. To better delimit the dissolved plume, a well was then installed midway between the outer two wells. Nearly 300 intervals in the Lockport Dolomite were sampled by this procedure. The wells were grouted after testing.

This program revealed extensive subsurface migration of DNAPL chemicals from the landfill. Hyde Park chemicals were detected in seeps emanating from the dolomite along the Niagara Gorge in 1984. Dissolved chemical and DNAPL plumes determined from these studies are delineated in Figure 11-21.

Immiscible flow modeling was conducted to conceptually examine planned hydraulic containment and DNAPL recovery options (Faust et al., 1989). The modeling analyses demonstrate that:

(1) Viscosity should be measured to provide better predictions of potential DNAPL recovery;

(2) Wells should be operated at pumping rates that do not significantly dewater the aquifer adjacent to the wells;

(3) Wells should be open primarily to the permeable zones containing the highest saturations of DNAPL; and,

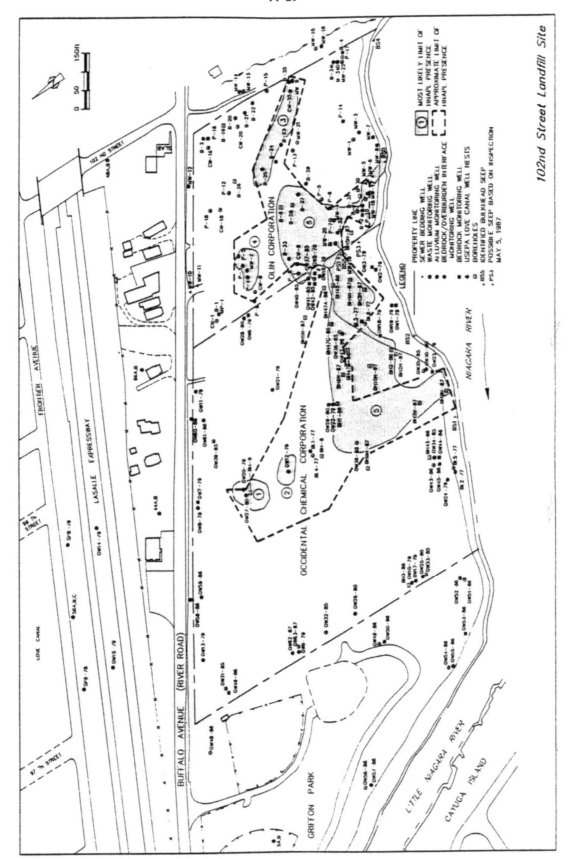

Figure 11-18. Approximate horizontal extent of DNAPL in fill and alluvium at the 102nd Street Landfill (from OCC/Olin, 1990).

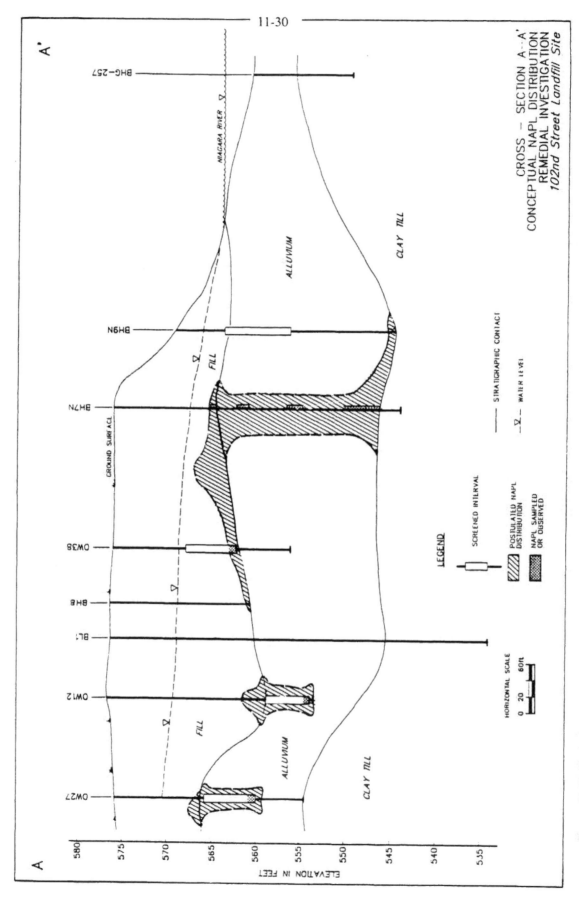

CROSS – SECTION A–A'
CONCEPTUAL NAPL DISTRIBUTION
REMEDIAL INVESTIGATION
102nd Street Landfill Site

Figure 11-19. Typical conceptual DNAPL distribution along a cross-section at the 102nd Street Landfill (from OCC/Olin, 1990).

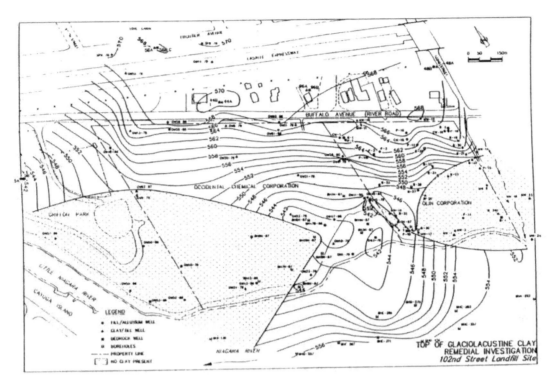

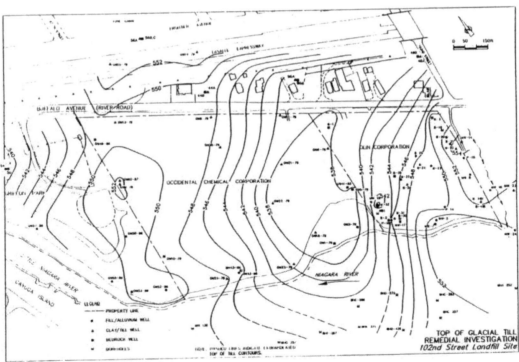

Figure 11-20.   The surface of the silty clay and glacial till capillary barrier layers appear to form a stratigraphic trap beneath the south-central portion of the site (from OCC/Olin, 1990).

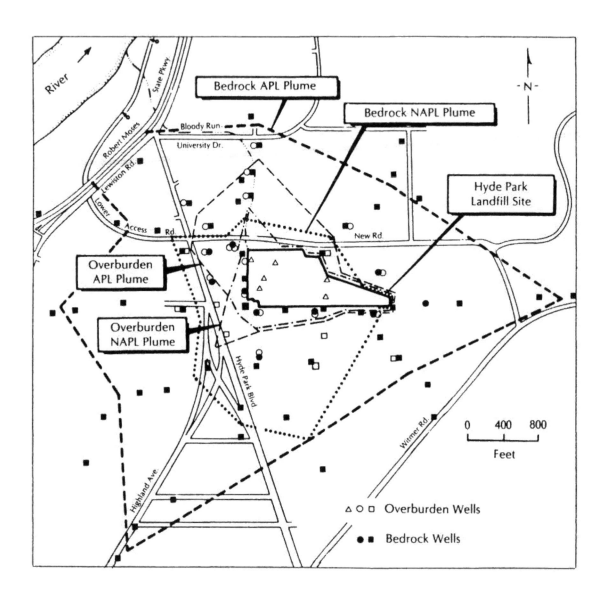

Figure 11-21. Boundaries of dissolved chemical and DNAPL plumes emanating from the Hyde Park Landfill (from Cohen et al., 1987).

(4) DNAPL recovery rates should be monitored frequently on a well-by-well basis to aid prediction of DNAPL recovery rates and to decide when to delete and/or add new extraction wells.

11.3.4 S-Area Landfill

The S-Area is located within the OCC manufacturing plant in Niagara Falls (Figure 11-11). The S-Area property was partially reclaimed from the Niagara River by dumping fill materials between 1938 and 1947. Hooker buried an estimated 63,000 tons of liquid and solid chemical residues at the site between the late 1940s and 1961. Wastes were disposed of in 15 to 18 ft deep parallel trenches, and some wastes were buried in tank cars.

The discovery of contaminated sludge in water supply tunnels cut into the Lockport Dolomite that convey river water to the City of Niagara Falls water treatment plant adjacent to the S-Area site (Figure 11-22) aroused concern regarding chemical migration from the landfill in 1978. Subsequent investigations revealed that the clay-till aquitard that is present at the 102nd Street landfill is absent beneath a portion of the site. Whereas the clay-till surface appears to form a stratigraphic trap for DNAPL accumulation at the 102nd Street landfill, discontinuity in the clay-till surface beneath the S-Area site provides a conduit for DNAPL migration to the Lockport Dolomite.

DNAPL and dissolved chemicals have migrated significant distances from the S-Area landfill in overburden and bedrock. Extensive drilling and testing surveys conducted by OCC between 1986 and 1988 generally delineated the extent of contamination at the site (Conestoga-Rovers and Associates, 1988a,b). Protocols for drilling in bedrock at S-Area are described in Chapter 9.4, and observations of DNAPL in bedrock wells is shown in Figure 9-7. Conestoga-Rovers and Associates (1988) noted an increase in DNAPL density and decrease in DNAPL viscosity with depth in the Lockport Dolomite beneath S-Area.

11.4 SUMMARY

The case studies illustrate several critical aspects of DNAPL site evaluation. Assessing DNAPL presence based on historic information, field observations, and monitoring data is needed to guide site characterization and remediation activities. Knowledge or suspicion of DNAPL presence requires that special precautions be taken during field work to minimize the potential for inducing unwanted DNAPL migration by drilling or pumping. Delineation of subsurface geologic conditions is crucial to site evaluation because DNAPL movement can be largely controlled by the capillary properties of subsurface media. It is particularly important to determine, if practicable, the spatial distribution of fine-grained capillary barriers and preferential DNAPL pathways (e.g., fractures and coarse-grained strata). Finally, the case studies evidence that it will usually be necessary to control (i.e., contain or remove) DNAPL zone contamination in order to attempt restoration of the aquifer downgradient from the DNAPL source. Failure to adequately consider DNAPL presence can lead to a flawed assessment of remedial alternatives.

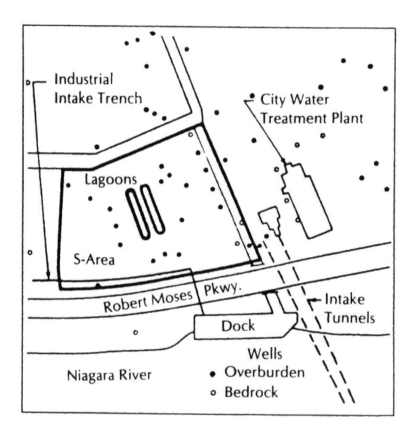

Figure 11-22.   Proximity of the City of Niagara Falls Water Treatment Plant water-supply intake
tunnels to the S-Area Landfill (from Cohen et al., 1987).

12 RESEARCH NEEDS

As awareness of DNAPL contamination increased in the 1980s, research was conducted to better understand the behavior of DNAPL in the subsurface. Much of this research was an expansion of the investigations performed by Schwille (1988). DNAPL research is currently focusing on remediation (National Center for Groundwater Research, 1992). Through this progression of DNAPL research, relatively little effort has been expended on developing new site characterization tools or methods for DNAPL sites.

What has generally occurred at DNAPL sites is that tools and techniques utilized at contamination sites in general have been applied with varying degrees of success. Additionally, some new tools and methods have been developed and others have been adapted to better satisfy the requirements of a DNAPL site investigation. Site characterization strategies have also evolved to more closely match the special concerns and risks posed by DNAPL presence.

Despite substantial progress, additional research on DNAPL site characterization tools and methods is warranted utilizing a variety of venues: laboratories, controlled field sites with emplaced DNAPL, and uncontrolled contamination sites. Additional research and technology transfer efforts should focus on:

(1) Well drilling techniques to demonstrate the isolation of DNAPL zones through the use of double-cased wells or other techniques;

(2) Well and boring abandonment techniques to demonstrate the efficacy of different grouting mixtures and methods to prevent preferential vertical fluid migration;

(3) The utility of surface and borehole geophysical methods to better characterize DNAPL presence and distribution, and stratigraphic controls on DNAPL movement;

(4) The utility of soil gas surveying to better characterize NAPL presence and related chemical migration;

(5) Methods to determine in-situ NAPL saturation (e.g., borehole geophysics, simple quantitative sample analysis);

(6) Techniques to determine field-scale constitutive relationships between saturation, capillary pressure, and relative permeability;

(7) Practical field or laboratory techniques to delineate mobile DNAPL from DNAPL in stratigraphic traps from DNAPL at residual saturation;

(8) Additional cost-effective methods to determine NAPL presence, composition, and properties;

(9) Techniques to better define site stratigraphy, heterogeneity, and fracture distributions;

(10) The long-term capacity of capillary barriers (e.g., clayey soil layers) to prevent DNAPL movement, including methods for determining barrier continuity and time-dependent aspects of DNAPL-mineral structure and wettability interactions;

(11) Identifying the limited characterization efforts required to determine and implement appropriate remedial measures at DNAPL contamination sites;

(12) Further optimization of characterization strategies given different source, hydrogeologic, risk, and remedy considerations; and,

(13) Refinement of pilot test designs, protocols, and monitoring requirements to determine the feasibility and/or technical impracticality of alternative remedial measures.

13 REFERENCES

Abdul, A.S., T.L. Gibson and D.N. Rai, 1990. Laboratory studies of the flow of some organic solvents and their aqueous solutions through bentonite and kaolin clays, *Ground Water*, 28(4):424-533.

Abdul, A.S., S.F. Kia and T.L. Gibson, 1989. Limitations of monitoring wells for the detection and quantification of petroleum products in soils and aquifers, *Ground Water Monitoring Review*, 9(2):90-99.

Abriola, L.M., 1983. Mathematical modeling of the multiphase migration of organic compounds in a porous medium, Ph.D. Dissertation, Department of Civil Engineering, Princeton University, Princeton, New Jersey.

Abriola, L.M., 1988. Multiphase flow and transport models for organic chemicals: A review and assessment, Electric Power Research Institute, Palo Alto, California.

Abriola, L.M. and G.F. Pinder, 1985a. A multiphase approach to the modeling of porous media contamination by organic compounds, 1. Equation development, *Water Resources Research*, 21(1):11-18.

Abriola, L.M. and G.F. Pinder, 1985b. A multiphase approach to the modeling of porous media contamination by organic compounds, 2. Numerical simulation, *Water Resources Research*, 21(1):19-26.

ACGIH, 1990. Threshold limit values for chemical substances and physical agents and biological exposure indices, American Conference of Government Industrial Hygienists, 122 pp.

Acher, A.J., P. Boderie and B. Yanon, 1989. Soil pollution by petroleum products, I: Multiphase migration of kerosene components in soil columns, *Journal of Contaminant Hydrology*, 4, pp. 333-345.

Acker, W.L., III, 1974. *Basic Procedures for Soil Sampling and Core Drilling*, Acker Drill Co., Scranton, Pennsylvania, 246 pp.

Adams, T.V. and D.R. Hampton, 1990. Effects of capillarity on DNAPL thickness in wells and adjacent sands, Presented at the International Association of Hydrogeologists Conference on Subsurface Contamination by Immiscible Fluids, Calgary, Alberta, April 18-20.

Adamson, A.W., 1982. *Physical Chemistry of Surfaces*, John Wiley and Sons, Inc., New York, pp. 332-368.

Addison, R.F., 1983. PCB replacements in dielectric fluids, *Environmental Science and Technology*, 17(10):486A-494A, ACS.

Agrelot, J.C., J.J. Malot and M.J. Visser, 1985. Vacuum: Defense system for ground water VOC contamination, *Proceedings of the Fifth National Symposium and Exposition on Aquifer Restoration and Ground Water Monitoring*, National Water Well Association, Columbus, Ohio, pp. 485-494.

Akstinat, M.H., 1981. Surfactants for enhanced oil recovery processes in high salinity systems - Product selection and evaluation, in *Enhanced Oil Recovery*, F.J. Fayers, ed., Elsevier, New York, pp. 43-80.

Alford-Stevens, A.L., 1986. Analyzing PCBs, *Environmental Science and Technology*, 20(12):1194-1199.

Aller, L., T.W. Bennett, G. Hackett, R.J. Petty, J.H. Lehr, H. Sedoris and D.M. Nielsen, 1989. *Handbook of Suggested Practices for the Design and Installation of Ground-Water Monitor Wells*, National Water Well Association, Dublin, Ohio, 398 pp.

Althoff, W.F., R.W. Cleary and P.H. Roux, 1981. Aquifer decontamination for volatile organics: A case history, *Ground Water*, 19(5):495-504.

Amott, E., 1959. Observations relating to the wettability of porous rock, *Trans. AIME*, 216:156-162.

Amyx, J.W., D.M. Bass, Jr. and R.L. Whiting, 1960. *Petroleum Reservoir Engineering*, McGraw-Hill Book Co. New York, 610 pp.

Anderson, D.C., K.W. Brown and J. Green, 1981. Organic leachate effects on the permeability of clay liners, *Proceedings of the National Conference on Management of Uncontrolled Hazardous Waste Sites*, Hazardous Materials Control Research Institute, Silver Spring, Maryland, pp. 223-229.

Anderson, M.R., 1988. The dissolution and transport of dense non-aqueous phase liquids in saturated porous media, Ph.D. Dissertation, Oregon Graduate Center, Beaverton, Oregon.

Anderson, M.R., R.L. Johnson and J.F. Pankow, 1987. The dissolution of residual dense non-aqueous phase liquid (DNAPL) from a saturated porous medium, *Proceedings of the NWWA/API Conference on Petroleum Hydrocarbons and Organic Chemicals in Ground Water: Prevention, Detection, and Restoration*, National Water Well Association-American Petroleum Institute, Houston, Texas, pp. 409-428.

Anderson, M.R., R.L. Johnson and J.F. Pankow, 1992a. Dissolution of dense immiscible solvents into groundwater: Laboratory experiments involving a well-defined residual source, *Ground Water*, 30(2):250-256.

Anderson, M.R., R.L. Johnson and J.F. Pankow, 1992b. Dissolution of dense chlorinated solvents into groundwater: 3. Modeling contaminant plumes from fingers and pools of solvent, *Environmental Science and Technology*, 26(5):901-908.

Anderson, M.R. and J.F. Pankow, 1986. A case study of a chemical spill: Polychlorinated biphenyls (PCBs) 3. PCB sorption and retardation in soil underlying the site, *Water Resources Research*, 22(7):1051-1057.

Anderson, W.G., 1986a. Wettability literature survey--part 1: Rock/oil/brine interactions, and the effects of core handling on wettability; *Journal of Petroleum Technology*, October, pp. 1125-1149.

Anderson, W.G., 1986c. Wettability literature survey--part 3: The effects of wettability on the electrical properties of porous media, *Journal of Petroleum Technology*, December, pp. 1371-1378.

Anderson, W.G., 1987a. Wettability literature survey--part 4: The effects of wettability on capillary pressure, *Journal of Petroleum Technology*, October, pp. 1283-1300.

Anderson, W.G., 1987b. Wettability literature survey--part 5: The effects of wettability on relative permeability, *Journal of Petroleum Technology*, November, pp. 1453-1468.

Anderson, W.G., 1987c. Wettability literature survey--part 6: The effects of wettability on waterflooding, *Journal of Petroleum Technology*, December, pp. 1605-1622.

Annan, A.P., P. Bauman, J.P. Greenhouse and J.D. Redman, 1991. Geophysics and DNAPLs, *Proceedings of the Fifth National Outdoor Action Conference on Aquifer Restoration, Ground Water Monitoring and Geophysical Methods*, Las Vegas, Nevada, NGWA, Dublin, Ohio, pp. 963-977.

API, 1980. *Underground Spill Cleanup Manual*, American Petroleum Institute Publication No. 1628, Washington, D.C., 34 pp.

API, 1989. A guide to the assessment and remediation of underground petroleum releases, American Petroleum Institute Publication No. 1628, 2nd ed., Washington, D.C., 81 pp.

Ardito, C.P. and J.F. Billings, 1990. Alternative remediation strategies: The subsurface volatiliztion and ventilation system, *Proceedings of Petroleum Hydrocarbons and Organic Chemicals in Ground Water: Prevention, Detection, and Restoration*, National Water Well Association-American Petroleum Institute, Houston, Texas, pp. 281-308.

Arthur D. Little, Inc., 1981. The role of capillary pressure in the S-Area landfill, Report to Wald, Harkrader, & Ross, Washington, D.C.

Arthur D. Little, Inc., 1982. Capillary pressure considerations for the S-Area landfill: A critical review of the literature, Report to Wald, Harkrader, and Ross, Washington, D.C.

Arthur D. Little, Inc., 1983. S-Area two phase flow model, Report to Wald, Harkrader, and Ross, Washington, D.C.

ASTM, 1990a. Standard method for penetration test and split-barrel sampling of soils, in *Annual Book of ASTM Standards*, Vol. 04.08, D1586-84, American Society for Testing and Materials, Philadelphia, Pennsylvania.

ASTM, 1990b. Standard practice for the design and installation of ground water monitoring wells in aquifers, American Society for Testing and Materials, D5092-90, Philadelphia, Pennsylvania.

Atwater, J.W., 1984. A case study of a chemical spill: Polychlorinated biphenyls (PCBs) revisited, *Water Resources Research*, 20:317-319.

Austin, G.T., 1984. *Shreve's Chemical Process Industries*, Fifth ed., McGraw-Hill Book Co., New York, 859 pp.

Azbel, D., 1981. *Two-Phase Flows in Chemical Engineering*, Cambridge University Press, Cambridge.

Baechler, F.E. and D.S. MacFarlane, 1990. Sydney tar ponds clean up: Hydrogeological assessment - Coke ovens complex, Presented at the International Association of Hydrogeologists Conference on Subsurface Contamination by Immiscible Fluids, Calgary, Alberta, April 18-20.

Baehr, A.L., 1987. Selective transport of hydrocarbons in the unsaturated zone due to aqueous and vapor phase partitioning, *Water Resources Research*, 23(10):1926-1938.

Baehr, A.L. and M.Y. Corapcioglu, 1987. A compositional multiphase model for groundwater contamination by petroleum products, 2. Numerical solution, *Water Resources Research*, 23(1):201-214.

Baehr, A.L., G.E. Hoag and M.C. Marley, 1989. Removing volatile contaminants from the unsaturated zone by inducing advective air-phase transport, *Journal of Contaminant Hydrology*, 4(1):2-6.

Banerjee, S., 1984. Solubility of organic mixtures in water, *Environmental Science and Technology*, 18(8):587-591.

Barber, C., G.B. Davis, D. Briegel and J.K. Ward, 1990. Factors controlling the concentration of methane and other volatiles in groundwater and soil-gas around a waste site, *Journal of Contaminant Hydrology*, 5(2):155-169.

Barcelona, M.J., J.P. Gibb, J.A. Helfrich and E.E. Garske, 1985. Practical guide for ground-water sampling, USEPA/600/2-85-104, 169 pp.

Barcelona, M.J., J.P. Gibb and R.A. Miller, 1983. A guide to the selection of materials for monitoring well construction and ground-water sampling, Illinois State Water Survey Contract Report 327 to USEPA Robert S. Kerr Environmental Research Laboratory, USEPA Contract CR-809966-01, 78 pp.

Barcelona, M.J., J.F. Keely, W.A. Pettyjohn and A. Wehrmann, 1987. Handbook ground water, USEPA/625/6-87/016, 212 pp.

Bear, J., 1988. *Dynamics of Fluids in Porous Media*, Dover Publishers, New York, 1988. pp. 444-449.

Bear, J., 1979. *Hydraulics of Groundwater*, McGraw-Hill Book Co., New York, 569 pp.

Bedient, P.B., A.C. Rodgers, T.C. Bouvette, M.B. Tomson and T.H. Wang, 1984. Ground-water quality at a creosote waste site, *Ground Water*, 22(3):318-379.

Begor, K.F., M.A. Miller and R.W. Sutch, 1989. Creation of an artificially produced fracture zone to prevent contaminated ground-water migration, *Ground Water*, 27(1):57-65.

Beikirch, M.G., 1991. Experimental evaluation of surfactant flushing for aquifer restoration, M.S. Thesis, Geology Department, University of New York at Buffalo, Buffalo, New York.

Belanger, D.W., A.R. Lotimer and R.B. Whiffen, 1990. The migration of coal tar in a number of hydrogeologic environments, Presented at the International Association of Hydrogeologists Conference on Subsurface Contamination by Immiscible Fluids, Calgary, Alberta, April 18-20.

Benner, F.C. and F.E. Bartell, 1941. The effect of polar impurities upon capillary and surface phenomena in petroleum production, *Drill Prod. Pract.*, pp. 341-348.

Benson, R.C., 1988. Surface and downhole geophysical techniques for hazardous waste site investigation, *Hazardous Waste Control*, 1(2).

Benson, R.C., 1991. Remote sensing and geophysical methods for evaluation of subsurface conditions, in *Practical Handbook of Ground-Water Monitoring*, D.M. Nielsen, ed., Lewis Publishers, Chelsea, Michigan, pp. 143-194.

Benson, R.C. and R.A. Glaccum, 1979. Radar surveys for geotechnical site assessments, in *Geophysical Methods in Geotechnical Engineering*, Specialty Session, ASCE, Atlanta, Georgia, pp. 161-178.

Benson, R.C., R.A. Glaccum and M.R. Noel, 1982. Geophysical techniques for sensing buried wastes and waste migration, USEPA Environmental Monitoring Systems Laboratory, Las Vegas, Nevada, 236 pp.

Benson, R.C. and L. Yuhr, 1987. Assessment and long term monitoring of localized subsidence using ground penetrating radar, *Proceedings of the Second Multidisciplinary Conference on sinkholes and the Environmental Impact of Karst*, Orlando, Florida.

Berg, R.R., 1975. Capillary pressures in stratigraphic traps, *The American Association of Petroleum Geologists Bulletin*, 59(6):939-956.

Bishop, P.K., M.W. Burston, D.N. Lerner and P.R. Eastwood, 1990. Soil gas surveying of chlorinated solvents in relation to groundwater pollution studies, *Quarterly Journal of Engineering Geology*, London, 23:255-265.

Black, W.H., H.R. Smith and F.D. Patton, 1986. Multiple-level ground water monitoring with the MP system, *Proceedings of the NWWA-AGU Conference on Surface and Borehole Geophysical Methods and Ground Water Instrumentation*, National Water Well Association, Worthington, Ohio, pp. 41-61.

Blackwell, R.J., 1981. Miscible displacement: Its status and potential for enhanced oil recovery, in *Enhanced Oil Recovery*, F.J. Fayers, ed., Elsevier, New York, pp. 237-245.

Blake, S.B. and M.M. Gates, 1986. Vacuum enhanced hydrocarbon recovery: a case study, *Proceedings of Petroleum Hydrocarbons and Organic Chemicals in Ground Water: Prevention, Detection, and Restoration*, National Water Well Association-American Petroleum Institute, Houston, Texas.

Blake, S.B., B. Hockman and M. Martin, 1990. Applications of vacuum dewatering techniques to hydrocarbon remediation, *Proceedings of Petroleum Hydrocarbons and Organic Chemicals in Ground Water: Prevention, Detection, and Restoration*, National Water Well Association-American Petroleum Institute, Houston, Texas, pp. 211-226.

Blevins, T.R., J.R. Duerksen and J.W. Ault, 1984. Light-oil steamflooding -- An emerging technology, *Journal of Petroleum Technology*, 36:1115-1122.

BNA, 1992. *BNA's Environmental Due Diligence Guide*, The Bureau of National Affairs, Inc., Washington, D.C.

Boberg, T.C., 1988. *Thermal Methods of Oil Recovery*, Exxon Monograph, John Wiley & Sons, New York, 411 pp.

Breiner, S., 1973. Applications manual for portable magnetometers, GeoMetrics, Sunnyvale, California, 58 pp.

Breit, V.S., E.H. Mayer and J.D. Charmichael, 1981. Caustic flooding in the Wilmington Field, California: Laboratory, modeling, and field results, in *Enhanced Oil Recovery*, F.J. Fayers, ed., Elsevier, New York pp. 223-236.

Brooks, R.H. and A.T. Corey, 1964. Hydraulic properties of porous media, Hydrology Paper no. 3, Colorado State University, Fort Collins, CO, 27 pp.

Brooks, R.H. and A.T. Corey, 1966. Properties of porous media affecting fluid flow, *Journal of the Irrigation and Drainage Division*, ASCE, IR2:61-88.

Brown, H.W., 1951. Capillary pressure investigations, *Trans. AIME*, 192:67-74.

Brown, K.W., J.C. Thomas and J.W. Green, 1984. Permeability of compacted soils to solvents mixtures and petroleum products, *Land Disposal of Hazardous Waste Proceedings of the Tenth Annual Research Symposium*, USEPA/600/9-84-007.

Brusseau, M., 1991. Transport of organic chemicals by gas advection in structured or heteroegeneous porous media: Development of a model and application of column experiments, *Water Resources Research*, 27(12):3189-3199.

Brutsaert, W., 1973. Numerical solution of multiphase well flow, *Proceedings American Society of Civil Engineers Journal of Hydraulics Division*, 99:1981-2001.

Caenn, R., D.B. Burnett and G.V. Chilingarian, 1989. Polymer flooding, in *Enhanced Oil Recovery, II, Processes and Operations*, E.C. Donaldson, G.V. Chilingarian, and T.F. Yen, eds., Elsevier, New York, pp. 157-187.

Cambell, M.D. and J.H. Lehr, 1984. *Water Well Technology*, McGraw-Hill Book Co., New York, 681 pp.

Camp, Dresser and McKee, 1987. Identification and review of multiphase codes for application to UST release detection, USEPA Office of Underground Storage Tanks.

Campbell, T.C., 1981. The role of alkaline chemicals in oil displacement processes, in *Surface Phenomena in Enhanced Oil Recovery*, D.O. Shah, ed., Plenum Press, New York, pp. 293-306.

Carsel, R.F. and R.S. Parrish, 1988. Developing joint probability distributions of soil and water retention characteristic, *Water Resources Research*, 24(5):755-769.

Carslaw, H.S. and J.C. Jaeger, 1959. *Conduction of Heat in Solids*, 2nd Edition, Oxford University Press, 510 pp.

Cartwright, K. and M. McComas, 1968. Geophysical surveys in the vicinity of sanitary landfills in Northeastern Illinois, *Ground Water* 6(1):23-30.

Cary, J.W., J.F. McBride and C.S. Simmons, 1989a. Trichloroethylene residuals in the capillary fringe as affected by air-entry pressure, *Journal of Environmental Quality*, 18:72-77.

Cary, J.W., J.F. McBride and C.S. Simmons, 1991. Assay of organic liquid contents in predominantly water-wet unconsolidated porous media, *Journal of Contaminant Hydrology*, no.8, pp. 135-142.

Cary, J.W., C.S. Simmons and J.F. McBride, 1989b. Predicting oil infiltration and redistribution in unsaturated soils, *Soil Science Society of America Journal*, 53(2):335-342.

Castor, T.P., W.H. Somerton and J.F. Kelly, 1981. Recovery mechanisms of alkaline flooding, in *Surface Phenomena in Enhanced Oil Recovery*, D.O. Shah, ed., Plenum Press, New York, pp. 249-291.

CH2M-Hill, 1989. Evaluation of ground-water extraction remedies, USEPA/540/2-89/054.

Chatzis, I., M.S. Kuntamukklua and N.R. Morrow, 1988. Effect of capillary number on the microstructure of residual oil in strongly water-wet sandstones, *SPE Reservoir Engineering*, August, pp. 902-912.

Chatzis, I., N.R. Morrow and H.T. Lim, 1983. Magnitude and detailed structure of residual oil saturation, *Society of Petroleum Engineers Journal*, April, pp. 311-326.

Chauveteau, G. and A. Zaitoun, 1981. Basic rheological behavior of xanthan polysaccharide solutions in porous media: Effects of pore size and polymer concentration, in *Enhanced Oil Recovery*, F.J. Fayers, ed., Elsevier, New York, pp. 197-212.

Cherry, J.A. and S. Feenstra, 1991. Identification of DNAPL sites: An eleven point approach, draft document in Dense Immiscible Phase Liquid Contaminants in Porous and Fractured Media, short course notes, Waterloo Centre for Ground Water Research, Kitchener, Ontario.

Cherry, J.A., S. Feenstra, B.H. Kueper and D.W. McWhorter, 1990. Status of in situ technologies for cleanup of aquifers contaminated by DNAPLs below the water table, in *International Specialty Conference on How Clean is Clean? Cleanup Criteria for Contaminated Soil and Groundwater*, Air and Waste Management Association, pp. 1-18.

Cherry, J.A. and P.E. Johnson, 1982. A multi-level device for monitoring in fractured rock, *Ground Water*, 2(3):41-44.

Chiang, C.Y., K.R. Loos and R.A. Klopp, 1992. Field determination of geological/chemical properties of an aquifer by cone penetrometry and headspace analysis, *Ground Water*, 30(3):428-436.

Chiou, C.T., R.L. Malcolm, T.I. Brinton and D.E. Kile, 1986. Water solubility enhancement of some organic pollutants and pesticides by dissolved humic and fulvic acids, *Environmental Science and Technology*, 20(5):502-508.

Chouke, R.L., P. Van Meurs and C. Van der Poel, 1959. The instability of slow, immiscible, viscous liquid-liquid displacements in permeable media, *Petrol. Trans. AIME*, 216:188-194.

Christy, T.M. and S.C. Spradlin, 1992. The use of small diameter probing equipment for contaminated site investigation, Geoprobe Systems, Salina, Kansas, 15 pp.

Clark, L., 1988a. *The Field Guide to Water Wells and Boreholes*, Geological Society of London Professional Handbook Series, John Wiley and Sons, New York, 155 pp.

Clark, R.R., 1988b. A new continuous-sampling, wireline system for acquisition of uncontaminated, minimally disturbed soil samples, *Ground Water Monitoring Review*, 8(4):66-72.

Clay, D.R., 1992. Considerations in ground-water remediation at Superfund sites and RCRA facilities -- Update, USEPA memorandum, 13 pp.

Cohen, R.M., A.P. Bryda, S.T. Shaw and C.P. Spalding, 1992. Evaluation of visual methods to detect NAPL in soil and water, *Ground Water Monitoring Review*, 12(4):132-141.

Cohen, R.M., R.R. Rabold, C.R. Faust, J.O. Rumbaugh, III and J.R. Bridge, 1987. Investigation and hydraulic containment of chemical migration: Four landfills in Niagara Falls, *Civil Engineering Practice*, Journal of the Boston Society of Civil Engineers Section/ASCE, 2(1):33-58.

Conestoga-Rovers and Associates, 1978. Phase I pollution abatement plan -- Upper groundwater regime, Love Canal chemical landfill, Niagara Falls, N.Y., Report to the City of Niagara Falls.

Conestoga-Rovers and Associates, 1986. Appendix B, Plans, specifications, and protocols for the subsurface investigation, S-Area/Water Treatment Plant, S-Area Remedial Program, Report prepared for Occidental Chemical Corporation, Niagara Falls, New York, 33 pp.

Conestoga-Rovers and Associates, 1988a. Assessment of the extent of APL/NAPL migration from the S-Area in the Lockport Bedrock, S-Area Remedial Program, Report prepared for Occidental Chemical Corporation, Niagara Falls, New York, 70 pp.

Conestoga-Rovers and Associates, 1988b. Assessment of the geological and hydrogeological characteristics of the overburden below the S-Area and Northern Area, S-Area Remedial Program, Report to Occidental Chemical Corporation, Niagara Falls, New York.

Connor, J.A., C.J. Newell and D.K. Wilson, 1989. Assessment, field testing, and conceptual design for managing dense non-aqueous phase liquids (DNAPL) at a Superfund site, *Proceedings of Petroleum Hydrocarbons and Organic Chemicals in Ground Water: Prevention, Detection, and Restoration*, National Water Well Association-American Petroleum Institute, Houston, Texas, pp. 519-533.

Convery, M.P., 1979. The behavior and movement of petroleum products in unconsolidated surficial deposits, M.S. Thesis, University of Minnesota.

Conway, H.L., E.J. Quinn and T.N. Wasielewski, 1985. Coal tar disposal site investigation Wallingford and Norwalk, Northeast Utilities Service Co., Connecticut.

Corapcioglu, M.Y. and A.L. Baehr, 1987. A compositional multiphase model for groundwater contamination by petroleum products, 1: Theoretical considerations, *Water Resources Research*, 23(1):191-200.

Corey, A.T., 1986. *Mechanics of Immiscible Fluids in Porous Media*, Water Resources Publications, Littleton, Colorado, 255 pp.

Corey, A.T., C.H. Rathjens, J.H. Henderson and M.R.J. Wyllie, 1956. Three-phase relative permeability, *Society of Petroleum Engineering Journal*, 207, pp. 349-351.

Craig, F.F., Jr., 1971. *The Reservoir Engineering Aspects of Waterflooding Monograph*, Vol. 3, SPE of AIME, Henry L. Doherty Series, Dallas, Texas.

Critchlow, H.B., 1977. *Modern Reservoir Engineering -- A Simulation Approach*, Prentice-Hall, Englewood Cliffs, New Jersey, 354 pp.

Crow, W.L., E.R. Anderson and E. Minugh, 1985. Subsurface venting of hydrocarbon vapors from an underground aquifer, American Petroleum Institute, Washington, D.C.

Crow, W.L., E.R. Anderson and E. Minugh, 1987. Subsurface venting of hydrocarbons emanating from hydrocarbon product on groundwater, *Ground Water Monitoring Review*, 7(1):51-57.

Cullinane, M.J., Jr., L.W. Jones and P.G. Malone, 1986. Handbook for stabilization-solidification of hazardous waste, EPA/540/2-86/001, USEPA Hazardous Waste Engineering Research Laboratory, Cincinnati, Ohio.

Dakin, R.A. and A.T. Holmes, 1987. Monitoring, migration, and control of an ethylene dichloride contaminant plume in a gravel aquifer, *Proceedings of Geotechnique in Resource Development*, Canadien Geotechnical Society, Regina, Saskatchewan, pp. 375-387.

Dalton, M.G., B.E. Huntsman and K. Bradbury, 1991. Acquisition and interpretation of water-level data, in *Practical Handbook of Ground-Water Monitoring*, D.M. Nielsen, ed., Lewis Publishers, Chelsea, Michigan, pp. 367-396.

Dames and Moore, 1985. Groundwater quality assessment report for hazardous waste management facility, Utah Power & Light pole treatment yard, Idaho Falls, Idaho.

Davis, H., J.L. Jehn and S. Smith, 1991. Monitoring well drilling, soil sampling, rock coring, and borehole logging, in *Practical Handbook of Ground-Water Monitoring*, D.M. Nielsen, ed., Lewis Publishers, Chelsea, Michigan, pp. 195-238.

Davis, J.O., 1991. Depth zoning and specializing processing methods for electromagnetic geophysical surveys to remote sense hydrocarbon type groundwater contaminants, *Proceedings of the Fifth National Outdoor Action Conference on Aquifer Restoration, Ground Water Monitoring, and Geophysical Methods*, Las Vegas, Nevada, pp. 905-913.

Davis, S.N., D.J. Campbell, H.W. Bentley and T.J. Flynn, 1985. Ground water tracers, Robert S. Kerr Environmental Research Laboratory, Cooperative Agreement CR-910036, Ada, Oklahoma, 200 pp.

Dean, J.A., ed., 1973. *Lange's Handbook of Chemistry*, Eleventh Edition, McGraw-Hill Book Co., New York.

DeHoog, F.R., J.H. Knight and A.N. Stokes, 1982. An improved method for numerical inversion of Laplace transforms, *Siam Journal of Sci. Stat. Comp.*, 3(3):357-366.

Delshad, M. and G.A. Pope, 1989. Comparison of the three-phase oil relative permeability models, *Transport in Porous Media*, 4(1):59-83.

Demond, A.H. and P.V. Roberts, 1987. An examination of relative permeability relations for two-phase flow in porous media, *Water Resources Bulletin*, 23(4):617-628.

dePastrovich, T.L., Y. Baradat, R. Barthel, A. Chiarelli and D.R. Fussell, 1979. Protection of groundwater from oil pollution, CONCAWE, The Hague, 61 pp.

Derks, R., 1990. PCB transformers, *Hazmat World*, 3(2):72.

Dev, H., 1986. Radio frequency enhanced in situ decontamination of soils contaminated with halogenated hydrocarbons, *Proceedings of EPA Conference on Land Disposal, Remedial Action, Incineration and Treatment of Hazardous Wastes*, USEPA, Cincinnati, Ohio.

Dev, H., P. Conderelli, J.E. Bridges, C. Rogers and D. Downey, 1988. In situ radio frequency heating process for decontamination of soils, American Chemical Society Symposium on Solving Hazardous Waste Problems, ACS Symposium Series 7.

Dev, H. and D. Downey, 1989. In situ soil decontamination by radio-frequency heating - Field test, IIT Research Institute, Chicago, Illinois.

Devinny, J.S., L.G. Everett, J.C.S. Lu and R.L. Stollar, 1990. *Subsurface Migration of Hazardous Wastes*, Van Nostrand Reinhold, New York, 387 pp.

Devitt, D.A., R.B. Evans, W.A. Jury, T.H. Starks, B. Eklund and A. Gholson, 1987. Soil gas sensing for detection and mapping of volatile organics, EPA/600/8-87/036, USEPA Environmental Monitoring Systems Laboratory, Las Vegas, Nevada, 281 pp.

Dietrich, J.K. and P.B. Bonder, 1976. Three-phase oil relative permeability problem in reservoir simulation, SPE 6044, Presented at the 51st Annual Meeting of the SPE, New Orleans, October 3-6.

DiGiulio, D.C. and J.S. Cho, 1990. Conducting field tests for evaluation of soil vacuum extraction application, *Proceedings of the Fourth National Outdoor Action Conference on Aquifer Restoration, Ground Water Monitoring, and Geophysical Methods*, National Water Well Association, Las Vegas, Nevada, pp. 587-601.

Donaldson, E.C., G.V. Chiligarian and T.F. Yen, eds., 1989. *Enhanced Oil Recovery, II, Processes and Operations*, Elsevier, New York, 604 pp.

Donaldson, E.C., R.D. Thomas and P.B. Lorenz, 1969. Wettability determination and its effect on recovery efficiency, *Society of Petroleum Engineering Journal*, March, pp. 13-20.

Doscher, T.M. and F. Ghassemi, 1981. Steam drive - The successful enhanced oil recovery technology, in *Enhanced Oil Recovery*, F.J. Fayers, ed., Elsevier, New York, pp. 549-563.

Downey, D.C. and M.G. Elliot, 1990. Performance of selected in situ soil decontamination technologies: An Air Force perspective, *Environmental Progress*, 9(3):169-173.

Dracos, T., 1978. Theoretical considerations and practical implications on the infiltration of hydrocarbons in aquifers, Presented at the International Association of Hydrogeologists International Symposium on Groundwater Pollution by Oil Hydrocarbons, Prague, pp. 127-137.

Driscoll, F.G., 1986. *Groundwater and Wells*, Johnson Division, St. Paul, Minnesota, 1089 pp.

Dunlap, L.E., 1984. Abatement of hydrocarbon vapors in buildings, in *Petroleum Hydrocarbons and Organic Chemicals in Ground Water: Prevention, Detection, and Restoration*, National Water Well Association-American Petroleum Institute, Houston, Texax, pp. 504-518.

E.C. Jordan Co., 1987. Love Canal remedial project task V-C, Implementation of a long-term monitoring program, Prepared for the New York State Department of Environmental Conservation, Albany, New York.

Earth Dimensions, 1979. Soil investigation Love Canal -- southern section, Niagara Falls, New York.

Earth Dimensions, 1980. Litigation soil sampling boring logs, Prepared for the New York State Department of Health, Albany, New York.

Eastcott, L., W.Y. Shiu and D. Mackay, 1988. Environmentally relevant physical-chemical properties of hydrocarbons: A review of data and development of simple correlations, *Oil and Chemical Pollution*, 4:191-216.

Edge, R.W. and K. Cordry, 1989. The Hydropunch(tm): An in situ sampling tool for collecting ground water from unconsolidated sediments, *Ground Water Monitoring Review*, 9(3):177-183.

Edil, T.B., M.M.K. Chang, L.T. Lan and T.V. Riewe, 1992. Sealing characteristics of selected grouts for water wells, *Ground Water*, 30(3):351-361.

Edison Electric Institute, 1984. Handbook of manufactured gas plant sites.

Eganhouse, R.P. and J.A. Calder, 1973. The solubility of medium molecular weight aromatic hydrocarbons and the effects of hydrocarbon co-solutes and salinity, *Geochim. Cosmochim. Acta*, 37.

Ehrlich, G.G., D.F. Goerlitz, E.M. Godsy and M.F. Hult, 1982. Degradation of phenolic contaminants in ground water by anaerobic bacteria: St. Louis Park, Minnesota, *Ground Water*, 20:703-715.

Ellis, W.D., J.R. Payne and G.D. McNabb, 1985. Treatment of contaminated solid with aqueous surfactants, EPA/600/2-85/129, USEPA, Cincinnati, Ohio.

EPRI, 1985. Field measurement methods for hydrogeologic investigations: A critical review of the literature, EPRI Report EA-4301, Electric Power Research Institute, Palo Alto, California.

EPRI, 1989. Techniques to develop data for hydrogeochemical models, Electric Power Research Institute Report EN-6637, Palo Alto, California.

Evans, J.C., H.Y. Fang and I.J. Kugelman, 1985. Organic fluid effects on the permeability of soil-bentonite slurry walls, *Proceedings of the National Conference on Hazardous Waste and Environmental Emergencies*, Hazardous Waste Control Research Institute, Silver Spring, Maryland.

Falta, R.W., I. Javandel, K. Pruess and P.A. Witherspoon, 1989. Density-drive flow of gas in the unsaturated zone due to evaporation of volatile organic chemicals, *Water Resources Research*, 25(10):2159-2169.

Farr, A.M., R.J. Houghtalen and D.B. McWhorter, 1990. Volume estimates of light nonaqueous phase liquids in porous media, *Ground Water*, 28(1):48-56.

Faust, C.R., 1984. Affidavit re S-Area, U.S., N.Y. v. Hooker Chemicals and Plastics Corp. et al., Civil Action No. 79-988, Federal District Court, Buffalo, New York.

Faust, C.R., 1985a. Affidavit re Hyde Park, U.S., N.Y. v. Hooker Chemicals and Plastics Corp. et al., Civil Action No. 79-989, Federal District Court, Buffalo, New York.

Faust, C.R., 1985b. Transport of immiscible fluids within and below the unsaturated zone: A numerical model, *Water Resources Research*, 21(4):587-596.

Faust, C.R., J.H. Guswa and J.W. Mercer, 1989. Simulation of three-dimensional flow of immiscible fluids within and below the unsaturated zone, *Water Resources Research*, 25(12):2449-2464.

Faust, C.R., S.J. Wamback and C.P. Spalding, 1990. Characteristics of the migration of immiscible fluids in glacial deposits and dolomites in Niagara Falls, New York, Presented at the International Association of Hydrogeologists Conference on Subsurface Contamination by Immiscible Fluids, Calgary, Alberta, April 18-20.

Fayers, F.J. and J.P. Matthews, 1984. Evaluation of normalized Stone's methods for estimating three-phase relative permeabilities, *SPE Journal*, 24, pp. 224-232.

Federal Register, July 27, 1990. USEPA proposed rules, Volume 55, No. 145.

Feenstra, S., 1989. A conceptual framework for the evaluation of polychlorinated biphenyl (PCB) in ground water.

Feenstra, S., 1990. Evaluation of multi-component DNAPL sources by monitoring of dissolved-phase concentrations, Presented at the International Association of Hydrogeologists Conference on Subsurface Contamination by Immiscible Fluids, Calgary, Alberta, April 18-20.

Feenstra, S. and J.A. Cherry, 1988. Subsurface contamination by dense non-aqueous phase liquids (DNAPL) chemicals, Paper presented at International Groundwater Symposium, International Association of Hydrogeologists, Halifax, Nova Scotia, May 1-4.

Feenstra, S. and J.A. Cherry, 1990. Groundwater contamination by creosote, Paper presented at the Eleventh Annual Meeting of the Canadien Wood Preserving Association, Toronto, Ontario.

Feenstra, S., D.M. Mackay and J.A. Cherry, 1991. A method for assessing residual NAPL based on organic chemical concentrations in soil samples, *Groundwater Monitoring Review*, 11(2):128-136.

Ferrand, L.A., P.C.D. Milly and G.F. Pinder, 1989. Experimental determination of three-fluid saturation profiles in porous media, *Journal of Contaminant Hydrology*, 4(4):373-395.

Ferry, J.P., P.J. Dougherty, J.B. Moser and R.M. Schuller, 1986. Occurrence and recovery of a DNAPL in a low-yielding bedrock aquifer, *Proceedings of Petroleum Hydrocarbons and Organic Chemicals in Ground Water: Prevention, Detection, and Restoration*, National Water Well Association-American Petroleum Institute, Houston, Texas,, pp. 722-733.

Fiedler, F.R., 1989. An investigation of the relationship between actual and apparent gasoline thickness in a uniform sand aquifer, M.S. Thesis, University of New Hampshire, Durham, New Hampshire.

Fitzpatrick, V.F., C.L. Timmerman and J.L. Buelt, 1986. In situ vitrification - A candidate process for in situ destruction of hazardous waste, *Proceedings of the Seventh Superfund Conference*, Hazardous Materials Control Research Institute, Washington, D.C.

Flumerfelt, R.W., A.B. Catalano and C-H. Tong, 1981. On the coalescence characteristics of low tension oil-water-surfactant systems, in *Surface Phenomena in Enhanced Oil Recovery*, D.O. Shah, ed., Plenum Press, New York, pp. 571-594.

Fountain, J.C., 1991. In-situ extraction of DNAPL by surfactant flushing: Theoretical background and results of a field test, Presented at the USEPA DNAPL Workshop, Dallas, Texas, April 1991.

Franks, B.J., ed., 1987. Movement and fate of creosote waste in ground water near an abandoned wood-preserving plant near Pensacola, Florida, *Proceedings of the Third Technical Meeting U.S. Geological Survey Program on Toxic Waste -- Ground Water Contamination*, USGS Open-File Report 87-109.

Frick, T., ed., 1962. *Petroleum Production Handbook, 2nd ed.*, Society of Petroleum Engineers of AIME, Dallas, Texas.

Fried, J.J., P. Muntzer and L. Zilliox, 1979. Ground-water pollution by transfer of oil hydrocarbons, *Ground Water*, 17(6):586-594.

Fussell, D.R., H. Godjen, P. Hayward, R.H. Lilie, A. Marco and C. Panisi, 1981. Revised inland oil spill clean-up manual, CONCAWE Report No. 7/81, Den Haag, 150 pp.

Gas Research Institute, 1987. Management of manufactured gas plant sites, GRI-87/0260.

GeoTrans, 1988. Love Canal NAPL investigation report, Prepared for the New York State Department of Law.

GeoTrans, 1989. Groundwater monitoring manual for the electric utility industry, Edison Electric Institute, Washington, D.C.

Ghiorse, W.C., K. Malachowsky, E.L. Madsen and J.L. Sinclair, 1990. Microbial degradation of coal-tar derived organic compounds at a town-gas site, *Proceedings: Environmental Research Conference on Groundwater Quality and Waste Disposal*, EPRI Report EN-6749, 31 pp.

Gierke, J.S. N.J. Hutzler and D.B. McKenzie, 1990. Experimental and model studies of hte mechanisms influencing vapor extraction performance, *Proceedings of Petroleum Hydrocarbons and Organic Chemicals in Ground Water: Prevention, Detection, and Restoration*, National Water Well Association-American Petroleum Institute, Houston, Texas, pp. 325-338.

Gilkeson, R.H., P.C. Heigold and D.E. Layman, 1986. Practical application of theoretical models to magnetometer surveys on hazardous waste disposal sites -- A case history, *Ground Water Monitoring Review*, 6(1):54-61.

Glaccum, R., M. Noel, R. Evans and L. McMillion, 1983. Correlation of geophysical and organic vapor analyzer data over a conductive plume containing volatile organics, *Proceedings of the 3rd National Symposium on Aquifer Restoration and Ground Water Monitoring*, National Water Well Association, Dublin, Ohio, pp. 421-427.

Goerlitz, D.F., D.E. Troutman, E.M. Godsy and B.J. Franks, 1985. Migration of wood-preserving chemicals in contaminated groundwater in a sand aquifer at Pensacola, Florida, *Environmental Science and Technology*, 19:955-961.

Gogarty, W.B., 1983. Enhanced oil recovery by the use of chemicals, *Journal of Petroleum Technology*, 35:1581-1590.

Gould, R.F., ed, 1964. *Contact Angle Wettability and Adhesion*, Advances in Chemistry Series, American Chemical Society.

Goyal, K.L. and S. Kumar, 1989. Steamflooding for enhanced oil recovery, in *Enhanced Oil Recovery, II, Processes and Operations*, E.C. Donaldson, G.V. Chlingarian, and T.F. Yen, eds., Elsevier, New York, pp. 317-349.

Grady, S.J. and F.P. Haeni, 1984. Application of electromagnetic techniques in determining distribution and extent of ground water contamination at a sanitary landfill, Farmington, Connecticut, *Proceedings of the NWWA/EPA Conference on Surface and Borehole Geophysical Methods in Ground Water Investigations*, February 7-9, San Antonio, Texas, pp. 338-367.

Greenhouse, J.P. and D.D. Slaine, 1983. The use of reconnaissance electromagnetic methods to map contaminant migration, *Ground Water Monitoring Review*, 3(2):47-59.

Greenhouse, J.P. and M. Monier-Williams, 1985. Geophysical monitoring of ground water contamination ground waste disposal sites, *Ground Water Monitoring Review* 5(4):63-69.

Gretsky, P., R. Barbour and G.S. Asimenios, 1990. Geophysics, pit surveys reduce uncertainty, *Pollution Engineering*, 22(6):102-108.

Griffin, R.A. and E.S.K. Chian, 1980. Attenuation of water-soluble polychlorinated biphenyls by earth materials, EPA-600/2-80-87, USEPA, 92 pp.

Griffith, D.H. and R.F. King, 1969. *Applied Geophysics for Engineers and Geologists*, Pergamon Press.

Groves, F.R., Jr., 1988. Effect of cosolvents on the solubility of hydrocarbons in water, *Environmental Science and Technology*, 22(3):282-286.

Guarnaccia, J.F., P.T. Imhoff, B.C. Missildine, M. Oostrom, M.A. Celia, J.H. Dane, P.R. Jaffe and G.F. Pinder, 1992. Multiphase chemical transport in porous media, USEPA Environmental Research Brief, EPA/600/S-92/002, Robert S. Kerr Environmental Research Laboratory, Ada, Oklahoma, 19 pp.

Guswa, J.H., 1985. Application of multi-phase flow theory at a chemical waste landfill, Niagara Falls, New York, *Proceedings of the Second International Conference on Groundwater Quality Research*, National Center for Ground Water Research, Oklahoma State University, Stillwater, Oklahoma, pp. 108-111.

Guyod, H., 1972. Application of borehole geophysics to the investigation and development of groundwater resources, *Water Resources Bulletin*, 8(1):161-174.

Hackett, G., 1987. Drilling and constructing monitoring wells with hollow-stem augers, part I: Drilling considerations, *Ground Water Monitoring Review*, 7(4):51-62.

Hackett, G., 1988. Drilling and constructing monitoring wells with hollow-stem augers, part II: Monitoring well installation, *Ground Water Monitoring Review*, 8(1):60-68.

Haeni, P., 1986. Application of seismic-refraction techniques to hydrologic studies, USGS Open File Report No. 84-746, 144 pp.

Haley, J.L., B. Hanson, and J.P.E. des Rosiers, 1990. Remedial actions for Superfund sites with PCB contamination, *Proceedings of HMCRI's 11th Annual National Conference and Exhibition Superfund '90*, Hazardous Materials Control Research Institute, Silver Spring, Maryland, pp. 575-579.

Halliburton Co., 1981. *Practical Subsurface. Evaluation*, Houston, Texas, pp. 56-58.

Hampton, D.R., 1988. Laboratory investigation of the relationship between actual and apparent product thickness in sands, extended abstract, American Association of Petroleum Geologists Symposium on Environmental Concerns in the Petroleum Industry, Palm Springs, California, May 10, 1988.

Hampton, D.R. and P.D.G. Miller, 1988. Laboratory investigation of the relationship between actual and apparent product thickness in sands, *Proceedings of the Conference on Petroleum Hydrocarbons and Organic Chemicals in Ground Water: Prevention, Detection and Restoration*, National Water Well Association-American Petroleum Institute, Houston, Texax, pp. 157-181.

Hedgcoxe, H.R. and W.S. Stevens, 1991. Hydraulic control of vertical DNAPL migration, *Proceedings of the Petroleum Hydrocarbons and Organic Chemicals in Ground Water: Prevention, Detection, and Restoration*, National Water Well Association-American Petroleum Institute, Houston, Texas, pp. 327-338.

Hendry, M.J., 1988. Do isotopes have a place in ground-water studies, *Ground Water*, 26(4):410-415.

Herman, G., 1989. Analysis of Love Canal NAPLs, Occidental Chemical Corporation report.

Herrling, B. and W. Buermann, 1990. A new method for in-situ remediation of volatile contaminants in groundwater - Numerical simulation of the flow regime, *Proceedings of the VIII International Conference on Computational Methods in Water Resources*, Venice, July 11-15.

Herrling, B., W. Buermann and J. Stamm, 1990. In-situ remediation of volatile contaminants in groundwater by a new system of 'Underpressure-Vaporizer-Wells', Presented at the International Association of Hydrogeologists Conference on Subsurface Contamination by Immiscible Fluids, Calgary, Alberta, April 18-20.

Herzog, B.L., J.D. Pennino and G.L. Nielsen, 1991. Groundwater sampling, in *Practical Handbook of Ground-Water Monitoring*, D.M. Nielsen, ed., Lewis Publishers, Chelsea, Michigan, pp. 449-500.

Hess, K.M., W.N. Herkelrath and H.I. Essaid, 1992. Determination of subsurface fluid contents at a crude-oil spill site, *Journal of Contaminant Hydrology*, 10(1):75-96.

Hesselink, F.Th. and M.J. Faber, 1981. Polymer-surfactant interaction and its effect on the mobilization of capillary-trapped oil, in *Surface Phenomena in Enhanced Oil Recovery*, D.O. Shah, ed., Plenum Press, New York, pp. 861-869.

HEW, 1972. PCBs and the environment, NTIS publication COM-72-10419, Springfield, Virginia.

Hinchee, R.E., 1989. Enhanced biodegradation through soil venting, *Proceedings of the Workshop on Soil Vacuum Extraction*, Robert S. Kerr Environmental Research Laboratory, Ada, Oklahoma, April 27-28.

Hinchee, R.E., D.C. Downey, R.R. Dupont, P. Aggarwal and R.N. Miller, 1990. Enhancing biodegradation of petroleum hydrocabon through soil venting, *Journal of Hazardous Materials* (accepted).

Hinchee, R.E. and H.J. Reisinger, 1987. A practical application of multiphase transport theory to ground water contamination problems, *Ground Water Monitoring Review*, 7(4):84-92.

Hoag, G.E. and M.C. Marley, 1986. Gasoline residual saturation in unsaturated uniform aquifer materials, *Journal of Environmental Engineering*, ASCE, 112(3):586-604.

Hochmuth, D.P. and D.K. Sunada, 1985. Groundwater model of two-phase immiscible flow in coarse material, *Ground Water*, 23(5):617-626.

Hoekstra, P. and B. Hoekstra, 1991. Geophysics applied to environmental, engineering, and ground water investigations, short course notes, Blackhawk Geosciences, Inc., Bowie, Maryland.

Hoffmann, B, 1969. Uber die ausbreitung geloster kohlenwasserstoffe im grundwasserleiter, Mitteilungen aus dem Institut fur Waserwirtschaft und Landwirtschaftlichen Wasserbau der Tech Hochschule, Hannover, 16.

Hoffmann, B., 1970. Dispersion of soluble hydrocarbons in groundwater stream, *Adv. Water Poll Res.*, Pergamon, Oxford, England, 2HA-7b, pp. 1-8.

Holmes, D.B. and K.W. Cambell, 1990. Contaminant stratification at a deeply penetrating, multiple component DNAPL site, *Proceedings of the 11th Annual National Conference and Exhibition Superfund '90*, Hazardous Materials Control Research Institute, Washington, D.C., pp. 492-497.

Holzer, T.L., 1976. Application of groundwater flow theory to a subsurface oil spill, *Ground Water*, 14(3):138-145.

Homsy, G.M., 1987. Viscous fingering in porous media, *Annual Review of Fluid Mechanics*, (19):271-311.

Honarpour, M., L. Koederitz and A.H. Harvey, 1986. *Relative Permeability of Petroleum Reservoirs*, CRC Press, Inc., Boca Raton, FL, 143 pp.

Houthoofd, J.M., J.H. McCready and M.H. Roulier, 1991. Soil heating technologies for in situ tratment: A review, in *Remedial Action, Treatment, and Disposal of Hazardous Waste, Proceedings of the Seventeenth Annual RREL Hazardous Waste Research Symposium*, EPA/600/9-91/002, USEPA, pp. 190-203.

Howard, P.H., R.S. Boethling, W.F. Jarvis, W.M. Meylan and E.M. Michalenko, 1991. *Handbook of Environmental Degradation Rates*, Lewis Publishers, Chelsea, Michigan, 725 pp.

Hubbert, M.K., 1953. Entrapment of petroleum under hydrostatic conditions, *The Bulletin of the American Association of Petroleum Geologists*, 37(8):1954-2026.

Hughes, B.M., R.W. Gillham and C.A. Mendoza, 1990a. Transport of trichloroethylene vapors in the unsaturated zone: A field experiment, Presented at the International Association of Hydrogeologists Conference on Subsurface Contamination by Immiscible Fluids, Calgary, Alberta, April 18-20.

Hughes, B.M., R.D. McClellan and R.W. Gillham, 1990b. Application of soil-gas sampling technology to studies of trichloroethylene vapour transport in the unsaturated zone, Waterloo Centre for Groundwater Research.

Huling, S.G. and J.W. Weaver, 1991. Dense nonaqueous phase liquids, USEPA Groundwater Issue Paper, EPA/540/4-91, 21 pp.

Hult, M.F. and M.E. Schoenberg, 1984. Preliminary evaluation of ground-water contamination by coal-tar derivatives, St. Louis Park area, Minnesota, USGS Water-Supply Paper 2211, 53 pp.

Hunt, J.R., N. Sitar and K.S. Udell, 1988a. Nonaqueous phase liquid transport and cleanup, 1: Analysis of mechanisms, *Water Resources Research*, 24(8):1247-1258.

Hunt, J.R., N. Sitar and K.S. Udell, 1988b. Nonaqueous phase liquid transport and cleanup, 2: Experimental studies, *Water Resources Research*, 24(8):1259-1269.

Hunter, J.A., R.A. Burns, R.L. Good, H.A. MacAulay and R.M. Cagne, 1982. Optimum field techniques for bedrock reflection mapping with the multichannel engineering seismograph, in Current research, Part B, Geological Survey of Canada, Paper 82-1b, pp. 125-129.

Hutzinger, O., S. Safe and V. Zitko, 1974. *The Chemistry of PCBs*, CRC Press, Cleveland, Ohio.

Hutzler, N.F., B.E. Murphy and J.S. Gierke, 1989. State of technology review: Soil vapor extraction systems, Cooperative Agreement CR-814319-01-1, Hazardous Waste Engineering Research Laboratory, USEPA, Cincinnati, Ohio, 36 pp.

Interagency Task Force, 1979. Draft report on hazardous waste disposal in Erie and Niagara Counties, New York.

Jackson, R.E. and R.J. Patterson, 1989. A remedial investigation of an organically polluted outwash aquifer, *Ground Water Monitoring Review*, 9(3):119-125.

Jackson, R.E., R.J. Patterson, B.W. Graham, J. Bahr, D. Belanger, J. Lockwood and M.W. Priddle, 1985. Contaminant hydrogeology of toxic organic chemicals at a disposal site, Gloucester, Ontario, 1. Chemical concepts and site assessment, IWD Scientific Series No. 141, Environment Canada, Ottawa, Ontario, 114 pp.

Jackson, R.E., M.W. Priddle and S. Lesage, 1990. Transport and fate of CFC-113 in ground water, *Proceedings of Petroleum Hydrocarbons and Organic Chemicals in Ground Water: Prevention, Detection, and Restoration*, National Water Well Association-American Petroleum Institute, Houston, Texas, pp. 129-142.

Janssen-Van Rosmalen, R. and F.Th. Hesselink, 1981. Hot caustic flooding, in *Enhanced Oil Recovery*, F.J. Fayers, ed., Elsevier, New York, pp. 573-586.

JBF Scientific Corporation, 1981. The interaction of S-Area soils and liquids: Review and supplementary laboratory studies, Report submitted to USEPA.

Jhaveri, V. and A.J. Mazzacca, 1983. Bio-reclamation of ground and groundwater, Presented at the National Conference on Management of Uncontrolled Hazardous Waste Sites, Washington, D.C., October 31-November 2.

Johnson, P.C., M.W. Kemblowski and J.D. Colthart, 1988. Practical screening models for soil venting applications, *Proceedings of Petroleum Hydrocarbons and Organic Chemicals in Ground Water: Prevention, Detection, and Restoration*, National Water Well Association-American Petroleum Institute, Houston, Texas, pp. 521-546.

Johnson, P.C., M.W. Kemblowski and J.D. Colthart, 1990a. Quantitative analysis for the cleanup of hydrocarbon contaminated soils by in-situ soil venting, *Ground Water*, 28(3):403-412.

Johnson, P.C., C.C. Stanely, M.W. Kemblowski, D.L. Byers and J.D. Colthart, 1990b. A practical approach to the design, operation, and monitoring of in situ soil-venting systems, *Ground Water Monitoring Review*, 10:159-178.

Johnson, R.L., 1991. The dissolution of dense immiscible solvents into groundwater: Implications for site characterization and remediation, *Groundwater Quality and Analysis at Hazardous Waste Sites (in press)*, S. Lesage and R.E. Jackson, eds., 24 pp.

Johnson, R.L. and F.D. Guffey, 1990. Contained recovery of oily wastes (CROW), Draft Final Report, USEPA, Cincinnati, Ohio, 97 pp.

Johnson, R.L. and J.F. Pankow, 1992. Dissolution of dense immiscible solvents in groundwater: 2.Dissolution from pools of solvent and implications for the remediation of solvent-contaminated sites, *Environmental Science & Technology*, 26(5)896-901, ACS.

J.R. Kolmer and Associates, 1990. Love Canal sampling and analysis report, Prepared for Piper and Marbury, Washington, D.C.

Jury, W.A., D. Russo, G. Streile and H. El Abd, 1990. Evaluation of volatilization by organic chemicals residing below the soil surface, *Water Resources Research*, 26(1):13-20.

Keely, J., 1989. Performance evaluation of pump-and-treat remediations, USEPA/540/4-89-005, Robert S. Kerr Environmental Research Laboratory, Ada, Oklahoma.

Kelley, W.E., 1976. Geoelectric sounding for delineating ground-water contamination, *Ground Water*, 14(1):6-10.

Kemblowski, M.W. and C.Y. Chiang, 1990. Hydrocarbon thickness fluctuations in monitoring wells, *Ground Water*, 28(2):244-252.

Kenaga, E.E. and C.A.I. Goring, 1980. Relationship between water solubility, soil sorption, octanol-water partitioning, and concentration of chemicals in biota, *Aquatic Toxicology*, ASTM STP 707, J.G. Eaton, P.R. Parrish and A.C. Hendricks, eds., American Society for Testing and Materials, Philadelphia, Pennsylvania, pp. 78-115.

Kerfoot, H.B., 1987. Soil-gas measurement for detection of groundwater contamination by volatile organic compounds, *Environmental Science and Technology*, 21(10):1022-1024, ACS.

Kerfoot, H.B., 1988. Is soil-gas analysis an effective means of tracking contaminant plumes in ground water? What are the limitations of the technology currently employed?, *Ground Water Monitoring Review*, 8(2):54-57.

Kerfoot, H.B. and L.J. Barrows, 1987. Soil-gas measurement for detection of subsurface organic contamination, Lockheed Co., USEPA Report, Contract No. 68-03-3245.

Keys, W.S., 1988. Borehole geophysics applied to ground-water investigations, USGS Open-File Report 87-539, 303 pp.

Keys, W.S. and L.M. MacCary, 1976. Application of borehole geophysics to water-resources investigations, Techniques of Water-Resources Investigations of the United States Geological Survey, Chapter E1.

Koerner, R.M., A.E. Lord, Jr. and J.J. Bowders, 1981. Utilitization and assessment of a pulsed RF system to monitor subsurface liquids, *Proceedings of the National Conference on Management of Uncontrolled Hazardous Waste Sites*, Hazardous Materials Control Research Institute, Silver Spring, Maryland, pp. 165-170.

Konstantinova-Shlezinger, M.A., ed. 1961. *Fluorimetric Analysis*, translated from Russian by Israel Program for Scientific Translations Ltd. in 1965, Jerusalem, 376 pp.

Korte, N.E. and P.M. Kearl, 1991. The utility of multiple-completion monitoring wells for describing a solvent plume, *Ground Water Monitoring Review*, 11(2):153-156.

Kraemer, C.A., J.A. Shultz and J.W. Ashley, 1991. Monitoring well post-installation considerations, in *Practical Handbook of Ground-Water Monitoring*, D.M. Nielsen, ed., Lewis Publishers, Chelsea, Michigan, pp. 333-366.

Kueper, B.H., W. Abbot and G. Farquhar, 1989. Experimental observations of multiphase flow in heterogeneous porous media, *Journal of Contaminant Hydrology*, 5:83-95.

Kueper, B.H. and E.O. Frind, 1988. An overview of immiscible fingering in porous media, *Journal of Contaminant Hydrology*, 2, pp. 95-110.

Kueper, B.H. and E.O. Frind, 1991a. Two-phase flow in heterogeneous porous media, 1. Model development, *Water Resources Research*, 27(6):1049-1058.

Kueper, B.H. and E.O. Frind, 1991b. Two-phase flow in heterogeneous porous media, 2. Model application, *Water Resources Research*, 27(6):1059-1070.

Kueper, B.H., C.S. Haase and H.L. King, 1991. Consideration of DNAPL in the operation and monitoring of waste disposal ponds constructed in fractured rock and clay, *Proceedings of First Canadien Conference on Environmental Geotechnics*, Canadien Geotechnical Society, Montreal, Quebec.

Kueper, B.H. and D.B. McWhorter, 1991. The behavior of dense, nonaqueous phase liquids in fractured clay and rock, *Ground Water*, 29(5):716-728.

Kuhn, E.P., P.J. Colberg, J.L. Schnoor, O. Wanner, A.J.B. Zehnder and R.P. Schwarzenbach, 1985. Microbial transformation of substituted benzenes during infiltration of river water to groundwater: Laboratory column studies, *Environmental Science and Technology*, 19:961-968.

Kumar, S., T.F. Yen, G.V. Chilingarian and E.C. Donaldson, 1989. Alkaline flooding, in *Enhanced Oil Recovery, II, Processes and Operations*, E.C. Donaldson, G.V. Chilingarian and T.F. Yen, eds., Elsevier, pp. 219-254.

Kuppusamy, T., J. Sheng, J.C. Parker and R.J. Lenhard, 1987. Finite-element analysis of multiphase immiscible flow through soils, *Water Resources Research*, 23(4):625-632.

Kurt, C.E. and R.C. Johnson, Jr., 1982. Permeability of grout seals surrounding thermoplastic well casing, *Ground Water*, 20(4):415-419.

Labaste, A. and L. Vio, 1981. The Chateaurenard (France) polymer flood field test, in *Enhanced Oil Recovery*, F.J. Fayers, ed., Elsevier, New York, pp. 213-222.

Ladwig, K.J., 1983. Electromagnetic induction methods for monitoring acid mine drainage, *Ground Water Monitoring Review*, 3(1):46-51.

Lankston, R.W. and M.M. Lankston, 1983. An introduction to the utilization of the shallow or engineering seismic reflection method, Geo-Compu-Graph, Inc.

Lappalla, E.G. and G.M. Thompson, 1983. Detection of groundwater contamination by shallow soil gas sampling in the vadose zone, *Proceedings of the Characterization and Monitoring of the Vadose Zone Conference*, National Water Well Association, Las Vegas, Nevada, pp. 659-679.

Lattman, L.H., 1958. Technique of mapping geologic fracture traces and lineaments on aerial photographs, *Photogrammetric Engineering*, 84:568-576.

Lattman, L.H. and R.R. Parizek, 1964. Relationship between fracture traces and the occurrence of ground-water in carbonate rocks, *Journal of Hydrology*, 2:73-91.

Lavigne, D., 1990. Accurate, on-site analysis of PCBs in soil - A low cost approach, *Proceedings of HMCRI's 11th Annual National Conference and Exhibition Superfund '90*, Hazardous Materials Control Research Institute, Washington, D.C., pp. 273-276.

Leach, R.O., O.R. Wagner, H.W. Wood and C.F. Harpke, 1962. A laboratory and field study of wettability adjustment in waterflooding, *Journal of Petroleum Technology*, 44:206.

Lee, M.D., J.M. Thomas, R.C. Borden, P.B. Bedient, J.T. Wilson and C.H. Ward, 1988. Biorestoration of aquifers contaminated with organic compounds, *CRC Critical Reviews in Environmental Control*, 18(1):29-89.

Lee, M.D. and C.H. Ward, 1984. Reclamation of contaminated aquifers: Biological techniques, *Proceedings of Hazardous Material Spills Conference (April 9-12)*, Nashville, Tennessee, pp. 98-103.

Leinonen, P.J. and D. Mackay, 1973. The multicomponent solubility of hydrocarbons in water, *Canadien Journal of Chemical Engineering*, 51:230-233.

Lenhard, R.J., 1992. Measurement and modeling of three-phase saturation-pressure hysteresis, *Journal of Contaminant Hydrology*, 9(1992):243-269.

Lenhard, R.L. and J.C. Parker, 1987a. A model for hysteretic constitutive relations governing multiphase flow: 2. Permeability-saturation relations, *Water Resources Research*, 23(12):2197-2206.

Lenhard, R.J. and J.C. Parker, 1987b. Measurement and prediction of saturation-pressure relationships in three phase porous media systems, *Journal of Contaminant Hydrology*, 1(1987):407-424.

Lenhard, R.J. and J.C. Parker, 1990. Estimation of free hydrocarbon volume from fluid levels in monitoring wells, *Ground Water*, 28(1):57-67.

Lesage, S., R.E. Jackson, M.W. Priddle and P.G. Riemann, 1990. Occurrence and fate of organic solvent residues in anoxic groundwater at the Gloucester Landfill, Canada, *Environmental Science and Technology*, 24(4):559-566.

Leuschner, A.P. and L.A. Johnson, Jr., 1990. In situ physical and biological treatment of coal tar contaminated soil, *Proceedings of Petroleum Hydrocarbons and Organic Chemicals in Ground Water: Prevention, Detection, and Restoration*, National Water Well Association-American Petroleum Institute, Houston, Texas, pp. 427-441.

Leverett, M.C., 1941. Capillary behavior in porous solids, *Trans. AIME*, Petroleum Engineering Division, 142, pp. 152-169.

Lin, C., G.F. Pinder and E.F. Wood, 1982. Water resources program report 83-WR-2, October, Water Resources Program, Princeton University, Princeton, New Jersey.

Lin, F.J., G.J. Besserer and M.J. Pitts, 1987. Laboratory evaluation of crosslinked polymer and alkaline-polymer-surfactant flood, *Journal of Canadien Petroleum Technology*, 26:54-65.

Lindorff, D.E. and K. Cartwright, 1977. Ground-water contamination: Problems and remedial actions, Environmental Geology Notes No. 81, Illinois State Geological Survey, Urbana, Illinois, 58 pp..

Litherland, S.T. and D.W. Anderson, 1990. The trouble with DNAPLs, *Proceedings of HMCRI's 11th Annual National Conference and Exhibition Superfund '90*, Hazardous Materials Control Research Institute, Washington, D.C., pp. 565-574.

Littman, W., 1988. *Polymer Flooding*, Elsevier, New York, 212 pp.

Lokke, H., 1984. Leaching of ethylene glycol and ethanol in subsoils, *Water, Air and Soil Pollution*, 22:373-387.

Lord, A.E., Jr., D.E. Hullings, R.M. Koerner and J.E. Brugger, 1989. Laboratory studies of vacuum-assisted steam stripping of organic contaminants from soil, *Proceedings of the Fifteenth Annual Research Symposium on Land Disposal, Remedial Action, Incineration and Treatment of Hazardous Waste*, USEPA/600/9-90-006.

Lord, A.E., Jr., R.M. Koerner, V.P. Murphy and J.E. Brugger, 1987. In-situ, vacuum-assisted steam stripping of contaminants from soil, *Proceedings of Superfund '87, 8th National Conference on Management of Uncontrolled Hazardous Waste Sites*, Hazardous Materials Control Research Institute, Silver Spring, Maryland, pp. 390-395.

Lord, A.E., Jr., R.M. Koerner, V.P. Murphy and J.E. Brugger, 1988. Laboratory studies of vacuum-assisted steam stripping of organic contaminants from soil, *Proceedings of the Fourteenth Annual Research Symposium on Land Disposal, Remedial Action, Incineration and Treatment of Hazardous Wastes*, USEPA/600/9-88/021.

Lord, A.E., Jr., L.J. Sansone and R.M. Koerner, 1991. Vacuum-assisted steam stripping to remove pollutants from contaminated soil - A laboratory study, in *Remedial Action, Treatment, and Disposal of Hazardous Waste, Proceedings of the Seventeenth Annual RREL Hazardous Waste Research Symposium*, USEPA/600/9-91/002, pp. 329-352

Lucius, J.E., G.R. Olhoeft, P.L. Hill and S.K. Duke, 1990. Properties and hazards of 108 selected substances, USGS Open-File Report 90-408, 559 pp.

Luckner, C.A., M.T. van Genuchten and D.R. Nielsen, 1989. A consistent set of parametric models for two-phase flow of immiscible fluids in the subsurface, *Water Resources Research*, 25(10):2187-2193.

Lurk, P.W., S.S. Copper, P.G. Malone and S.H. Lieberman, 1990. Development of innovative penetrometer systems for the detection and delineation of contaminated groundwater and soil, *Proceedings of Superfund 1990*, Hazardous Materials Control Research Institute, Silver Spring, Maryland, pp. 297-299.

Lyman, W.J., W.F. Reehl and D.H. Rosenblatt, 1982. *Handbook of Chemical Property Estimation Methods, Environmental Behavior of Organic Compounds*, McGraw-Hill Book Co., New York.

Mackay, D.M., M.B. Bloes and K.M. Rathfelder, 1990. Laboratory studies of vapor extraction for remediation of contaminated soil, Presented at the International Association of Hydrogeologists Conference on Subsurface Contamination by Immiscible Fluids, Calgary, Alberta, April 18-20.

Mackay, D.M. and J.A. Cherry, 1989. Groundwater contamination: Pump-and-treat remediation, *Environmental Science and Technology*, 23(6):620-636, ACS.

Mackay, D.M., P.V. Roberts and J.A. Cherry, 1985. Transport of organic contaminants in groundwater, *Environmental Science and Technology*, 19(5):384-392.

Mackay, D., W.Y. Shiu, A. Maijanen and S. Feenstra, 1991. Dissolution of non-aqueous phase liquids in groundwater, *Journal of Contaminant Hydrology*, 8(1):23-42.

Mandl, G. and C.W. Volek, 1969. Heat and mass transport in steam-drive processes, *Society of Petroleum Engineers Journal*, 9(1):59-79.

Manji, K.H. and B.W. Stasiuk, 1988. Design considerations for Dome's David alkali/polymer flood, *Journal of Canadien Petroleum Technology*, May-June, pp. 49-54.

Marburg Associates and W.P. Parkin, 1991. *Site Auditing: Environmental Assessment of Property*, Specialty Technical Publishers, Inc., Vancouver, British Columbia.

Marley, M.C. and G.E. Hoag, 1984. Induced soil venting for recovery/restoration of gasoline hydrocarbons in the vadose zone, *Proceedings of Petroleum Hydrocarbons and Organic Chemicals in Ground Water*, National Water Well Association-American Petroleum Institute, Houston, Texas, pp. 473-503.

Marrin, 1988. Soil-gas sampling and misinterpretation, *Ground Water Monitoring Review*, 7(2):51-54.

Marrin, D.L. and H.B. Kerfoot, 1988. Soil gas surveying techniques, *Environmental Science and Technology*, 22(7):740-745, ACS.

Marrin, D.L. and G.M. Thompson, 1984. Remote detection of volatile organic contaminants in ground water via shallow soil gas sampling, *Proceedings of Petroleum Hydrocarbons and Organic Chemicals in Ground Water: Prevention, Detection, and Restoration*, National Water Well Association-American Petroleum Institute, Houston, Texas, pp. 172-187.

Marrin, D.L. and G.M. Thompson, 1987. Gaseous behavior of TCE overlying a contaminated aquifer, *Ground Water*, 25(1):21-27.

Massmann, J.W., 1989. Applying groundwater flow models in vapor extraction system design, *Journal of Environmental Engineering*, 115(1):129-149.

Mattraw, H.C. and B.J. Franks, eds., 1984. Movement and fate of creosote waste in groundwater, Pensacola, Florida, USGS Open-File Report 84-466, 93 pp.

Mayer, E.H., R.L. Berg, J.D. Charmichael and R.M. Weinbrandt, 1983. Alkaline injection for enhanced oil recovery -- A status report, *Journal of Petroleum Technology*, 35(1):209-221.

McCaffery, F.G. and J.P. Batycky, 1983. Flow of immiscible liquids through porous media, in *Handbook of*

Fluids in Motion, N.P. Cheremisinoff and R. Gupta, eds., pp. 1027-1048.

McClellan, R.D. and R.W. Gillham, 1990. Vacuum extraction of trichloroethene from the vadose zone, Presented at the International Association of Hydrogeologists Conference on Subsurface Contamination by Immiscible Fluids, Calgary, Alberta, April 18-20.

McElwee, C.D., J.J. Butler, Jr. and J.M. Healey, 1991. A new sampling system for obtaining relatively undisturbed samples of unconsolidated coarse sand and gravel, *Ground Water Monitoring Review*, 11(3):182-191.

McGinnis, G.D., 1989. Overview of the wood-preserving industry, *Proceedings Technical Assistance to USEPA Region IX: Forum on Remediation of Wood Preserving Sites*, San Francisco, California.

McGinnis, G.D., H. Borazjani, D.F. Pope, D.A. Strobel and L.K. McFarland, 1991. On-site treatment of creosote and pentachlorophenol sludges and contaminated soil, USEPA/600/2-91/019, 241 pp.

McIelwain, T.A., D.W. Jackman and P. Beukema, 1989. Characterization and remedial assessment of DNAPL PCB oil in fractured bedrock: A case study of the Smithville, Ontario site.

McNeill, J.D., 1980. Electromagnetic resistivity mapping of contaminant plumes, *Proceedings of National Conference on Management of Uncontrolled Hazardous Waste Sites*, Hazardous Materials Control Research Institute, Silver Spring, Maryland, pp. 1-6.

McWhorter, D.B., 1991. The flow of DNAPL to wells, drains, and sumps, Presented at the USEPA DNAPL Workshop, Dallas, Texas, April 1991.

Mehdizadeh, A., G.L. Langnes, J.O. Robertson, Jr., T.F. Yen, E.C. Donaldson and G.V. Chilingarian, 1989. Miscible flooding, in *Enhanced Oil Recovery, II, Processes and Operations*, E.C. Donaldson, G.V. Chilingarian and T.F Yen, eds., Elsevier, New York, pp. 107-128.

Meiser and Earl, 1982. Use of fracture traces in water well location: A handbook, U.S. Department of the Interior, Office of Water Research and Technology Report OWRT TT/82 1, 55 pp.

Melrose, J.C. and C.F. Brandner, 1974. Role of capillary forces in determination of microscopic displacement efficiency for oil recovery by water flooding, *Journal of Can. Petroleum Technology*, 10:54.

Mendoza, C.A. and E.O. Frind, 1990a. Advective-dispersive transport of dense organic vapors in the unsaturated zone, 1. Model development, *Water Resources Research*, 26(3):379-387.

Mendoza, C.A. and E.O. Frind, 1990b. Advective-dispersive transport of dense organic vapours in the unsaturated

zone, 2. Sensitivity analysis, *Water Resources Research*, 26(3):388-398.

Mendoza, C.A. and T.A. McAlary, 1990. Modeling of groundwater contamination caused by organic solvent vapors, *Ground Water*, 28(2):199-206.

Menegus, D.K. and K.S. Udell, 1985. A study of steam injection into water saturated capillary porous media, in *Heat Transfer in Porous Media and Particulate Flows*, ASME, 46:151-157.

Mercer, J.W. and R.M. Cohen, 1990. A review of immiscible fluids in the subsurface: Properties, models, characterization and remediation, *Journal of Contaminant Hydrology*, 6:107-163.

Mercer, J.W., D.C. Skipp and D. Giffin, 1990. Basics of pump-and-treat ground-water remediation technology, USEPA-600/8-90/003, Robert S. Kerr Environmental Research Laboratory, Ada, Oklahoma, 31 pp.

Michalski, A., 1989. Application of temperature and electrical conductivity logging in ground water monitoring, *Ground Water Monitoring Review*, 9(3):112-118.

Miller, C.A., 1975. Stability of moving surfaces in fluid systems with heat and mass transport, III. Stability of displacement fronts in porous media, *AI ChemEng Journal*, 21(3):474-479.

Miller, C.T., M.M. Poirier-McNeill and A.S. Mayer, 1990. Dissolution of trapped nonaqueous phase liquids: Mass transfer characteristics, *Water Resources Research*, 26(11):2783-2796.

Miller, M.M., S.P. Wasik, G.L. Huang, W.Y. Shiu and D. Mackay, 1985. Relationships between octanol-water partition coefficient and aqueous solubility, *Environmental Science and Technology*, 19:522-529.

Miller, R.N., R.E. Hinchee, C.M. Vogel, R.R. DuPont and D.C. Downey, 1990. A field scale investigation of enhanced petroleum hydrocarbon biodegradation in the vadose zone at Tyndall AFB, Florida, *Proceedings of Petroleum Hydrocarbons and Organic Chemicals in Ground Water: Prevention, Detection, and Restoration*, National Water Well Association-American Petroleum Institute, Houston, Texas, pp. 339-351.

Miller, S., 1982. The PCB imbroglio, *Environmental Science and Technology*, 17(1):11-14.

Millington, R.J. and J.P. Quirk, 1961. Permeability of porous solids, *Transactions of the Faraday Society*, 57:1200-1207.

Moein, G.J., A.J. Smith, Jr., K.E. Biglane, B. Loy and T. Bennett, 1976. Follow-up study of the distribution and fate of polychlorinated biphenyls and benzenes in soil and ground water samples after an accidental spill of transformer fluid, USEPA, Atlanta, Georgia, 145 pp.

Mohanty, K.K., H.T. Davis, L.E. Scriven, 1987. Physics of oil entrapment in water-wet rock, *SPE Reservoir Engineering*, February, pp. 113-128.

Monsanto, undated. Aroclor plasticizers, Technical Bulletin O/PL-306, 51 pp.

Monsanto, 1988. Polychlorinated biphenyls material safety data sheets.

Montgomery, J.H., 1991. *Groundwater Chemicals Desk Reference, Volume II*, Lewis Publishers, Chelsea, Michigan, 944 pp.

Montgomery, J.H. and L.M. Welkom, 1990. *Groundwater Chemicals Desk Reference*, Lewis Publishers, Chelsea, Michigan, 640 pp.

Mooney, H.M., 1975. Bison Instruments earth resistivity meters instruction manual, Bison Instruments, Inc., Minneapolis, Minnesota, 24 pp.

Mooney, H.M., 1980. Handbook of engineering geophysics, Bison Instruments, Inc., Minneapolis, Minnesota.

Moore, D.R.J. and S.L. Walker, 1991. Canadien water quality guidelines for polychlorinated biphenyls in coastal and estuarine waters, Environment Canada, Scientific Series No. 186, 61 pp.

Morgan, J.H., 1992. Horizontal drilling applications of petroleum technologies for environmental purposes, *Ground Water Monitoring Review*, 12(3):98-102.

Morrow, N.R., I. Chatzis and J.J. Taber, 1988. Entrapment and mobilization of residual oil in bead packs, *SPE Reservoir Engineering*, August, pp. 927-934.

Mualem, Y., 1976. A new model for predicting the hydraulic conductivity of unsaturated porous media, *Water Resources Research*, 12(3):513-522.

Mueller, J.G., P.J. Chapman and P.H. Pritchard, 1989. Creosote-contaminated sites, *Environmental Science and Technology*, 23(10):1197-1201.

Mull, R., 1971. Migration of oil products in the subsoil with regard to groundwater pollution by oil, *Advances in Water Pollution Research*, Pergamon, Oxford, pp. 1-8.

Mull, R., 1978. Calculations and experimental investigations of the migration of hydrocarbons in natural soils, Presented at the International Association of Hydrogeologists International Symposium on Groundwater Pollution by Oil Hydrocarbons, Prague, pp. 167-181.

Murarka, I.P., 1990. Manufactured gas plant site investigations, *EPRI Journal*, September, pp. 1-7.

Nash, J., 1987. Field studies of in situ soil washing, USEPA Report EPA-600/S2-87/110.

Nash, J. and R.P. Traver, 1986. Field evaluation of in situ washing of contaminated soils with water/surfactants, *Proceedings of the Twelve Annual Research Symposium*, USEPA/600/9-86/022.

National Center for Groundwater Research, 1992. Extended abstracts, *Proceedings of the Subsurface Restoration Third International Conference on Ground Water Quality Research*, Rice University, Houston, Texas, 343 pp.

National Research Council, 1990. *Ground Water Models: Scientific and regulatory applications*, Water Science and Technology Board, National Academy Press, Washington, D.C., 303 pp.

National Research Council of Canada (NRCC), 1980. A case study of a spill of industrial chemicals - Polychlorinated biphenyls and chlorinated benzenes, NRCC Report 17586, Ottawa, Ontario.

Nelson, R.C., J.B. Lawson, D.R. Thigpen and G.L. Stegemeier, 1984. Cosurfactant-enhanced alkaline flooding, SPE/DOE Paper 12672, Presented at the SPE/DOE Fourth Symposium on Enhanced Oil Recovery, Tulsa, Oklahoma, April 15-18.

Neustadter, E.L., 1984. Surfactants in enhanced oil recovery, in *Surfactants*, T.F. Tadros, ed., Academic Press, New York.

New York State Department of Health (NYSDOH), 1980. Love Canal litigation soil sampling, Report prepared for the New York State Department of Law, Albany, New York.

Newell, C.J., J.A. Connor, D.K. Wilson and T.E. McHugh, 1991. Impact of dissolution of dense non-aqueous phase liquids (DNAPLs) on groundwater remediation, *Proceedings of Petroleum Hydrocarbons and Organic Chemicals in Ground Water: Prevention, Detection, and Restoration*, National Water Well Association-American Petroleum Institute, Houston, Texas, pp. 301-315.

Newell, C.J. and R.R. Ross, 1992. Estimating potential for occurrence of DNAPL at Superfund sites, USEPA Quick Reference Fact Sheet, Robert S. Kerr Environmental Research Laboratory, Ada, Oklahoma.

Newman, W., J.M. Armstrong and M. Ettenhofer, 1988. an improved soil gas survey method using adsorbent tubes for sample collection, *Proceedings of the 2nd National Outdoor Action Conference on Ground Water Restoration, Ground Water Monitoring, and Geophysical Methods*, National Water Well Association, Dublin, Ohio, pp. 1033-1049.

Nielsen, D.M. and R. Schalla, 1991. Design and installation of ground-water monitoring wells, in *Practical Handbook of Ground-Water Monitoring*, D.M. Nielsen, ed., Lewis Publishers, Chelsea, Michigan, pp. 239-332.

NIOSH, 1990. *NIOSH Pocket Guide to Chemical Hazards*, National Institute for Occupational Safety and Health, U.S. Department of Health and Human Services, 245 pp.

Nirmalakhandan, N.N. and R.E. Speece, 1988. Prediction of aqueous solubility of organic chemicals based on molecular structure, *Environmental Science and Technology*, 22(3):328-338.

NJDEP, 11/14/91. Basis and background for the ground water quality standards, New Jersey Department of Environmental Protection and Energy.

Noggle, J., 1985. *Physical Chemistry*, Little, Brown and Co., Boston.

Norris, S.E., 1972. The use of gamma logs in determining the character of unconsolidated sediments and well construction features, *Ground Water*, 10(6):14-21.

Novak, J.T., C.D. Goldsmith, R.E. Benoit and J.H. O'Brien, 1984. Biodegradation of alcohols in subsurface systems, Degradation, retention and dispersion of pollutants in groundwater, Specialized seminar, September 12-14, Copenhagen, Denmark, pp. 61-75.

Novosad, J., 1981. Experimental study and interpretation of surfactant retention in porous media, in *Enhanced Oil Recovery*, F.J. Fayers, ed., Elsevier, New York, pp. 101-121.

Nutting, P.G., 1934. Some physical and chemical properties of reservoir rocks bearing on the accumulation and discharge of oil, *Problems of Petroleum Geology*, W.E. Wrather and F.H. Lahee, eds., AAPG, pp. 825-832.

NWWA, 1981. *Water Well Specifications*, National Water Well Association Committee on Water Well Standards, Premier Press, Berkeley, California, 156 pp.

O'Brien and Gere Engineers, Inc., 1988. *Hazardous Waste Site Remediation*, Van Nostrand Reinhold, New York, 422 pp.

O'Connor, M.J., J.G. Agar and R.D. King, 1984. Practical experience in the management of hydrocarbon vapors in the subsurface, in *Petroleum Hydrocarbons and Organic Chemicals in Ground Water: Prevention, Detection, and Restoration*, National Water Well Association-American Petroleum Institute, Houston, Texas, pp. 519-533.

OCC/Olin, 1990. Remedial investigation final report, Volume 1 -- Text, 102nd Street Landfill site, Niagara Falls, New York.

Offeringa, J., R. Barthel and J. Weijdema, 1981. The interplay between research and field operatiosn in the development of thermal recovery methods, in *Enhanced Oil Recovery*, F.J. Fayers, ed., Elsevier, 1981, pp. 527-541.

Olhoeft, G.R., 1984. Applications and limitations of ground penetrating radar, Expanded abstract, 54th Annual International Meeting of the Society of Exploration Geophysicists, December 2-6, Atlanta, Georgia, pp. 147-148.

Olhoeft, G.R., 1986. Direct detection of hydrocarbons and organic chemicals with ground penetrating radar and complex resistivity, *Proceedings of Petroleum*

Hydrocarbons and Organic Chemicals in Ground Water: Prevention, Detection, and Restoration, National Water Well Association-American Petroleum Institute, Houston, Texas, pp. 284-305.

Olhoeft, G.R., 1992. *Geophysics Advisor Expert System*, USGS Circular 1022.

Orellana, E. and H.M. Mooney, 1966. Master tables and curves for vertical electrical sounding over layered structures, Interciencia, Madrid, Spain, available from U.S. Army Engineer Waterways Experiment Station, Vicksburg, Mississippi.

Osborne, M. and J. Sykes, 1986. Numerical modeling of immiscible organic transport at the Hyde Park landfill, *Water Resources Research*, 22(1):25-33.

Osoba, J.S., J.G. Richardson, J.K. Kerver, J.A. Hafford and P.M. Blair, 1951. Laboratory determinations of relative permeability, *Trans. AIME*, 192:47.

Ostendorf, D.W., L.E. Leach, E.S. Hinlein and Y. Xie, 1991. Field sampling of residual aviation gasoline in sandy soil, *Ground Water Monitoring Review*, 11(2):107-120.

Palmer, D., 1980. *The Generalized Reciprocal Method of Seismic Refraction Interpretation*, K.B.S. Burke, ed., Department of Geology, Univ. of New Brunswick, Frederickton, New Brunswick, Canada, 104 pp.

Palombo, D.A. and J.H. Jacobs, 1982. Monitoring chlorinated hydrocarbons in groundwater, *Proceedings of the National Conference on Management of Uncontrolled Hazardous Waste Sites*, Hazardous Materials Control Research Institute, Silver Spring, Maryland, pp. 165-168.

Parizek, R.R., 1976. On the nature and significance of fracture traces and lineaments in carbonate and other terranes, in *Karst Hydrology and Water Resources, Proceedings of the U.S.-Yugoslavian Symposium*, Water Resources Publications, Fort Collins, Colorado.

Parizek, R.R., 1987. Groundwater monitoring in fracture flow dominated rocks, in *Pollution, Risk Assessment and Remediation in Groundwater Systems*, Khanbilvardi and Filos, eds., Scientific Publications, Washington, D.C., pp. 89-148.

Parker, J.C., 1989. Multiphase flow and transport in porous media, *Review Geophysics AGU*, 27(3):311-328.

Parker, J.C. and R.J. Lenhard, 1987. A model for hysteretic constitutive relations governing multiphase flow: 1. Saturation-pressure relations, *Water Resources Research*, 23(12):2187-2196.

Parker, J.C., R.J. Lenhard and T. Kuppusamy, 1987. A parametric model for constitutive properties governing multiphase fluid flow in porous media, *Water Resources Research*, 23(4):618-624.

Peaceman, D.W., 1977. *Fundamentals of Numerical Reservoir Simulation*, Elsevier, Amsterdam, 176 pp.

Pedersen, T.A. and J.T. Curtis, 1991. Soil vapor extraction technology reference handbook, EPA/540/2-91/003, USEPA Office of Research and Development, Cincinnati, Ohio, 316 pp.

Pfannkuch, H., 1984. Mass-exchange processes at the petroleum-water interface, Paper presented at the Toxic-Waste Technical Meeting, Tucson, AZ (March 20-22), M.F. Hult, ed., USGS Water-Resources Investigations Report 84-4188.

Phillipson, W.R. and D.A. Sangrey, 1977. Aerial detection techniques for landfill pollutants, in *Management of Gas and Leachate in Landfills, Proceedings of the Third Annual Municipal Solid Waste Research Symposium*, USEPA/600/9-77-026, St. Louis, Missouri, pp. 104-114.

Pinder, G.F., R.M. Cohen and J. Feld, 1990. DNAPL Migration from the Love Canal landfill, Niagara Falls, New York, Presented at the International Association of Hydrogeologists Conference on Subsurface Contamination by Immiscible Fluids, Calgary, Alberta, April 18-20.

Piontek, K., T. Sale and T. Simpkin, 1989. Bioremediation of subsurface wood preserving contamination, Paper presented at the Mississippi Forest Product Laboratory Forum on Bioremediation of Wood-Treating Waste, 19 pp.

Pitchford, A.M., A.T. Mazzella and K.R. Scarbrough, 1989. Soil-gas and geophysical techniques for detection of subsurface organic contamination, USEPA Environmental Monitoring Systems Laboratory, Las Vegas, Nevada.

Pitts, M.J., S.R. Clark and S.M. Smith, 1989. West Kiehl field alkaline-surfactant-polymer oil recovery system design and application, *Proceedings of the International Symposium on Enhanced Oil Recovery*, pp. 273-290.

Plumb, R.H., Jr. and A.M. Pitchford, 1985. Volatile organic scans: Implications for ground water monitoring, *Proceedings of Petroleum Hydrocarbons and Organic Chemicals in Ground Water: Prevention, Detection, and Restoration*, National Water Well Association-American Petroleum Institute, Houston, Texas, pp. 207-222.

Poulsen, M.M. and B.H. Kueper, 1992. A field experiment to study the behavior of tetrachloroethylene in unsaturated porous media, *Environmental Science and Technology*, 26(5):889-895, ACS.

Powers, S.E., L.M. Abriola and W.J. Weber, Jr., 1992. An experimental investigation of nonaqueous phase liquid dissolution in saturated subsurface systems: Steady state mass transfer rates, *Water Resources Research*, 28(10):2691-2705.

Powers, S.E., C.O. Loureiro, L.M. Abriola and W.J. Weber, Jr., 1991. Theoretical study of the significance of nonequilibrium dissolution of nonaqueous phase liquids in subsurface systems, *Water Resources Research*, 27(4):463-477.

Prats, M., 1989. Operational aspects of steam injection processes, in *Enhanced Oil Recovery*, F.J. Fayers, ed., Elsevier, New York.

Purcell, W.R., 1949. Capillary pressures -- Their measurement using mercury and the calculation of permeability therefrom, *Trans. AIME*, 186:39-48.

Quinn, E.J., T.N. Wasielewski and H.L. Conway, 1985. Assessment of coal tar constituents migration: Impacts on soils, ground water and surface water.

Rannie, E.H. and R.L. Nadon, 1988. An inexpensive, multi-use, dedicated pump for ground water monitoring wells, *Ground Water Monitoring Review*, 8(4):100-107.

Rao, P.S.C., L.S. Lee and A.L. Wood, 1991. Solubility, sorption and transport of hydrophobic organic chemicals in complex mixtures, USEPA Environmental Research Brief, EPA/600/M-91/009, Robert S. Kerr Environmental Research Laboratory, Ada, Oklahoma, 14 pp.

Rathfelder, K., 1989. Numerical simulation of soil vapor extraction systems, Ph.D. Dissertation, Department of Civil Engineering, University of California, Los Angeles, California.

Rathfelder, K., W.W-G. Yeh and D. Mackay, 1991. Mathematical simulation of soil vapor extraction systems: Model development and numerical examples, *Journal of Contaminant Hydrology*, 8(1991):263-297.

Rathmell, J.J., P.H. Braun and T.K. Perkins, 1973. Reservoir waterflood residual oil saturation from laboratory tests, *Journal of Petroleum Technology*, pp. 175-185.

Raven, K.G. and P. Beck, 1990. Case studies of coal tar and creosote contamination in Ontario, Presented at the International Association of Hydrogeologists Conference on Subsurface Contamination by Immiscible Fluids, Calgary, Alberta, April 18-20.

Raymond, R.L., 1974. Reclamation of hydrocarbon contaminated groundwaters, U.S. Patent Office, 3,846,290, patented November 5.

Redman, J.D., B.H. Kueper and, A.P. Annan, 1991. Dielectric stratigraphy of a DNAPL spill and implications for detection with ground penetrating radar, *Proceedings of the Fifth National Outdoor Action Conference on Aquifer Restoration, Ground Water Monitoring, and Geophysical Methods*, May 13-16, 1991, Las Vegas, Nevada, NGWA, pp. 1017-1030.

Redpath, B.B., 1973. Seismic refraction exploration for engineering site investigations, Technical Report E-73-4, U.S. Army Engineer Waterways Experiment Station, Explosive Excavation Research Laboratory, Livermore, California, 51 pp.

Regalbuto, D.P., J.A. Barrera and J.B. Lisiecki, 1988. In-situ removal of VOCs by means of enhanced volatilization, *Proceedings of Petroleum Hydrocarbons and Organic Chemicals in Ground Water: Prevention, Detection, and Restoration*, National Water Well Association-American Petroleum Institute, Houston, Texas, pp. 571-590.

Rhodes, E.O., 1979. The history of coal tar and light oil, in *Bituminous Materials: Asphalts, tars, and pitches, Volume III: Coal tars and pitches*, A.J. Hoiberg ed.,. R.E. Krieger Publishing Co., Huntington, New York, pp. 1-31.

Ridgeway, W.R. and D. Larssen, 1990. A comparison of two multiple-level ground-water monitoring systems, in *Ground Water and Vadose Zone Monitoring*, D.M. Nielsen and A.I. Johnson, eds., ASTM, Philadelphia, Pennsylvania, pp. 213-237.

Riggs, C.O. and A.W. Hatheway, 1988. Groundwater monitoring field practice -- An overview, in *Ground-Water Contamination Field Methods*, ASTM Publication code number 04-963000-38, Philadelphia, Pennsylvania, pp. 121-136.

Ritchey, J.D., 1986. Electronic sensing devices used for in situ ground-water monitoring, *Ground Water Monitoring Review*, 16(2):108-113.

Rivett, M.O. and J.A. Cherry, 1991. The effectiveness of soil gas surveys in delineation of groundwater contamination: Controlled experiments at the Borden field site, *Proceedings of the Petroleum Hydrocarbons and Organic Chemicals in Ground Water: Prevention, Detection, and Restoration*, National Water Well Association-American Petroleum Institute, Houston, Texas, pp. 107-124.

Rivett, M.O., S. Feenstra and J.A. Cherry, 1991. Field experimental studies of residual solvent source emplaced in the groundwater zone, *Proceedings of the Petroleum Hydrocarbons and Organic Chemicals in Ground Water: Prevention, Detection, and Restoration*, National Water Well Association-American Petroleum Institute, Houston, Texas, pp. 283-300.

Rivett, M.O., D.N. Lerner and J.W. Lloyd, 1990. Temporal variations of chlorinated solvents in abstraction wells, *Ground Water Monitoring Review*, 10(4):127-133.

Roberts, J.R., J.A. Cherry and F.W. Schwartz, 1982. A case study of a chemical spill: Polychlorinated biphenyls (PCBs) 1. History, distribution, and surface translocation, *Water Resources Research*, 18(3):525-534.

Robertson, C.G., 1992. Groundwater extraction system case history, IBM Corporation, Dayton, New Jersey, Presentation to Committee on Groundwater Extraction Systems, National Research Council, March 24, Washington, D.C.

Robertson, P.K. and R.G. Campanella, 1986. *Guidelines for Use and Interpretation of the Electronic Cone Penetration Test*, 3rd ed., The University of British Columbia.

Rodgers, R.B. and W.F. Kean, 1980. Monitoring groundwater contamination at a fly-ash disposal site using surface electrical resistivity methods, *Ground Water*, 18(5):472-478.

Rosenfeld, J.K. and R.H. Plumb, Jr., 1991. Groundwater contamination at wood treatment facilities, *Ground Water Monitoring Review*, 11(1):133-140.

Rossi, S.S. and W.H. Thomas, 1981. Solubility behavior of three aromatic hydrocarbons in distilled water and natural seawater, *Environmental Science and Technology*, 15:715-716.

Roux, P.H. and W.F. Althoff, 1980. Investigation of organic contamination of ground water in South Brunswick Township, New Jersey, *Ground Water*, 18(5):464-471

Rumbaugh, J.O., J.A. Caldwell and S.T. Shaw, 1987. A geophysical ground water monitoring program for a sanitary landfill: Implementation and preliminary analysis, *Proceedings of the First National Outdoor Action Conference on Aquifer Restoration, Ground Water Monitoring, and Geophysical Methods*, Las Vegas, Nevada, National Water Well Association, May 18-21.

Ruth, J.H., 1986. Odor thresholds and irritation levels of several chemical substances: A review, *American Industrial Hygiene Association Journal*, 47:142-151.

Salager, J.L., J.C. Morgan, R.S. Schecter, W.H. Wade and E. Vasquex, 1979. Optimum formulation of surfactant/water/oil systems for minimum interfacial tension or phase behavior, *Society of Petroleum Engineers Journal*, pp. 107-115.

Salathiel, R.A., 1973. Oil recovery by surface film drainage in mixed wettability rocks, *Journal of Petroleum Technology*, 25:1216-1224.

Sale, T.C. and K. Piontek, 1988. In situ removal of waste wood-treating oils from subsurface materials, *Proceedings of USEPA Forum on Remediation of Wood-Preserving Sites*, San Francisco, 24 pp.

Sale, T.C., K. Piontek and M. Pitts, 1989. Chemically enhanced in situ soil washing, *Proceedings of Petroleum Hydrocarbons and Organic Chemicals in Ground Water: Prevention, Detection, and Restoration*, National Water Well Association-American Petroleum Institute, Houston, Texas, pp. 487-503.

Sale, T.C., D. Stieb, K.R. Piontek and B.C. Kuhn, 1988. Recovery of wood-treating oil from an alluvial aquifer using dual drainlines, *Proceedings of Petroleum Hydrocarbons and Organic Chemicals in Ground Water: Prevention, Detection, and Restoration*, National Water

Well Association-American Petroleum Institute, Houston, Texas, pp. 419-422.

Sanders, P.J., 1984. New tape for ground-water measurements, *Ground Water Monitoring Review*, 4(1):39-42.

Saraf, D.N. and F.G. McCaffery, 1982. Two- and three-phase relative permeabilities: A review, Petroleum Recovery Institute, Report No. 81-8.

Sayegh, S.G. and F.G. McCaffery, 1981. Laboratory testing procedures for miscible floods, in *Enhanced Oil Recovery*, F.J. Fayers, ed., Elsevier, New York, pp. 285-298.

Schaumburg, F.D., 1990. Banning trichloroethylene: Responsible reaction or overkill?, *Environmental Science and Technology*, 24(1):17-22.

Scheigg, H.O., 1985. Considerations on water, oil, and air in porous media, *Water Science Technology*, 23(4/5):467-476.

Schmelling, S.G., 1992. Personal communication, USEPA Robert S. Kerr Environmental Research Laboratory, Ada, Oklahoma.

Schmidtke, K., E. McBean and F. Rovers, 1987. Drawdown impacts in dense non-aqueous phase liquids, in *NWWA Ground Water Monitoring Symposium*, National Water Well Association, Las Vegas, Nevada, pp. 39-51.

Schowalter, T.T., 1979. Mechanics of secondary hydrocarbon migration and entrapment, *The American Association of Petroleum Geologists Bulletin*, 63(5):723-760.

Schuller, R.M., W.W. Beck, Jr. and D.R. Price, 1982. Case study of contaminant reversal and groundwater restoration in a fractured bedrock, *Proceedings of National Conference on Management of Uncontrolled Hazardous Waste Sites*, Hazardous Materials Control Research Institute, Bethesda, Maryland, pp. 94-96.

Schwartz, F.W., J.A. Cherry and J.R. Roberts, 1982. A case study of a chemical spill: Polychlorinated biphenyls (PCBs), 2. Hydrogeological conditions and contaminant migration, *Water Resources Research*, 18(3):535-545.

Schwille, F., 1988. *Dense Chlorinated Solvents in Porous and Fractured Media*, Lewis Publishers, Chelsea, Michigan, 146 pp.

Senn, R.B. and M.S. Johnson, 1987. Interpretation of gas chromatographic data in subsurface hydrocarbon investigations, *Ground Water Monitoring Review*, 7(1):58-63.

Shah, D.O., 1981. Fundamental aspects of surfactant-polymer flooding process, in *Enhanced Oil Recovery*, F.J. Fayers, ed., Elsevier, New York, pp. 1-41.

Shangraw, T.C., D.P. Michaud and T.M. Murphy, 1988. Verification of the utility of a Photovac gas chromatograph for conduct of soil gas surveys, *Proceedings of the 2nd National Outdoor Action Conference on Aquifer Restoration, Ground Water Monitoring, and Geophysical Methods*, National Water Well Association, Dublin, Ohio, pp. 1089-1108.

Sharma, M.K. and D.O. Shah, 1989. Use of surfactants in oil recovery, in *Enhanced Oil Recover, II, Processes and Operations*, E.C. Donaldson, G.V. Chilingarian and T.F. Yen, eds., Elsevier, New York, pp. 255-315.

Sharma, M.M. and R.W. Wunderlich, 1985. The alteration of rock properties due to interactions with drilling fluid components, SPE Annual Technical Conference paper SPE 14302, Las Vegas, Nevada, September 22-25.

Shifrin, N., 1986. Affidavit re Hyde Park, U.S. N.Y. v. Hooker Chemicals and Plastics Corp. et al., Civil Action No. 79-989.

Shiu, W.Y., A. Maijanen, A.L.Y. Ng and D. Mackay, 1988. Preparation of aqueous solutions of sparing soluble organic substances, II. Multicomponent systems -- Hydrocarbon mixtures and petroleum products, *Environmental Toxicology and Chemistry*, 7:125-137.

Shugar, G.J. and J.T. Ballinger, 1990. *Chemical Technicians' Ready Reference Handbook*, McGraw-Hill Book Co., New York, 890 pp.

Siddiqui, S.H. and R.R. Parizek, 1971. Hydrologic factors influencing well yields in folded and faulted carbonate rocks in Central Pennsylvania, *Water Resources Research*, 7(5):1295-1312.

Silka, L.R., 1986. Simulation of the movement of volatile organic vapor through the unsaturated zone as it pertains to soil-gas surveys, *Proceedings of the NWWA/API Conference on Petroleum Hydrocarbons and Organic Chemicals in Ground Water: Prevention, Detection, and Restoration*, National Water Well Association-American Petroleum Institute, Houston, Texas, pp. 204-226.

Silka, L.R. and D.L. Jordan, 1993. Vapor analysis/ extraction, in *Geotechnical Practice for Waste Disposal*, D.E. Daniel, ed., Chapman & Hall, London, pp. 378-429.

Sims, R.C., 1988. Onsite bioremediation of wood preserving contaminants in soils, *Proceedings Technical Assistance to U.S. EPA Region IX: Forum on Remediation of Wood Preserving Sites*, October 24-25, San Fransisco, California.

Sims, R.C., 1990. Soil remediation techniques at uncontrolled hazardous waste sites: A critical review, *Journal of the Air & Waste Management Association*, 40:(5)704-732.

Sitar, N., J.R. Hunt and J.T. Geller, 1990. Practical aspects of multiphase equilibria in evaluating the degree of contamination, Presented at the International Association of Hydrogeologists Conference on Subsurface

Contamination by Immiscible Fluids, Calgary, Alberta, April 18-20.

Slaine, D.D. and J.P. Greenhouse, 1982. Case studies of geophysical contaminant mapping at several waste disposal sites, *Proceedings of the Second National Symposium on Aquifer Restoration and Ground Water Monitoring*, National Water Well Association, Worthington, Ohio, pp. 299-315.

Sleep, B.E. and J.F. Sykes, 1989. Modeling the transport of volatile organics in variably saturated media, *Water Resources Research*, 25(1):81-92.

Slobod, R.L. and A. Chambers, 1951. Use of centrifuge for determining connate water, residual water, and capillary pressure curves of small core samples, *Trans. AIME*, 192:127-134.

Smith, D.A., 1966. Theoretical considerations of sealing and non-sealing faults, *AAPG Bulletin*, 50:363-374.

Smolley, M. and J.C. Kappmeyer, 1991. Cone penetrometer tests and Hydropunch™ sampling: A screening technique for plume definition, *Ground Water Monitoring Review*, 11(2):101-106.

Snell, R.W., 1962. Three-phase relative permeability in an unconsolidated sand, *Journal of Inst. Petroleum*, 84:80-88.

Spittler, T.M., W.S. Clifford and L.G. Fitch, 1985. A new method for detection of organic vapors in the vadose zone, *Proceedings of the 2nd Conference on Characterization and Monitoring in the Vadose Zone*, National Water Well Association, Dublin, Ohio, pp. 236-246.

Sresty, G.C., H. Dev, R.H. Snow and J.E. Bridges, 1986. Recovery of bitumen from tar sand deposits with the radio frequency process, *Society of Petroleum Engineering Journal*, January, pp. 85-94.

Starr, R.C. and R.A. Ingleton, 1992. A new method for collecting core samples without a drilling rig, *Ground Water Monitoring Review*, 12(1):91-95.

Steeples, D.W., 1984. High resolution seismic reflections at 200 Hz, *Oil and Gas Journal*, December 3, pp. 86-92.

Stephanatos, B.N., 1988. Modeling the transport of gasoline vapors by an advective-diffusion unsaturated zone model, *Proceedings of Petroleum Hydrocarbons and Organic Chemicals in Ground Water: Prevention, Detection, and Restoration*, National Water Well Association-American Petroleum Institute, Houston, Texas, pp. 591-611.

Stewart, M.T., 1982. Evaluation of electromagnetic methods for rapid mapping of salt-water interfaces in coastal aquifers, *Ground Water*, 20(5):538-545.

Stipp, D., 1991. Super waste? Throwing good money at bad water yields scant improvement, *The Wall Street Journal*, May 15, p. 1.

Stollar, R.L. and P. Roux, 1975. Earth resistivity surveys -- A method for defining ground-water contamination, *Ground Water*, 13(2):145-150.

Stone, H.L., 1970. Probability model for estimating three-phase relative permeability, *Journal of Petroleum Technology*, 20:214-218.

Stone, H.L., 1973. Estimation of three-phase relative permeability and residual oil data, *Journal of Can. Petroleum Technology*, 12(4):53-61.

Stoner, E.R. and M.F. Baumgardner, 1979. Data acquisition through remote sensing, in *Planning the Uses and Management of Land*, M.T. Beatty, G.W. Petersen, and L.D. Swindale, eds., Soil Science Society of America, Madison, Wisconsin, pp. 159-186.

Sudicky, E.A., 1986. A natural gradient experiment on solute transport in a sand aquifer: Spatial variability of hydraulic conductivity and its role in the dispersion process, *Water Resources Research*, 22(13):2069-2082.

Sudicky, E.A. and P.S. Huyakorn, 1991. Contaminant migration in imperfectly known heterogeneous groundwater systems, Ninth U.S. National Committee Report to the IUGG General Assembly, *Reviews of Geophysics*, pp. 240-253.

Suflita, J.M. and G.D. Miller, 1985. Microbial metabolism of chlorophenolic compounds in ground water aquifers, *Environmental Toxicology and Chemistry*, 4:751-758.

Surkalo, 'H., 1990. Enhanced alkaline flooding, *Journal of Petroleum Technology*, 42(1):6-7.

Surkalo, H., M.J. Pitts, B. Sloat and D. Larsen, 1986. Polyacramide vertical conformance process improved sweep efficiency and oil recovery in the OK field, Society of Petroleum Engineers Paper 14115.

Sutter, J.L., J. Glass and K. Davies, 1991. Superfund groundwater extraction evaluation case studies and recommendations, *Groundwater Remediation*, pp. 236-239.

Sverdrup, K.A., 1986. Shallow seismic refraction survey of near-surface ground water flow, *Ground Water Monitoring Review*, 6(1):80-83.

Swanson, R.G., 1981. *Sample Examination Manual*, Methods in Exploration Series, The American Association of Petroleum Geologists, Tulsa, Oklahoma, pp. 20-24.

Sweeney, J.T., 1984. Comparison of electrical resistivity methods for investigation of ground water conditions of a landfill site, *Ground Water Monitoring Review*, 4(1):52-59.

Taber, J.J., 1981. Research on enhanced oil recovery, past, present, and future, in *Surface Phenomena in Enhanced Oil Recovery*, D.O. Shah, ed., Plenum Publishing Corp.

Technos, Inc., 1980. Geophysical investigation results, Love Canal, New York, Report to GCA Corporation and USEPA, Technos, Inc., Miami, Florida.

Telford, W.M., L.P. Geldar, R.E. Sheriff and D.A. Keys, 1982. *Applied Geophysics*, Cambridge University Press.

Testa, S.M. and M.T. Paczkowski, 1989. Volume determination and recoverability of free hydrocarbon, *Ground Water Monitoring Review*, 9(1):120-128.

Tetra Tech, Inc., 1988. Chemical data for predicting the fate of organic chemicals in water, Volume 2: Database, EPRI EA-5818, Volume 2, Electric Power Research Institute, Palo Alto, California, 411 pp.

Texas Research Institute, 1984. Forced venting to remove gasoline vapors from a large-scale model aquifer, American Petroleum Institute, Washington, D.C., 60 pp.

Thomas, G.W., 1982. *Principles of Hydrocarbon Reservoir Simulation*, International Human Resources Development Corporation, Boston, 207 pp.

Thompson, G.M. and D.L. Marrin, 1987. Soil gas contaminant investigations: A dynamic approach, *Ground Water Monitoring Review*, 7(3):88-93.

Thornton, J.S. and W.L. Wootan, 1982. Venting for the removal of hydrocarbon vapors from gasoline contaminated soil, *Journal of Environmental Science and Health*, A17(1):31-44.

Tillman, N., K. Ranlet and T.J. Meyer, 1989a. Soil gas surveys: Part I, *Pollution Engineering*, 21(7):86-89.

Tillman, N., K. Ranlet and T.J. Meyer, 1989b. Soil gas surveys: Part II, Procedures, *Pollution Engineering*, 21(8):79-84.

Treiber, L.E., D.L. Archer and W.W. Owens, 1972. A laboratory evaluation of the wettability of fifty oil producing reservoirs, *Journal of the Society of Petroleum Engineers*, 12(6):531-540.

Troutman, D.E., E.M. Godsy, D.F. Goerlitz and G.G. Ehrlich, 1984. Phenolic contamination in the sand-and-gravel aquifer from a surface impoundment of wood treatment wastes, Pensacola, Florida, USGS Water-Resources Investigations Report 84-4230, 36 pp.

Tuck, D.M., P.R. Jaffe, D.A. Crerar and R.T. Mueller, 1988. Enhanced recovery of immobile residual non-wetting hydrocarbons from the unsaturated zone using surfactant solutions, *Proceedings of Petroleum Hydrocarbons and Organic Chemicals in Ground Water: Prevention, Detection, and Restoration*, National Water Well Association-American Petroleum Institute, Houston, Texas, pp. 457-479.

Udell, K.S. and L.D. Stewart, 1989. Field study on in situ steam injection and vacuum extraction for recovery of volatile organic solvents, University of California Berkeley, Sanitary Engineering and Environmental Health Research Laboratory, UCB-SHEEHRL Report No. 89-2.

Udell, K.S. and L.D. Stewart, 1990. Combined steam injection and vacuum extraction for aquifer cleanup, Presented at the International Association of Hydrogeologists Conference on Subsurface Contamination by Immiscible Fluids, Calgary, Alberta, April 18-20.

Uhlman, K., 1992. Groundwater dating locates VOC source, *Environmental Protection*, 3(9):56-62.

Urish, D.W., 1983. The practical application of surface electrical resistivity to detection of ground-water pollution, *Ground Water*, 21:144-152.

USDA, 1980. The biologic and economic assessment of pentachlorophenol, inorganic arsenicals, creosote, Volume I: Wood preservatives, U.S. Department of Agriculture Technical Bulletin 1658-I, 435 pp.

USDI, 1974. *Earth Manual*, U.S. Department of the Interior Water and Power Resources Service, U.S. Government Printing Office, Washington, D.C., 810 pp.

USEPA, 1978. Electrical resistivity evaluations of solid waste disposal facilities, 94 pp.

USEPA, 1980. Procedures manual for ground water monitoring at solid waste disposal facilities, EPA/SW-611, 269 pp.

USEPA, 1983. The PCB regulations under TSCA: Over 100 questions and answers to help you meet these requirements, revised edition No. 3.

USEPA, 1984a. Case studies 1-23: Remedial response at hazardous waste sites, EPA/540/2-84-002b, Cincinnati, Ohio.

USEPA, 1984b. Summary report: Remedial response at hazardous waste sites, EPA/540/2-84-002a, Cincinnati, Ohio.

USEPA, 1986. RCRA ground-water monitoring technical enforcement guidance document, EPA OSWER-9950.1, 317 pp.

USEPA, 1987. A compendium of Superfund field operations methods, EPA/540/P-87/001, 644 pp.

USEPA, 1988. Guidance for conducting remedial investigations and feasibility studies under CERCLA, USEPA/540/G-89/004.

USEPA, 1989a. Interim final RCRA facility investigation (RFI) guidance, USEPA/530/SW-89-031.

USEPA, 1989b. Risk assessment guidance for Superfund, Volume I, Human health evaluation manual (Part A), Interim final, USEPA/540/1-89/002.

USEPA, 1989c. Evaluation of ground-water extraction remedies: Vol. 2, case studies 1-19, interim final, EPA/540/2-89/054b, Washington, D.C.

USEPA, 1989d. Proceedings technical assistance to USEPA Region IX: Forum on remediation of wood preserving sites, October 24-25, 1988, San Francisco, California.

USEPA, 1989e. Terra-Vac in situ vacuum extraction system: Applications analysis report, EPA/540/A5-89-003.

USEPA, 1990a. Handbook on in situ treatment of hazardous waste-contaminated soils, EPA/540/2-90/002.

USEPA, 1990b. The Superfund innovative technology evaluation program: Technology profiles, EPA/540/5-90/006, 170 pp.

USEPA, 1990c. Guidance on remedial actions for Superfund sites with PCB contamination, EPA/540/G-90/007.

USEPA, 1990d. Approaches for remediation of uncontrolled wood preserving sites, EPA/625/7-90/011, 21 pp.

USEPA, 10/17/91. Revised priority list of hazardous substances, Federal Register, 56(201):52165-52175.

USEPA, 1991a. Site characterization for subsurface remediation, EPA/625/4-91/026, 259 pp.

USEPA, 1991b. Evaluation of ground-water extraction remedies: Phase II, Vol. 1, summary report (November), OERR, Washington, D.C.

USEPA, 1992. Dense nonaqueous phase liquids -- A workshop summary, Dallas, Texas, April 17-18, 1991, EPA/600/R-92/030, Robert S. Kerr Environmental Research Laboratory, Ada, Oklahoma.

USEPA-ERT, 1988a. Standard operating procedure 2051: Charcoal tube sampling, November 7, 1988.

USEPA-ERT, 1988b. Standard operating procedure 2052: Tenax tube sampling, November 8, 1988.

USGS, 1985. Movement and fate of creosote waste in ground water near an abandoned wood-preserving plant near Pensacola, Florida, *Proceedings of the Second Technical Meeting U.S. Geological Survey Program on Toxic Waste -- Ground Water Contamination*, USGS Open-File Report 86-481.

U.S. International Trade Commission, 1991. Synthetic organic chemicals, United States production and sales, 1990, USITC Publication 2470, Washington, D.C.

Valocchi, A., 1985. Validity of the local equilibrium assumption for modeling sorbing solute transport through homogeneous soils, *Water Resources Research*, 21(6):808-820.

van Dam, J., 1967. The migration of hydrocarbons in a water bearing stratum, in *The Joint Problems of the Oil and Water Industries*, P. Hepple, ed., Elsevier, Amsterdam, pp. 55-96.

van der Waarden, M., A.L.A.M. Bridie and W.M. Groenewoud, 1971. Transport of mineral oil components to ground water, I. Model experiments on the transfer of hydrocarbons from a residual oil zone to trickling water, *Water Research*, 5:213-226.

van Genuchten, M. Th., 1980. A closed-form equation for predicting the hydraulic conductivity of unsaturated soils, *Soil Science Society Am. Journal*, 44:892-898.

Verschueren, K., 1983. *Handbook of Environmental Data on Organic Chemicals*, Van Nostrand Reinhold, New York, 1310 pp.

Villaume, J.F., 1984. Coal tar wastes: Their environmental fate and effects, in *Solid, Hazardous, and Radioactive Wastes: Management, Emergency Response, and Health Effects*, S.K. Majumdar and F.W. Miller, eds., Pennsylvania Academy of Science, Easton, Pennsylvania.

Villaume, J.F., 1985. Investigations at sites contaminated with DNAPLs, *Ground Water Monitoring Review*, 5(2):60-74.

Villaume, J.F., 1991. State of the practice: High-viscosity DNAPLs, Presented at the USEPA DNAPL Workshop, Dallas, Texas.

Villaume, J.F., P.C. Lowe and D.F. Unites, 1983. Recovery of coal gasification wastes: An innovative approach, *Proceedings of the Third National Symposium on Aquifer Restoration and Ground-Water Monitoring*, National Water Well Association, Dublin, Ohio, pp. 434-445.

Vogel, T.M., C.S. Criddle and P.L. McCarty, 1987. Transformations of halogenated aliphatic compounds, *Environmental Science and Technology*, 21(8):722-736.

Volek, C.W. and J.A. Pryor, 1972. Steam distillation drive -- Brea field, California, *Journal of Petroleum Technology*, 24:899-906.

Voorhees, K.J., J.C. Hickey and R.W. Klusman, 1984. Analysis of groundwater contamination by a new surface static trapping/mass spectrometry technique, *Analytical Chemistry*, 56:2602-2604.

Wagner, T.P., 1991. *Hazardous Waste Regulations*, 2nd edition, Van Nostrand Reinhold, New York, 488 pp.

Wang, F.H.L., 1988. Effect of wettability alteration on water/oil relative permeability, dispersion, and flowable saturation in porous media, *SPE Reservoir Engineering*, May, pp. 617-628.

Warner, D.L., 1969. Preliminary field studies using earth resistivity measurements for delineating zones of contaminated ground water, *Ground Water*, 7(1):9-16.

WCGR, 1991. Dense, Immiscible Phase Liquid Contaminants (DNAPLs) in Porous and Fractured Media, A Short Course, DNAPL Short Course notes, October 7-10, Kitchner Ontario, Canada, Waterloo Center for Groundwater Research, University of Waterloo.

Welch, S.J. and D.R. Lee, 1987. A multiple-packer/standpipe system for ground water monitoring in consolidated media, *Ground Water Monitoring Review*, 7(3):83-87.

Welge, H.J., 1949. Displacement of oil from porous media by water and gas, *Trans. AIME*, 186:133-145.

Welge, H.J. and W.A. Bruce, 1947. The restored state method for determination of oil in place and connate water, *Drilling and Production Practices*, American Petroleum Institute, 166 pp.

Willman, B.T., V.V. Valleroy, G.W. Runberg, A.J. Cornelius and L.W. Powers, 1961. Laboratory studies of oil recovery by steam injection, *Journal Petroleum Technology*, 13(7):681-690.

Wilson, A.R., 1990. *Environmental Risk: Identification and Management*, Lewis Publishers, Chelsea, Michigan, 336 pp.

Wilson, B.H. and J.F. Rees, 1985. Biotransformation of gasoline hydrocarbons in methanogenic aquifer material, *Proceedings of Petroleum Hydrocarbons and Organic Chemicals in Ground Water: Prevention, Detection, and Restoration*, National Water Well Association-American Petroleum Institute, Houston, Texas,.

Wilson, D.E., R.E. Montgomery and M.R. Sheller, 1987. A mathematical model for removing volatile subsurface hydrocarbons by miscible displacement, *Water, Air, and Soil Pollution*, 33(3-4):231-255.

Wilson, J.L. and S.H. Conrad, 1984. Is physical displacement of residual hydrocarbons a realistic possibility in aquifer restoration?, *Proceedings of the NWWA/API Conference on Petroleum Hydrocarbons and Organic Chemicals in Ground Water: Prevention, Detection, and Restoration*, National Water Well Association-American Petroleum Institute, Houston, Texas, pp. 274-298.

Wilson, J.L., S.H. Conrad, W.R. Mason, W. Peplinski and E. Hagen, 1990. Laboratory investigation of residual liquid organics, USEPA/600/6-90/004, Robert S. Kerr Environmental Research Laboratory, Ada, Oklahoma, 267 pp.

Wilson, J.T., L.E. Leach, M. Henson and J.N. Jones, 1986. In situ biorestoration as a groundwater remediation technique, *Ground Water Monitoring Review*, 6(4):56-64.

Wisniewski, G.M., G.P. Lennon, J.F. Villaume and C.L. Young, 1985. Response of a dense fluid under pumping stress, *Proceedings of the 17th Mid-Atlantic Industrial Waste Conference*, Lehigh University, pp. 226-237.

Wittmann, S.G., K.J. Quinn and R.D. Lee, 1985. Use of soil gas sampling techniques for assessment of groundwater contamination, *Proceedings of the NWWA/API Conference on Petroleum Hydrocarbons and Organic Chemicals in Ground Water -- Prevention, Detection, and Restoration*, National Water Well Association-American Petroleum Institute, Houston, Texas, pp. 291-309.

Working Group "Water and Petroleum", 1970. Evaluation and treatment of oil spill accidents on land with a view to the protection of water resources, Federal Ministry of Interior, Federal Republic of Germany, 2nd ed., Bonn, 138 pp.

Wright, D.L., G.R. Olhoeft and R.D. Watts, 1984. Ground-penetrating radar studies on Cape Cod, in *Surface and Borehole Geophysical Methods in Ground Water Investigations*, D.M. Nielsen, ed., National Water Well Association, Worthington, Ohio, pp. 666-680.

Yaniga, P.M. and J. Mulry, 1984. Accelerated aquifer restoration: Insitu applied techniques for enhanced free product recovery/adsorbed hydrocarbon reduction via bioreclamation, *Proceedings of Petroleum Hydrocarbons and Organic Chemicals in Ground Water: Prevention, Detection, and Restoration*, National Water Well Association-American Petroleum Institute, Houston, Texas, pp. 421-440.

Yen, W.S., A.T. Coscia and S.I. Kohen, 1989. Polyacrylamides, in *Enhanced Oil Recovery, II, Processes and Operations*, E.C. Donaldson, G.V. Chilingarian and T.F. Yen, eds., Elsevier, New York, pp. 189-218.

Yortsos, Y.C. and G.R. Gavalas, 1981. Analytical modelling of oil recovery by steam injection, II. Asymptotic and approximate solutions, *Society of Petroleum Engineers Journal*, 21(2):179-190.

Zalidis, G.C., M.D. Annable, R.B. Wallace, N.J. Hayden and T.C. Voice, 1991. A laboratory method for studying the aqueous phase transport of dissolved constituents from residually held NAPL in unsaturated soil columns, *Journal of Contaminant Hydrology*, 8(2):143-156.

Zapico, M.M., S. Vales and J.A. Cherry, 1987. A wireline piston core barrel for sampling cohesionaless sand and gravel below the water table, *Ground Water Monitoring Review*, 7(3):74-82.

Zilliox, L. and P. Muntzer, 1975. Effects of hydrodynamic processes on the development of groundwater pollution: Study of physical models in a saturated porous medium, *Prog. Water Technology*, 7(3/4):561-568.

Zilliox, L., P. Muntzer and J.J. Menanteau, 1973. Probleme de l'echange entre un produit petrolier immobile et l'eau en mouvement dans un milieu poreux, *Revue de l'Institut Francais du Petrole*, 28(2):185-200.

Zilliox, L., P. Muntzer and F. Schwille, 1974. Untersuchungen uber den stoffaustausch zwischen mineralol und wasser in porosen medien, *Deutsche Gewasserkundliche Mitteilungen*, 18, H2, April, pp. 35-37.

Zilliox, L., P. Muntzer and J.J. Fried, 1978. An estimate of the source of a phreatic aquifer pollution by hydrocarbons, oil-water contact and transfer of soluble substances in ground water, *Proceedings of the International Symposium on Groundwater Pollution by Oil-Hydrocarbons*, International Association of Hydrogeologists, Prague, pp. 209-227.

Zohdy, A.A., G.P. Eaton and D.R. Mabey, 1974. Application of surface geophysics to ground-water investigations, Techniques of Water-Resources Investigations of the USGS, Chapter D1, 116 pp.

Zytner, R.G., N. Biswas and J.K. Bewtra, 1989. PCE volatilized from stagnant water and soil, *ASCE Journal of Environmental Engineering*, 115(6):1199-1212.

APPENDIX A: DNAPL CHEMICAL DATA

A.1 DATA TABLE ENTRIES

Selected DNAPL chemical data are provided in Table A-1. Types of data listed are described below.

The *CAS #* is a unique identifier given by the American Chemical Society to chemicals included in the Chemical Abstracts Service Registry System. It is utilized to access several chemical databases and facilitates identification of compounds with multiple and ambiguous names.

Empirical formula identifies the carbon, hydrogen, and other elements within a particular compound.

Formula weight is the weight of one mole of a compound which contains Avogadro's number of molecules $(6.022045 \times 10^{23})$. It is calculated by summing the atomic weights given by the compound empirical formula.

Specific density is the density of a compound at a reference temperature, usually 20° C., divided by the density of water, usually at 4° C. All values are given using this convention, except where a different temperature associated with the density of the compound is given in the reference column.

Absolute viscosity, also known as dynamic viscosity, refers to the internal friction derived from molecular cohesion within a fluid that causes it to resist flow. It can be defined as (Lucius et al., 1990): "the ratio between the applied shear stress and the rate of shear" or the "force per unit area necessary to maintain a unit velocity gradient at right angles to the direction of flow between two parallel plates a unit distance apart." Kinematic viscosity equals the absolute viscosity divided by the liquid density. Absolute viscosity values are given in centipoise (cp) at 20° C. unless a different temperature is noted in parentheses in the references column.

Boiling point is the temperature at which the vapor pressure of a liquid equals or slightly exceeds atmospheric pressure. Values are given in degrees C.

Melting point is the temperature, usually measured in degrees C. at 1 atmosphere, at which the liquid and crystalline phases of a substance are in equilibrium.

Aqueous solubility is the saturated concentration of a compound in water at a given temperature and pressure (20° C. and 1 atmosphere unless otherwise noted in the references column). Representative values are listed for compounds having multiple values reported in the literature.

Vapor pressure is the partial pressure exerted by the vapor of the liquid (or solid) under equilibrium conditions at a given temperature, which is 20° C. unless otherwise noted in the references column. A relative measure of chemical volatility, vapor pressure is used to calculate air-water partition coefficients (i.e., Henry's Law Constants) and volatilization rate constants. Values are given in mm Hg (or Torr).

Henry's Law Constants are air-water partition coefficients that compare the concentration of a compound in the gas phase to its concentration in the water phase at equilibrium. Values are calculated by multiplying vapor pressure (atm) by formula weight (g/mole) and then dividing by aqueous solubility (g/m^3) and listed in atm-m^3/mole at 25° C.

K_{oc}, a soil/sediment organic carbon to water partition coefficient, is defined as:

$$K_{oc} = C_{oc} / C_w \qquad (A-1)$$

where C_{oc} is the equilibrium concentration of a chemical adsorbed to organic carbon in soil (mg chemical/kg organic carbon), and C_w is the equilibrium concentration of the same chemical dissolved in groundwater (mg chemical/L of solution). Log K_{oc} values are given in units of mL/g. Representative values are listed for compounds having multiple values reported in the literature.

K_{ow}, the n-octanol-water partition coefficient, defines the ratio of solute concentration in the water-saturated n-octanol phase to the solute concentration in the n-octanol saturated water phase (Montgomery, 1991). It reflects the relative affinity of a chemical for polar (water) or non-polar (hydrophobic organic) media. Values are unitless. Representative values are listed for compounds having multiple values reported in the literature.

Vapor density is the mass of chemical vapor per unit volume. It can be calculated as

$$VD = (P (FW)) / R K \qquad (A-2)$$

where VD is vapor density (g/L), P is pressure (atm), FW is formula weight (g/mol), R is the ideal gas constant (0.0820575 atm-L/mol-K), and K is the temperature (degrees Kelvin). Calculated values in g/L are provided

Table A-1. Selected data on DNAPL chemicals (refer to explanation in Appendix A).

DNAPL	Synonym	CAS #	Empirical Formula	Formula Weight (g)	Ref.	Specific Density (g/cc)	Ref.	Absolute Viscosity (cp)	Ref.
Aniline	Benzenamine	62-53-3	C6H7N	93.13	b	1.022	b	4.40	c
o-Anisidine	2-Methoxybenzenamine	90-04-0	C7H9NO	123.15	b	1.092	b		
Benzyl alcohol	Benzenemethanol	100-51-6	C7H8O	108.14	a	1.045	a	7.76	d(15)
Benzyl chloride	Chloromethylbenzene	100-44-7	C7H7Cl	126.59	a	1.100	a		
Bis(2-chloroethyl)ether	Bis(-chloroethyl)ether	111-44-4	C4H8Cl2O	143.01	b	1.220	b	2.14	c(25)
Bis(2-chloroisopropyl)ether	Bis(-chloroisopropyl)ether	108-60-1	C6H12Cl2O	171.07	a	1.103	a		
Bromobenzene	Phenyl bromide	108-86-1	C6H5BR	157.01	b	1.495	b	0.99	e(30)
Bromochloromethane	Chlorobromomethane	74-97-5	CH2BrCl	129.39	b	1.934	b	0.57	h
Bromodichloromethane	Dichlorobromomethane	75-27-4	CHBrCl2	163.83	a	1.980	a	1.71	e
Bromoethane	Ethyl bromide	74-96-4	C2H5Br	108.97	b	1.460	b	0.418	d(15)
Bromoform	Tribromomethane	75-25-2	CHBr3	252.73	a	2.890	a	2.02	c
Butyl benzyl phthalate	Benzyl butyl phthalate	85-68-7	C19H20O4	312.37	a	1.120	a		
Carbon disulfide	Carbon bisulfide	75-15-0	CS2	76.13	a	1.263	a	0.37	c
Carbon tetrachloride	Tetrachloromethane	56-23-5	CCl4	153.82	a	1.594	a	0.97	c
Chlorobenzene	Benzene chloride	108-90-7	C6H5Cl	112.56	a	1.106	a	0.80	c
2-Chloroethyl vinyl ether	(2-Chloroethoxy)ethene	110-75-8	C4H7ClO	106.55	a	1.048	a		
Chloroform	Trichloromethane	67-66-3	CHCl3	119.38	a	1.483	a	0.58	c
1-Chloro-1-nitropropane	Chloronitropropane	600-25-9	C3H6ClNO2	123.54	b	1.209	b		
2-Chlorophenol	o-Chlorophenol	95-57-8	C6H5ClO	128.56	a	1.263	a	2.25	e(45)
4-Chlorophenyl phenyl ether	p-Chlorodiphenyl ether	7005-72-3	C12H9ClO	204.66	a	1.203	a		
Chloropicrin	Trichloronitromethane	76-06-2	CCl3NO2	164.38	b	1.656	b		
m-Chlorotoluene		108-41-8	C6H4CH3Cl	126.59	f	1.072	f	0.75	h(38)
o-Chlorotoluene	2-Chloro-1-methylbenzene	95-45-8	C6H4CH3Cl	126.58	f	1.082	f	0.75	h(38)
p-Chlorotoluene		106-43-4	C6H4CH3Cl	126.59	f	1.066	f(25)		
Dibromochloromethane	Chlorodibromomethane	124-48-1	CHBr2Cl	208.28	a	2.451	a		
1,2-Dibromo-3-chloropropane	DPCP	96-12-8	C3H5Br2Cl	236.36	b	2.050	b		
Dibromodifluoromethane	Freon 12-B2	75-61-6	CBr2F2	209.82	b	2.297	b		
Dibutyl phthalate	Dibutyl-n-phthalate; DBP	84-74-2	C16H22O4	278.35	a	1.046	a	20.30	c
1,2-Dichlorobenzene	o-Dichlorobenzene	95-50-1	C6H4Cl2	147.00	a	1.305	a	1.32	c(25)
1,3-Dichlorobenzene	m-Dichlorobenzene	541-73-1	C6H4Cl2	147.00	a	1.288	a	1.04	c(25)
1,1-Dichloroethane	1,1-DCA	75-34-3	C2H4Cl2	98.96	a	1.176	a	0.44	c
1,2-Dichloroethane	Ethylene dichloride; 1,2-DCA	107-06-2	C2H4Cl2	98.96	a	1.235	a	0.80	c
1,1-Dichloroethene	Vinylidene chloride; 1,1-DCE	75-35-4	C2H2Cl2	96.94	a	1.218	a	0.36	c
trans-1,2-Dichloroethene	trans-1,2-DCE	156-60-5	C2H2Cl2	96.94	a	1.257	a	0.40	c
1,2-Dichloropropane	Propylene dichloride	78-87-5	C3H6Cl2	112.99	a	1.560	a	0.86	c
cis-1,3-Dichloropropene	cis-1,3-Dichloropropylene	10061-01-5	C3H4Cl2	110.97	a	1.224	a		
trans-1,3-Dichloropropene	trans-1,3-Dichloropropylene	10061-02-6	C3H4Cl2	110.97	a	1.182	a		
Dichlorvos	No-Pest Strip	62-73-7	C4H7Cl2)4P	220.98	b	1.415	b(25)		
Diethyl phthalate	DEP	84-66-2	C12H14O4	222.24	a	1.118	a	35.00	c
Dimethyl phthalate	DMP	131-11-3	C10H10O4	194.19	a	1.191	a	17.20	c(25)

Table A-1. Selected data on DNAPL chemicals (refer to explanation in Appendix A).

DNAPL	Synonym	CAS #	Empirical Formula	Formula Weight (g)	Ref.	Specific Density (g/cc)	Ref.	Absolute Viscosity (cp)	Ref.
Ethylene dibromide	1,2-Dibromoethane; EDB	106-93-4	C2H4Br2	187.86	b	2.179	b	1.72	c
Hexachlorobutadiene	HCBD	87-68-3	C4Cl6	260.76	a	1.554	a	2.45	c(38)
Hexachlorocyclopentadiene	HCCPD	77-47-4	C5Cl6	272.77	a	1.702	a		
Iodomethane	Methyl iodide	74-88-4	CH3I	141.94	b	2.279	b	0.52	d(15)
1-Iodopropane	Propyl iodide	107-08-4	C3H7I	169.99	b	1.749	b	0.84	d(15)
Malathion		121-75-5	C10H19O6PS2	330.36	b	1.230	b(25)		
Methylene chloride	Dichloromethane	75-09-2	CH2Cl2	84.93	a	1.327	a	0.43	c
Nitrobenzene	Nitrobenzol	98-95-3	C6H5NO2	123.11	a	1.204	a	2.01	c
Nitroethane	UN 2842	79-24-3	C2H5NO2	75.07	b	1.045	b(25)	0.66	d(25)
1-Nitropropane	UN 2608	108-03-2	C3H7NO2	89.09	b	1.008	b(24)	0.80	d(25)
2-Nitrotoluene	1-Methyl-2-nitrobenzene	88-72-2	C7H7NO2	137.14	b	1.163	b	2.37	d
3-Nitrotoluene	1-Methyl-3-nitrobenzene	99-08-1	C7H7NO2	137.14	b	1.157	b		
Parathion		56-38-2	C10H14NO5PS	291.27	b	1.260	b		
PCB-1016	Aroclor 1016	12674-11-2	varies	257.90	a	1.330	a(25)	19.3	g(38)
PCB-1221	Aroclor 1221	11104-28-2	varies	192.00	a	1.180	a(25)	4.8	g(38)
PCB-1232	Aroclor 1232	11141-16-5	varies	221.00	a	1.240	a(25)	8.2	g(38)
PCB-1242	Aroclor 1242	53469-21-9	varies	261.00	a	1.392	a(15)	24	g(38)
PCB-1248	Aroclor 1248	12672-29-6	varies	288.00	a	1.410	a(25)	65	g(38)
PCB-1254	Aroclor 1254	11097-69-1	varies	327.00	a	1.505	a(15)	700	g(38)
Pentachloroethane	Ethane pentachloride	76-01-7	C2HCl5	202.28	b	1.680	b	2.75	d(15)
1,1,2,2-Tetrabromoethane	Acetylene tetrabromide	79-27-6	C2H2Br4	345.65	b	2.875	b	9.79	d
1,1,2,2-Tetrachloroethane	Acetylene tetrachloride	79-34-5	C2H2Cl4	167.85	a	1.595	a	1.75	c
Tetrachloroethene	Perchloroethylene; PCE	127-18-4	C2Cl4	165.83	a	1.623	a	0.89	c
Thiophene	Thiacylopentadiene	110-02-1	C4H4S	84.14	b	1.065	b	0.65	d
1,2,4-Trichlorobenzene	1,2,4-TCB	120-82-1	C6H3Cl3	181.45	a	1.454	a	1.42	c
1,1,1-Trichloroethane	Methyl chloroform; 1,1,1-TCA	71-55-6	C2H3Cl3	133.40	a	1.339	a	1.20	c
1,1,2-Trichloroethane	1,1,2-TCA	79-00-5	C2H3Cl3	133.40	a	1.440	a	0.12	c
Trichloroethene	TCE	79-01-6	C2HCl3	131.39	a	1.464	a	0.57	c
1,1,2-Trichlorofluoromethane	Freon 11	75-69-4	CCl3F	137.37	a	1.487	a	0.42	c(25)
1,2,3-Trichloropropane	Allyl trichloride	96-18-4	C3H5Cl3	147.43	b	1.3889	b		
1,1,2-Trichlorotrifluoroethane	Freon 113	76-13-1	C2Cl3F3	187.38	b	1.564	b		
Tri-o-cresyl phosphate	o-Cresyl phosphate	78-30-8	C21H21O4P	368.37	b	1.955	b	80.00	d
Water	Ice	7732-18-5	H2O	18.02		1.000		1.00	

Table A-1. (continued)

DNAPL	Boiling Point (deg.C)	Ref.	Melting Point (deg.C)	Ref.	Aqueous Solubility (mg/L)	Ref.	Vapor Pressure (mm Hg)	Ref.	Henry's Law Constant (atm-m3/mol)	Ref.
Aniline	184	b	-6	b	3.50E+04	b	3.00E-01	b	1.36E-01	b
o-Anisidine	224	b	6	b	1.30E+04	b	<0.1	b	1.25E-06	b
Benzyl alcohol	205	a	-15	a	3.50E+04	a	<1	a		a
Benzyl chloride	179	a	-39	a	4.93E+02	a	9.00E-01	a	3.04E-04	a
Bis(2-chloroethyl)ether	179	b	-47	b	1.02E+04	b	7.10E-01	b	1.30E-05	b
Bis(2-chloroisopropyl)ether	187	a	-20	sax	1.70E+03	a	8.50E-01	a	1.10E-04	a
Bromobenzene	156	b	-31	b	5.00E+02	b	3.30E+00	b	2.40E-03	b
Bromochloromethane	68	b	-87	b	1.67E+04	b(25)	1.41E+00	b(25)	1.44E-03	b
Bromodichloromethane	90	a	-57	a	4.50E+03	a(0)	5.00E+01	a	2.12E-04	a
Bromoethane	38	b	-119	b	9.14E+03	b	3.75E+02	b	7.56E-03	b
Bromoform	149	a	8	a	3.01E+03	a	4.00E+00	a	5.32E-04	a
Butyl benzyl phthalate	370	a	-35	a	2.82E+00	a	8.60E-06	a	1.30E-06	a
Carbon disulfide	46	a	-112	a	2.10E+03	a	2.98E+02	a	1.33E-02	a
Carbon tetrachloride	77	a	-23	a	8.00E+02	a	9.00E+01	a	3.02E-02	a
Chlorobenzene	132	a	-46	a	5.00E+02	a	9.00E+00	a	4.45E-03	a
2-Chloroethyl vinyl ether	108	a	-70	a	1.50E+04	a	2.68E+01	a	2.50E-04	a
Chloroform	62	a	-63	a	8.00E+03	a	1.60E+02	a	3.20E-03	a
1-Chloro-1-nitropropane	142	b	<25	b	6.00E+00	b	5.80E+00	b(25)	1.57E-01	b
2-Chlorophenol	175	a	9	a	2.85E+04	a	1.42E+00	a(25)	8.28E-06	a
4-Chlorophenyl phenyl ether	284	a	-8	a	3.30E+00	a(25)	2.70E-03	a(25)	2.20E-04	a
Chloropicrin	112	b	-64	b	2.00E+03	b	2.00E+01	b	8.40E-02	b
m-Chlorotoluene	160	f	-48	f	4.80E+01	e	4.60E+00	e	1.60E-02	e
o-Chlorotoluene	159	f	-34	f	7.20E+01	e	2.70E+00	f	6.25E-03	e
p-Chlorotoluene	162	f	7	f	4.40E+01	e	4.50E+00	e	1.70E-02	e
Dibromochloromethane	117	a	-22	a	4.00E+03	a	7.60E+01	a	9.90E-04	a
1,2-Dibromo-3-chloropropane	196	b	6	b	1.00E+03	b	8.00E-01	b	2.49E-04	b
Dibromodifluoromethane	23	b	-141	b			6.88E+02	b		
Dibutyl phthalate	335	a	-35	a	1.01E+01	a	1.40E-05	a(25)	6.30E-05	a
1,2-Dichlorobenzene	180	a	-17	a	1.00E+02	a	1.00E+00	a	1.90E-03	a
1,3-Dichlorobenzene	173	a	-25	a	1.11E+02	a	2.30E+00	a(25)	3.60E-03	a
1,1-Dichloroethane	56	a	-97	a	5.50E+03	a	1.82E+02	a	4.30E-03	a
1,2-Dichloroethane	83	a	-35	a	8.69E+03	a	6.40E+01	a	9.10E-04	a
1,1-Dichloroethene	37	a	-122	a	4.00E+02	a	4.95E+02	a	2.10E-02	a
trans-1,2-Dichloroethene	47	a	-50	a	6.00E+02	a	2.65E+02	a	3.84E-01	a
1,2-Dichloropropane	96	a	-100	a	2.70E+03	a	4.20E+01	a	2.30E-03	a
cis-1,3-Dichloropropene	104	a	-84	a	2.70E+03	a	2.50E+01	a	1.30E-03	a
trans-1,3-Dichloropropene	112	a	-84	a	2.80E+03	a	2.50E+01	a	1.30E-03	a
Dichlorvos					1.00E+04	b	1.20E-02	b	5.00E-03	b
Diethyl phthalate	298	a	-40	a	9.28E+02	a	1.65E-03	a(25)	8.46E-07	a
Dimethyl phthalate	283	a	0	a	4.29E+03	a	1.65E-03	a(25)	4.20E-07	a

Table A-1. (continued)

DNAPL	Boiling Point (deg.C)	Ref.	Melting Point (deg.C)	Ref.	Aqueous Solubility (mg/L)	Ref.	Vapor Pressure (mm Hg)	Ref.	Henry's Law Constant (atm-m3/mol)	Ref.
Ethylene dibromide	131	b	10	b	4.32E+03	b	1.10E+01	b	7.06E-04	b
Hexachlorobutadiene	215	a	-21	a	2.55E+00	a	1.50E-01	a	2.60E-02	a
Hexachlorocyclopentadiene	237	a	-9	a	1.10E+00	a(22)	8.10E-02	a(25)	1.60E-02	a
Iodomethane	42.4	b	-66	b	1.40E+04	b	3.75E+02	b	5.48E-03	b
1-Iodopropane	102	b	-101	b	1.06E+03	b(23)	4.00E+01	b(24)	9.09E-03	b
Malathion			2.9	b	1.45E+02	b	1.25E-06	b	4.89E-09	b
Methylene chloride	40	a	-95	a	2.00E+04	a	3.49E+02	a	2.00E-03	a
Nitrobenzene	211	a	6	a	1.90E+03	a	1.50E-01	a	2.45E-05	a
Nitroethane	115	b	-50	b	4.50E+04	b	1.56E+01	b	4.66E-05	b
1-Nitropropane	130	b	-108	b	1.40E+04	b	7.50E+00	b	8.68E-05	b
2-Nitrotoluene	222	b	-3	b	6.00E+02	b	1.50E-01	b	4.51E-05	b
3-Nitrotoluene	233	b	16	b	5.00E+02	b	1.50E-01	b	5.41E-05	b
Parathion	375	b	6	b	1.20E+01	b	4.00E-04	b	8.56E-08	b
PCB-1016	325	a			2.30E-01	a	4.00E-04	a(25)		
PCB-1221	275	a	1	a	5.90E-01	a(24)	6.70E-03	a(25)	3.24E-04	a
PCB-1232	290	a	-35	a	1.45E+00	a(25)	4.60E-03	a(25)	4.64E+00	a
PCB-1242	325	a	-19	a	2.00E-01	a	1.00E-03	a	5.60E-04	a
PCB-1248	340	a	-7	a	5.00E-02	a	4.94E-04	a(25)	3.50E-03	a
PCB-1254	365	a	10	a	5.00E-02	a	6.00E-05	a	2.70E-03	a
Pentachloroethane	159	b	-22	b	5.00E+02	b	3.40E+00	b	2.45E-03	b
1,1,2,2-Tetrabromoethane	239	b	0	b	7.00E+02	b	1.00E-01	b	6.40E-05	b
1,1,2,2-Tetrachloroethane	146	a	-36	a	2.90E+03	a	5.00E+00	a	3.80E-04	a
Tetrachloroethene	121	a	-19	a	1.50E+02	a	1.40E+01	a	1.53E-02	a
Thiophene	84	b	-30	b	3.60E+03	b(18)	6.00E+01	b	2.93E-03	b
1,2,4-Trichlorobenzene	210	a	17	a	1.90E+01	a(22)	4.00E-01	a(25)	2.32E-03	a
1,1,1-Trichloroethane	74	a	-30	a	1.36E+03	a	1.00E+02	a	1.80E-02	a
1,1,2-Trichloroethane	114	a	-37	a	4.50E+03	a	1.90E+01	a	7.40E-04	a
Trichloroethene	87	a	-73	a	1.10E+03	a	5.78E+01	a	9.10E-03	a
1,1,2-Trichlorofluoromethane	24	a	-111	a	1.10E+03	a	6.87E+02	a	1.10E-01	a
1,2,3-Trichloropropane	142	b	-15	b			2.00E+00	b	3.18E-04	b
1,1,2-Trichlorotrifluoroethane	48	b	-35	b	2.00E+02	b	2.84E+02	b	3.33E-01	b
Tri-o-cresyl phosphate	410	b	-25	b	3.00E-01	b				
Water	100		0				1.75E+01			

Table A-1. (continued)

DNAPL	Log Koc (mL/g)	Ref.	Log Kow	Ref.	Vapor Density (g/L)	Ref.	Relative Vapor Density	Interfacial Liquid Tension (dyn/cm)	Ref.	Surface Tension (dyn/cm)	Ref.
Aniline	1.41	b	0.90	b	3.81	b	1.001	5.8		42.9	c
o-Anisidine			0.95	b	5.03	b					
Benzyl alcohol	1.98	a	1.10	a	4.42	a					
Benzyl chloride	2.28	a	2.30	a	5.17	a	1.004				
Bis(2-chloroethyl)ether	1.15	b	1.58	b	5.84	b	1.004			37.9	c
Bis(2-chloroisopropyl)ether	1.79	a	2.58	a	6.99	a	1.006				
Bromobenzene	2.33	b	3.01	b	6.42	b	1.019	39.8	j	35.8	e
Bromochloromethane	1.43	b	1.41	b	5.29	b	1.006			33.3	h
Bromodichloromethane	1.79	a	1.88	a	6.70	a	1.309				
Bromoethane	2.67	b	1.57	b	4.05	b	2.377			24.5	h
Bromoform	2.45	a	2.30	a	10.33	a	1.041			45.5	c
Butyl benzyl phthalate	2.32	a	4.78	a	12.76	a	1.000				
Carbon disulfide	2.47	a	1.84	a	3.11	a	1.646	48.4	j	32.3	c
Carbon tetrachloride	2.64	a	2.83	a	6.29	a	1.515	45.0	j	27.0	c
Chlorobenzene	1.68	a	2.84	a	4.60	a	1.035	37.4	j	33.2	c
2-Chloroethyl vinyl ether	0.82	a	1.28	a	4.36	a	1.095				
Chloroform	1.64	a	1.95	a	4.88	a	1.664	32.8	j	27.2	c
1-Chloro-1-nitropropane	3.34	b	4.25	b	5.05	b	1.025				
2-Chlorophenol	2.56	a	2.16	a	5.25	a	1.006			40.3	e
4-Chlorophenyl phenyl ether	3.60	a	4.08	a	8.36	a	1.000				
Chloropicrin	0.82	b	1.03	b	6.72	b	1.124				
m-Chlorotoluene	3.08	e	3.28	e			1.021			32.8	e
o-Chlorotoluene	3.20	e	3.42	f			1.012			32.9	h(25)
p-Chlorotoluene	3.08	e	3.3	e			1.020			34.6	h(25)
Dibromochloromethane	1.92	a	2.08	a	8.51	a	1.624				
1,2-Dibromo-3-chloropropane	2.11	b	2.63	b	9.66	b	1.008				
Dibromodifluoromethane					8.58	b	6.701				
Dibutyl phthalate	3.14	a	4.57	a	11.38	a	1.000			33.4	c
1,2-Dichlorobenzene	2.27	a	3.40	a	6.01	a	1.005	40.0	e	37.0	c
1,3-Dichlorobenzene	2.23	a	3.38	a	6.01	a	1.012			33.2	c
1,1-Dichloroethane	1.48	a	1.78	a	4.04	a	1.585			24.8	c
1,2-Dichloroethane	1.15	a	1.48	a	4.04	a	1.206	30.0	e(30)	32.2	c
1,1-Dichloroethene	1.81	a	2.13	a	3.96	a	2.545	37.0	e(23)	24.0	c(15)
trans-1,2-Dichloroethene	1.77	a	2.09	a	3.96	a	1.827	30.0	e	25.0	c
1,2-Dichloropropane	1.71	a	2.28	a	4.62	a	1.162			28.7	c
cis-1,3-Dichloropropene	1.68	a	1.41	a	4.54	a	1.094	23.8	e(27)	31.2	e
trans-1,3-Dichloropropene	1.68	a	1.41	a	4.54	a	1.094				
Dichlorvos	9.57	b	1.40	b	9.03	b	1.000				
Diethyl phthalate	1.84	a	2.35	a	9.08	a	1.000			37.5	c
Dimethyl phthalate	1.63	a	1.61	a	7.94	a	1.000				

Table A-1. (continued)

DNAPL	Log Koc (mL/g)	Ref.	Log Kow	Ref.	Vapor Density (g/L)	Ref.	Relative Vapor Density	Interfacial Liquid Tension (dyn/cm)	Ref.	Surface Tension (dyn/cm)	Ref.
Ethylene dibromide	1.64	b	1.76	b	7.68	b	1.080	36.5	e	38.7	c
Hexachlorobutadiene	3.67	a	4.78	a	10.66	a	1.002				
Hexachlorocyclopentadiene	3.63	a	5.04	a	11.15	a	1.001			37.5	e
Iodomethane	1.36	b	1.69	b	5.80	b	2.943			31.0	e
1-Iodopropane	2.16	b	2.49	b	6.95	b	1.259				
Malathion	2.46	b	2.89	b	13.50	b	1.000				
Methylene chloride	0.94	a	1.30	a	3.47	a	1.897	28.3	j	27.9	c
Nitrobenzene	2.01	a	1.95	a	5.03	a	1.001	25.7	j	43.0	c
Nitroethane			0.18	b	3.07	b	1.033				
1-Nitropropane			0.87	b	3.64	b	1.021				
2-Nitrotoluene			2.30	b	5.61	b	1.001				
3-Nitrotoluene			2.42	b	5.61	b	1.001				
Parathion	3.07	b	3.81	b	11.91	b	1.000				
PCB-1016	4.70	a	5.88	a			1.000				
PCB-1221	2.44	a	2.80	a			1.000				
PCB-1232	2.83	a	3.20	a	9.03	a	1.000				
PCB-1242	3.71	a	4.11	a	10.67	a	1.000				
PCB-1248	5.64	a	6.11	a			1.000				
PCB-1254	5.61	a	6.47	a	13.36	a	1.000				
Pentachloroethane	3.28	b	2.89	b	8.27	b	1.027			34.7	e
1,1,2,2-Tetrabromoethane	2.45	b	2.91	b	14.13	b	1.001				
1,1,2,2-Tetrachloroethane	2.07	a	2.56	a	6.86	a	1.032			36.0	c
Tetrachloroethene	2.42	a	2.60	a	6.78	a	1.088	44.4	e(25)	31.3	c
Thiophene	1.73	b	1.81	b	3.44	b	1.152				
1,2,4-Trichlorobenzene	3.98	a	4.02	a	7.42	a	1.003			39.1	c
1,1,1-Trichloroethane	2.18	a	2.47	a	5.45	a	1.479	45.0	e	25.4	c
1,1,2-Trichloroethane	1.75	a	2.18	a	5.45	a	1.091			34.0	c
Trichloroethene	2.10	a	2.53	a	5.37	a	1.272	34.5	e(24)	29.3	c
1,1,2-Trichlorofluoromethane	2.20	a	2.53	a	5.85	a	4.415			19.0	c
1,2,3-Trichloropropane							1.011				
1,1,2-Trichlorotrifluoroethane	2.59	b	2.57	b	7.66	b	3.062				
Tri-o-cresyl phosphate	3.37	b	5.11	b	15.06	b					

Table A-1. (continued)

DNAPL	Air Diffusion Coefficient (sq.cm./sec)	Ref.	Water Diffusion Coefficient (sq.cm/sec)	Ref.	Estimated Half-life in Soil (days)	Estimated Half-life in Groundwater (days)	RCRA or NJ Action Level Water (mg/L)	RCRA or NJ Action Level Soil (mg/kg)
Ethylene dibromide					28-180	20-120	4E-07	8E-03
Hexachlorobutadiene					28-180	56-360	4E-03	9E+01
Hexachlorocyclopentadiene					7-28	7-56	2E-01	6E+02
Iodomethane					7-28	14-56		
1-Iodopropane								
Malathion					3-7	8-103	2E-01 NJ	
Methylene chloride	1.02E-01	i	1.1E-06	c	7-28	14-56	5E-03	9E+01
Nitrobenzene	7.20E-02	h	7.6E-06	c	12-197	2-394	2E-02	4E+01
Nitroethane								
1-Nitropropane					28-180	56-360		
2-Nitrotoluene								
3-Nitrotoluene								
Parathion							2E-01	5E+02
PCB-1016							5E-06 *	9E-02 *
PCB-1221							5E-06 *	9E-02 *
PCB-1232							5E-06 *	9E-02 *
PCB-1242							5E-06 *	9E-02 *
PCB-1248							5E-06 *	9E-02 *
PCB-1254							5E-06 *	9E-02 *
Pentachloroethane								
1,1,2,2-Tetrabromoethane								
1,1,2,2-Tetrachloroethane					0.45-45	0.45-45	2E-03	4E+01
Tetrachloroethene	7.40E-02	i	7.5E-06	c	180-360	360-720	7E-04	1E+01
Thiophene								
1,2,4-Trichlorobenzene					28-180	56-360	7E-01	2E+03
1,1,1-Trichloroethane	7.96E-02	i	8E-06	h	140-273	140-546	3E+00	7E+03
1,1,2-Trichloroethane	7.90E-02	h	8E-06	h	136-360	136-720	6E-03	1E+02
Trichloroethene	8.11E-02	i	8.3E-06	c	180-360	321-1653	5E-03	6E+01
1,1,2-Trichlorofluoromethane					180-360	360-720	1E+01	2E+04
1,2,3-Trichloropropane					180-360	360-720	2E-01	5E+02
1,1,2-Trichlorotrifluoroethane					180-360	360-720		
Tri-o-cresyl phosphate								

Table A-1. (continued)

DNAPL	Air Diffusion Coefficient (sq.cm./sec)	Ref.	Water Diffusion Coefficient (sq.cm/sec)	Ref.	Estimated Half-life in Soil (days)	Estimated Half-life in Groundwater (days)	RCRA or NJ Action Level Water (mg/L)	RCRA or NJ Action Level Soil (mg/kg)
Aniline	7.50E-02	c(30)					6E-03	1E+02
o-Anisidine					28-180	56-360		
Benzyl alcohol							2E+00 NJ	
Benzyl chloride					0.62-12	0.62-12		
Bis(2-chloroethyl)ether					28-180	56-360	3E-03	5E+01
Bis(2-chloroisopropyl)ether					18-180	36-360	3E-01 NJ	
Bromobenzene								
Bromochloromethane								
Bromodichloromethane							3E-05	5E-01
Bromoethane								
Bromoform					28-180	56-360	7E-01	2E+03
Butyl benzyl phthalate					1-7	2-180	7E+00	2E+04
Carbon disulfide	8.92E-02	h	1.1E-05	h			4E+00	8E+03
Carbon tetrachloride	7.97E-02	i			180-360	7-360	3E-04	5E+00
Chlorobenzene	7.50E-02	c(30)	7.9E-06	h	68-150	136-300	7E-01	2E+03
2-Chloroethyl vinyl ether								
Chloroform	9.90E-02	i	9.1E-06	h	28-180	56-1800	6E-03	1E+02
1-Chloro-1-nitropropane								
2-Chlorophenol							2E-01	4E+02
4-Chlorophenyl phenyl ether								
Chloropicrin								
m-Chlorotoluene								
o-Chlorotoluene								
p-Chlorotoluene								
Dibromochloromethane					28-180	14-180	1E-02 NJ	
1,2-Dibromo-3-chloropropane					28-180	56-360	2E-06 NJ	
Dibromodifluoromethane								
Dibutyl phthalate	4.20E-02	c(25)	4.1E-05	c	2-23	2-23	4E+00	8E+03
1,2-Dichlorobenzene					28-180	56-360	6E-01 NJ	
1,3-Dichlorobenzene					28-180	56-360	6E-01 NJ	
1,1-Dichloroethane	8.90E-02	i			32-154	64-154	7E-02 NJ	
1,2-Dichloroethane	8.90E-02	i			100-180	100-360	5E-03	8E+00
1,1-Dichloroethene	9.11E-02	i	9.5E-06	h	28-180	56-132	7E-03	1E+01
trans-1,2-Dichloroethene	9.11E-02	i	9.5E-06	c			1E-01 NJ	
1,2-Dichloropropane					167-1289	334-2592	5E-04 NJ	
cis-1,3-Dichloropropene					5-11	5-11	1E-02	2E+01
trans-1,3-Dichloropropene					5-11	5-11	1E-02	2E+01
Dichlorvos								
Diethyl phthalate					3-56	6-112	3E+01	6E+04
Dimethyl phthalate					1-7	2-14	7E+00 NJ	

Table A-1. (continued)

DNAPL	Flash Point (deg.C)	Ref.	LEL (%)	Ref.	UEL (%)	Ref.	ACGIH TWA (ppm)	ACGIH STEL (ppm)	NIOSH IDLH (ppm)	Odor Low Threshold (ppm)	Odor High Threshold (ppm)
Aniline	70	b	1.3	b	11	b	Ca 2 (7.6)		Ca 100	5.25E-05	92
o-Anisidine	30	b(oc)					Ca 0.1 (0.50)		Ca 9.8		
Benzyl alcohol	93	a									
Benzyl chloride	60	b	1.1	b			Ca 1 (5.2)		10	4.54E-02	0.3
Bis(2-chloroethyl)ether	55	a									
Bis(2-chloroisopropyl)ether	85	a									
Bromobenzene	51	b									
Bromochloromethane	NC	b					200 (1060)		5000	3.17E+02	317
Bromodichloromethane											
Bromoethane	<-20	b	6.7	b	11.3	b	200 (891)	250 (1110)	3500	2.00E+02	200
Bromoform	NC	a					0.5 (5.2)			5.13E+02	513
Butyl benzyl phthalate	110	a									
Carbon disulfide	-30	a	1.3	a	50	a	10 (31)		500	7.80E-03	7
Carbon tetrachloride	NC	a					Ca 5 (31)		Ca 300	9.54E+00	238
Chlorobenzene	28	a	1.3	a	7.1	a	75 (345)		2400	2.13E-01	61
2-Chloroethyl vinyl ether	16	a									
Chloroform	NC	a					Ca 10 (49)		Ca 1000	5.12E+01	205
1-Chloro-1-nitropropane	62	b					2 (10)		2000		
2-Chlorophenol	64	a								3.59E-03	1
4-Chlorophenyl phenyl ether											
Chloropicrin	detonates	b					0.1 (0.67)		4	8.12E-01	1
m-Chlorotoluene											
o-Chlorotoluene											
p-Chlorotoluene											
Dibromochloromethane	NC	a									
1,2-Dibromo-3-chloropropane	77	b(oc)							Ca	9.98E-03	0
Dibromodifluoromethane	NC	b									
Dibutyl phthalate	157	a	0.5	a	2.5	a	(5)		803		
1,2-Dichlorobenzene	66	a	2.2	a	9.2	a	50 (301) C		1000	2.00E+00	50
1,3-Dichlorobenzene	63	a	2	a	9.2	a					
1,1-Dichloroethane	-6	a	5.6	a	16	a	200 (810)	250 (1010)	4000	1.10E+02	200
1,2-Dichloroethane	13	a	6.2	a	16	a	Ca 10 (4)		Ca 1000	5.93E+00	109
1,1-Dichloroethene	-15	a	6.5	a	15.5	a	Ca 5 (20)	Ca 20 (79)		5.04E+02	1009
trans-1,2-Dichloroethene	2	a	9.7	a	12.8	a	200 (793)		4000	8.47E-02	498
1,2-Dichloropropane	15.6	a	3.4	a	14.5	a	Ca 75 (347)	Ca 110 (508)	2000 ca	2.52E-01	131
cis-1,3-Dichloropropene	35	a	5.3	a	14.5	a	Ca 1 (4.5)				
trans-1,3-Dichloropropene	5.3	a	5.3	a	14.5	a	Ca 1 (4.5)				
Dichlorvos	NC	b					0.1 (0.90)		21		
Diethyl phthalate	140	a	0.7	a			(5)				
Dimethyl phthalate	146	a	1.2	a			(5)		1152		

Table A-1. (continued)

DNAPL	Flash Point (deg.C)	Ref.	LEL (%)	Ref.	UEL (%)	Ref.	ACGIH TWA (ppm)	ACGIH STEL (ppm)	NIOSH IDLH (ppm)	Odor Low Threshold (ppm)	Odor High Threshold (ppm)
Ethylene dibromide	NC	b					Ca		400 ca	1.00E+01	10
Hexachlorobutadiene	NC	a					Ca 0.02 (0.21)			1.13E+00	1
Hexachlorocyclopentadiene	NC	a					0.01 (0.11)			1.34E-01	0
Iodomethane	NC	b					Ca 2 (12)		800 ca		
1-Iodopropane											
Malathion	NC	b					(10)		364	9.99E-01	1
Methylene chloride	NC	c					Ca 50 (174)		5000 ca	1.55E+02	622
Nitrobenzene	88	a	1.8	a			1 (5)		200	4.67E-03	2
Nitroethane	28	b	3.4	b			100 (307)		1000	2.02E+02	202
1-Nitropropane	34	b	2.2	b			25 (91)		2300	2.96E+02	296
2-Nitrotoluene	106	b	2.2	b					200		
3-Nitrotoluene	101	b	1.6	b					200		
Parathion	NC	b					(0.1)		1.6	4.00E-02	0
PCB-1016	NC	a									
PCB-1221	141	a									
PCB-1232	152	a									
PCB-1242	176	a							Ca 0.9		
PCB-1248	193	a									
PCB-1254	222	a							Ca 0.3		
Pentachloroethane											
1,1,2,2-Tetrabromoethane	NC	a					1 (14)				
1,1,2,2-Tetrachloroethane	NC	a					Ca 1 (6.9)			3.06E+00	5
Tetrachloroethene	NC	a	NA		NA		Ca 50 (339)	Ca 200 (1357)	Ca 500	4.65E+00	69
Thiophene	-1.1	a									
1,2,4-Trichlorobenzene	105	a	2.5	a	6.6	a	5 (37) C			3.23E+00	3
1,1,1-Trichloroethane	NC	c					350 (1910)	450 (2460)	1000	9.95E+01	696
1,1,2-Trichloroethane	NC	c					Ca 10 (65)		Ca 500		
Trichloroethene	32.2	a	8	a	10.5	a	Ca 50 (269)	Ca 200 (1070)		2.10E-01	402
1,1,2-Trichlorofluoromethane	NC	a					1000 (5620) C		10000	4.98E+00	208
1,2,3-Trichloropropane	73.3	b	3.2	b	12.6	b	10 (60)		Ca 1000		
1,1,2-Trichlorotrifluoroethane	NC	b					1000 (7670)	1250 (9590)	4500	4.46E+01	134
Tri-o-cresyl phosphate	225	b					(0.1)		2.6		

for a reference temperature of 25° C. The vapor density of dry air at 25° C. and 1 atmosphere is 1.204 g/L.

Relative vapor density is calculated as the weighted mean formula weight of compound-saturated air relative to the mean formula weight of moist air (28.75 g/mol),

$$RVD = ((P_xFW/760)+(29.0(760-v_p)/760))/28.75 \quad (A-3)$$

where RVD is the relative vapor density (unitless), P_x is the vapor pressure in mm Hg, and FW is the compound formula weight (Schwille, 1988). Contaminated soil gas with a high RVD will tend to sink in the subsurface.

Interfacial liquid tension between DNAPL and water develops due to the difference between the greater mutual attraction of like molecules within each fluid and the lesser attraction of dissimilar molecules across the immiscible fluid interface (Chapter 4.2). Values are given in dynes per cm at 20° C. unless noted otherwise in the references column.

Surface tension refers to the interfacial tension between a liquid and its own vapor. Values are given in dynes per cm at 20° C. unless noted otherwise in the references column.

Air diffusion coefficients indicate the diffusivity of a chemical vapor in air. Values are given in cm^2/s at 20° C. unless noted otherwise in the references column

Water diffusion coefficients indicate the diffusivity of dilute solutes in water at 20° C. unless noted otherwise in the references column.

Low to high ranges of *estimated chemical half-lives in soil* are given in days. These ranges were developed by Howard et al. (1991) based on consideration of various degradation processes and limited available data.

Low to high ranges of *estimated chemical half-live in water* are given in days. These ranges were developed by Howard et al. (1991) based on consideration of various degradation processes and limited available data.

RCRA or NJ Action Levels for soil and water are proposed chemical concentrations in shallow soil and groundwater which would trigger conduct of a Corrective Measures Study under the Resource Conservation and Recovery Act (RCRA) (Federal Register, 7/27/90; NJDEP, 10/14/91). Values are given in ppm. A "*" by each PCB

Aroclor series indicates that the proposed action level is for total PCBs.

Flash point is the minimum temperature in degrees C. at which a liquid or solid emits ignitable flammable vapors given the presence of an ignition source such as a spark or flame. Given flash point temperatures are determined using the Tag closed cup (ASTM method D56) except where (oc) in the references column is used to denote open cup measurement (ASTM method D93). NC indicates that the DNAPL is non-combustible.

LEL, the lower explosive limit, refers to the minimum volumetric percent of a flammable gas or vapor in air at which ignition or explosion can occur in the presence of a spark or flame.

UEL, the upper explosive limit, refers to the maximum volumetric percent of a flammable gas or vapor in air at which ignition or explosion can occur in the presence of a spark or flame.

ACGIH TWA values represent the time-weighted average chemical concentration in breathing air, prescribed by the American Conference of Governmental Industrial Hygienists (ACGIH, 1990) to which nearly all workers can be exposed without adverse effect during a normal 8-hour work day and 40-hour work week. Values are given in ppm, and in mg/m^3 (in parentheses). Suspected or confirmed carcinogens are denoted by Ca.

AGCIH STEL values represent the 15-minute time-weighted average chemical concentration in breathing air, prescribed by the American Conference of Governmental Industrial Hygienists (ACGIH, 1990), to which nearly all workers can be exposed without adverse effect. These values should not be exceeded during any 15-minute period even if the TWA value is not exceeded. Values are given in ppm, and in mg/m^3 (in parentheses). Suspected or confirmed carcinogens are denoted by Ca.

NIOSH IDLH values represent chemical concentrations in breathing air specified by the National Institute of Occupational Safety and Health (NIOSH, 1990) to be Immediately Dangerous to Life and Health (IDLH). They are maximum concentrations to which one could be exposed for 30 minutes without suffering escape-impairing or irreversible health effects. Concentrations are given in ppm and in mg/m^3 (in parentheses).

Low and high odor threshold values represent a reported range of minimum chemical vapor concentrations detectable by the sense of smell as a noticeable change in the odor of the system (Ruth, 1986).

Coded *references* that appear in the reference columns include: a=Montgomery and Welkom (1990); b=Montgomery (1991); c=Lucius et al. (1990); d=Mercer et al. (1990); e=Mercer and Cohen (1990); f=Verschueren (1983); g=Monsanto (1988); h=Tetra Tech (1988); i=Mendoza and Frind (1990b); and, j=Dean (1973). Most of these references provide compilations of data reported by others and must be examined, therefore, to determine the original data source.

APPENDIX B: PARAMETERS AND CONVERSION FACTORS

Symbols and dimensions of selected parameters utilized in this document are listed in Table B-1. Conversion factors for length, area, volume, mass, time, density, velocity, force, and pressure are given in Tables B-2 to B-10, respectively.

Table B-1. Listing of selected parameters, symbols, and dimensions.

Parameter	Symbol	Dimensions
Angle of dip	θ	degrees
Concentration of chemical in soil gas	C_a C_g	moles/volume mass/volume mass/mass
Concentration of chemical at source	C_s	moles/volume mass/volume mass/mass
Concentration of chemical in water	C_w	moles/volume mass/volume mass/mass
Contact angle	ϕ	degrees
Contact area	L^2	length^2
Critical NAPL thickness or height	z_n	length
Density	ρ	mass/volume
Density, bulk	ρ_b	mass/volume
Density, NAPL	ρ_n	mass/volume
Density, water	ρ_w	mass/volume
Diffusion coefficient, air	D	length^2/time
Diffusion coefficient, effective	$D^\bullet$	length^2/time
Displacement by NAPL ratio	δ_n	dimensionless
Displacement by water ratio	δ_w	dimensionless
Fraction organic carbon content	f_{oc}	volume/volume dimensionless
Gradient, hydraulic	i	length/length dimensionless
Gradient, capillary pressure	i_{cp}	mass/time
Gradient, pressure due to gravity	i_g	density/density dimensionless
Gravity, acceleration	g	length/time^2
Head, capillary (capillary rise of water)	h_c	length

Table B-1. Listing of selected parameters, symbols, and dimensions.

Parameter	Symbol	Dimensions
Head, pressure head due to NAPL gravity	h_g	length
Head, hydraulic	h	length
Hydraulic Conductivity	K	length/time
Ideal Gas Constant	R	$R = 8.2057 \times 10^{-5}$ m³atm/(mol °K)
Interfacial tension (liquid and surface)	σ	force/length
Mass, dry sample	m_d	mass
Mass exchange coefficient	$m_x/L^2/t$	mass/length²/time
Mass exchange rate	m_x/t	mass/time
Mass, wet sample	m_w	mass
Mean soil particle diameter	d	length
Molecular weight	M	mass/moles
Mole fraction of compound A	X_A	dimensionless
Mole fraction of compound i	X_i	dimensionless
Partition coefficient, Henry's Law Constant	K_H	(mass • length²) / (time² • moles)
Partition coefficient, Henry's Law Constant, dimensionless	$K_{H'}$	dimensionless
Partition coefficient, octanol/water	K_{ow}	dimensionless
Partition coefficient, organic carbon	k_{oc}	volume/mass
Partition coefficient, sorption	k_d	volume/mass
Permeability, intrinsic	k	length²
Permeability, relative	k_r	dimensionless
Permeability, relative (air)	k_{ra}	dimensionless
Permeability, relative (NAPL)	k_{rn}	dimensionless
Permeability, relative (water)	k_{rw}	dimensionless
Pore size distribution index	λ	dimensionless
Porosity	n	volume/volume dimensionless
Porosity, air-filled	n_a	volume/volume dimensionless
Porosity, effective	n_e	volume/volume dimensionless
Porosity, total	n_t	volume/volume dimensionless

Table B-1. Listing of selected parameters, symbols, and dimensions.

Parameter	Symbol	Dimensions
Porosity, bulk water content	n_w	volume/volume dimensionless
Pressure	P	mass/(length*time^2)
Pressure, capillary	P_c	mass/(length*time^2)
Pressure due to gravity	P_g	mass/(length*time^2)
Pressure, NAPL	P_N	mass/(length*time^2)
Pressure, partial of chemical in gas phase	P	mass/(length*time^2)
Pressure, threshold entry (displacement)	P_d	mass/(length*time^2)
Pressure, vapor of pure solvent A	$P_A{}^o$	mass/(length*time^2)
Pressure, vapor of the solution containing solvent A	P_A	mass/(length*time^2)
Pressure, water	P_w	mass/(length*time^2)
Radial distance	r	length
Radius, pore	r	length
Radius, pore body	r_p	length
Radius, pore throat	r_t	length
Radius, source	r_s	length
Retardation factor, dissolved phase	R_f	dimensionless
Retardation factor, vapor phase	R_s	dimensionless
Saturation	s	volume/volume dimensionless
Saturation, effective nonwetting phase	s_{ne}	volume/volume dimensionless
Saturation, effective wetting phase	s_{we}	volume/volume dimensionless
Saturation, NAPL	s_n	volume/volume dimensionless
Saturation, residual	s_r	volume/volume dimensionless
Saturation, residual nonwetting phase	s_{nr}	volume/volume dimensionless
Saturation, residual wetting phase	s_{wr}	volume/volume dimensionless
Saturation, water	s_w	volume/volume dimensionless
Solubility, aqueous of compound i	S_i	mass/volume mass/mass

Table B-1. Listing of selected parameters, symbols, and dimensions.

Parameter	Symbol	Dimensions
Solubility, effective aqueous of compound i	S^e_i	mass/volume mass/mass
Time	t	time
Tortuosity factor	τ_a	dimensionless
Velocity, interstitial	v_i	length/time
Viscosity, absolute (also known as dynamic)	μ	mass/length/time
Viscosity, kinematic	ν	length^2/time
Volume, NAPL displaced by spontaneous imbibition	V_{nsp}	volume
Volume, NAPL displaced by spontaneous imbibition and forced displacement	V_{nt}	volume
Volume, sample pore volume	v_n	volume
Volume, water displaced by spontaneous imbibition	V_{wsp}	volume
Volume, water displaced by spontaneous imbibition and forced displacement	V_{wt}	volume
Volumetric retention capacity	R	volume/volume dimensionless

Table B-2. Length Conversion Factors (multiply by factor to convert row unit to column unit).

	mm	cm	m	km	in	ft	yd	mi
mm	1.0000E+00	1.0000E-01	1.0000E-03	1.0000E-06	3.9370E-02	3.2808E-03	1.0936E-03	6.2137E-07
cm	1.0000E+01	1.0000E+00	1.0000E-02	1.0000E-05	3.9370E-01	3.2808E-02	1.0936E-02	6.2137E-06
m	1.0000E+03	1.0000E+02	1.0000E+00	1.0000E-03	3.9370E+01	3.2808E+00	1.0936E+00	6.2137E-04
km	1.0000E+06	1.0000E+05	1.0000E+03	1.0000E+00	3.9370E+04	3.2808E+03	1.0936E+03	6.2137E-01
in	2.5400E+01	2.5400E+00	2.5400E-02	2.5400E-05	1.0000E+00	8.3333E-02	2.7778E-02	1.5783E-05
ft	3.0480E+02	3.0480E+01	3.0480E-01	3.0480E-04	1.2000E+01	1.0000E+00	3.3333E-01	1.8939E-04
yd	9.1440E+02	9.1440E+01	9.1440E-01	9.1440E-04	3.6000E+01	3.0000E+00	1.0000E+00	5.6818E-04
mi	1.6093E+06	1.6093E+05	1.6093E+03	1.6093E+00	6.3360E+04	5.2800E+03	1.7600E+03	1.0000E+00

Notes: mm=millimeters, cm=centimeters, m=meters, km=kilometers, in=inches, ft=feet, yd=yards, mi=miles; 1 micron (μm) = 0.001 mm.

Table B-3. Area Conversion Factors (multiply by factor to convert row unit to column unit).

	mm^2	cm^2	m^2	km^2	in^2	ft^2	yd^2	mi^2
mm^2	1.0000E+00	1.0000E-02	1.0000E-06	1.0000E-12	1.5500E-03	1.0764E-05	1.1960E-06	3.8610E-13
cm^2	1.0000E+02	1.0000E+00	1.0000E-04	1.0000E-10	1.5500E-01	1.0764E-03	1.1960E-04	3.8610E-11
m^2	1.0000E+06	1.0000E+04	1.0000E+00	1.0000E-06	1.5500E+03	1.0764E+01	1.1960E+00	3.8610E-07
km^2	1.0000E+12	1.0000E+10	1.0000E+06	1.0000E+00	1.5500E+09	1.0764E+07	1.1960E+06	3.8610E-01
in^2	6.4516E+02	6.4516E+00	6.4516E-04	6.4516E-10	1.0000E+00	6.9444E-03	7.7160E-04	2.4910E-10
ft^2	9.2903E+04	9.2903E+02	9.2903E-02	9.2903E-08	1.4400E+02	1.0000E+00	1.1111E-01	3.5870E-08
yd^2	8.3613E+05	8.3613E+03	8.3613E-01	8.3613E-07	1.2960E+03	9.0000E+00	1.0000E+00	3.2283E-07
mi^2	2.5900E+12	2.5900E+10	2.5900E+06	2.5900E+00	4.0145E+09	2.7878E+07	3.0976E+06	1.0000E+00

Notes: mm=millimeters, cm=centimeters, m=meters, km=kilometers, in=inches, ft=feet, yd=yards, mi=miles.

Table B-4. Volume Conversion Factors (multiply by factor to convert row unit to column unit).

	mm^3	cm^3	L	m^3	km^3	in^3	ft^3	yd^3
mm^3	1.0000E+00	1.0000E-03	1.0000E-06	1.0000E-09	1.0000E-18	6.1024E-05	3.5315E-08	1.3080E-09
cm^3	1.0000E+03	1.0000E+00	1.0000E-03	1.0000E-06	1.0000E-15	6.1024E-02	3.5315E-05	1.3080E-06
L	1.0000E+06	1.0000E+03	1.0000E+00	1.0000E-03	1.0000E-12	6.1024E+01	3.5315E-02	1.3080E-03
m^3	1.0000E+09	1.0000E+06	1.0000E+03	1.0000E+00	1.0000E-09	6.1024E+04	3.5315E+01	1.3080E+00
km^3	1.0000E+18	1.0000E+15	1.0000E+12	1.0000E+09	1.0000E+00	6.1024E+13	3.5315E+10	1.3080E+09
in^3	1.6387E+04	1.6387E+01	1.6387E-02	1.6387E-05	1.6387E-14	1.0000E+00	5.7870E-04	2.1433E-05
ft^3	2.8317E+07	2.8317E+04	2.8317E+01	2.8317E-02	2.8317E-11	1.7280E+03	1.0000E+00	3.7037E-02
yd^3	7.6455E+08	7.6455E+05	7.6455E+02	7.6455E-01	7.6455E-10	4.6656E+04	2.7000E+01	1.0000E+00

Notes: mm=millimeters, cm=centimeters, L=liters, m=meters, km=kilometers, in=inches, ft=feet, yd=yards.

Table B-5. Mass Conversion Factors (multiply by factor to convert row unit to column unit).

	mg	g	kg	oz (avoir.)	lb (avoir.)	ton (net)
mg	1.0000E+00	1.0000E-03	1.0000E-06	3.5274E-05	2.2046E-06	1.1023E-09
g	1.0000E+03	1.0000E+00	1.0000E-03	3.5274E-02	2.2046E-03	1.1023E-06
kg	1.0000E+06	1.0000E+03	1.0000E+00	3.5274E+01	2.2046E+00	1.1023E-03
oz (avoir.)	2.8350E+04	2.8350E+01	2.8350E-02	1.0000E+00	6.2500E-02	3.1250E-05
lb (avoir.)	4.5359E+05	4.5359E+02	4.5359E-01	1.6000E+01	1.0000E+00	5.0000E-04
ton (net)	9.0718E+08	9.0718E+05	9.0718E+02	3.2000E+04	2.0000E+03	1.0000E+00

Notes: mg=milligrams, g=grams, kg=kilograms, oz=ounces, lb=pounds.

Table B-6. Time Conversion Factors (multiply by factor to convert row unit to column unit.

time	s	min	hr	d	yr
s	1.0000E+00	1.6667E-02	2.7778E-04	1.1574E-05	3.1710E-08
min	6.0000E+01	1.0000E+00	1.6667E-02	6.9444E-04	1.9026E-06
hr	3.6000E+03	6.0000E+01	1.0000E+00	4.1667E-02	1.1416E-04
d	8.6400E+04	1.4400E+03	2.4000E+01	1.0000E+00	2.7397E-03
yr	3.1536E+07	5.2560E+05	8.7600E+03	3.6500E+02	1.0000E+00

Notes: s=seconds, min=minutes, hr=hours, d=days, yr=years.

Table B-7. Density Conversion Factors (multiply by factor to convert row unit to column unit).

	g/cm^3	kg/m^3	g/L	lbs/in^3	lbs/ft^3
g/cm^3	1.0000E+00	1.0000E+03	1.0000E+03	3.6127E-02	6.2428E+01
kg/m^3	1.0000E-03	1.0000E+00	1.0000E+00	3.6127E-05	6.2428E-02
g/L	1.0000E-03	1.0000E+00	1.0000E+00	3.6127E-05	6.2428E-02
lbs/in^3	2.7680E+01	2.7680E+04	2.7680E+04	1.0000E+00	1.7280E+03
lbs/ft^3	1.6018E-02	1.6018E+01	1.6018E+01	5.7871E-04	1.0000E+00

Notes: g/cm^3=grams per cubic centimeter, kg/m^3=kilograms per meter, g/L=grams per liter, lbs/in^3=pounds per cubic inch, lb/ft^3=pounds per cubic foot.

Table B-8. Velocity Conversion Factors (multiply by factor to convert row unit to column unit).

	cm/s	cm/d	m/s	m/d	m/yr	ft/s	ft/d	ft/yr
cm/s	1.0000E+00	8.6400E+04	1.0000E-02	8.6400E+02	3.1536E+05	3.2808E-02	2.8346E+03	1.0346E+06
cm/d	1.1574E-05	1.0000E+00	1.1574E-07	1.0000E-02	3.6500E+00	3.7973E-07	3.2808E-02	1.1975E+01
m/s	1.0000E+02	8.6400E+06	1.0000E+00	8.6400E+04	3.1536E+07	3.2808E+00	2.8346E+05	1.0346E+08
m/d	1.1574E-03	1.0000E+02	1.1574E-05	1.0000E+00	3.6500E+02	3.7973E-05	3.2808E+00	1.1975E+03
m/yr	3.1710E-06	2.7397E-01	3.1710E-08	2.7397E-03	1.0000E+00	1.0403E-07	8.9886E-03	3.2808E+00
ft/s	3.0480E+01	2.6335E+06	3.0480E-01	2.6335E+04	9.6122E+06	1.0000E+00	8.6400E+04	3.1536E+07
ft/d	3.5278E-04	3.0480E+01	3.5278E-06	3.0480E-01	1.1125E+02	1.1574E-05	1.0000E+00	3.6500E+02
ft/yr	9.6651E-07	8.3507E-02	9.6651E-09	8.3507E-04	3.0480E-01	3.1710E-08	2.7397E-03	1.0000E+00

Notes: cm/s=centimers per second, cm/d=centimeters per day, m/s=meters per second, m/d=meters per day, m/yr=meters per year, ft/s=feet per second, ft/d=feet per day, ft/yr=feet per year.

Table B-9. Force Conversion Factors (multiply by factor to convert row unit to column unit).

	dyne	kgF	N	lb	pdl
dyne	1.0000E+00	1.0200E-06	1.0000E-05	2.2480E-06	7.2330E-05
kgF	9.8070E+05	1.0000E+00	9.8070E+00	2.2050E+00	7.0930E+01
N	1.0000E+05	1.0200E-01	1.0000E+00	2.2480E-01	7.2330E+00
lb	4.4480E+05	4.5360E-01	4.4480E+00	1.0000E+00	3.2174E+01
pdl	1.3830E+04	1.4100E-02	1.3830E-01	3.1080E-02	1.0000E+00

Notes: kgF=kilogram force, N=Newton, lb=pound, pdl=poundal, $1\ N = 1\ kg \cdot m/s^2$.

Table B-10. Pressure Conversion Factors (multiply by factor to convert row unit to column unit).

	atm	bar	cm (water)	in (water)	mm (Hg)	in (Hg)	Pa	lbs/in^2
atm	1.0000E+00	1.0133E+00	1.0333E+03	4.0681E+02	7.6000E+02	2.9921E+01	1.0133E+05	1.4700E+01
bar	9.8692E-01	1.0000E+00	1.0198E+03	4.0149E+02	7.5006E+02	2.9530E+01	1.0000E+05	1.4508E+01
cm (water)	9.6780E-04	9.8062E-04	1.0000E+00	3.9371E-01	7.3553E-01	2.8958E-02	9.8062E+01	1.4227E-02
in (water)	2.4582E-03	2.4907E-03	2.5399E+00	1.0000E+00	1.8682E+00	7.3551E-02	2.4907E+02	3.6135E-02
mm Hg	1.3158E-03	1.3332E-03	1.3596E+00	5.3527E-01	1.0000E+00	3.9370E-02	1.3332E+02	1.9342E-02
in Hg	3.3421E-02	3.3864E-02	3.4533E+01	1.3596E+01	2.5400E+01	1.0000E+00	3.3864E+03	4.9129E-01
Pa	9.8692E-06	1.0000E-05	1.0198E-02	4.0149E-03	7.5006E-03	2.9530E-04	1.0000E+00	1.4508E-04
psi	6.8050E-02	6.8952E-02	7.0314E+01	2.7683E+01	5.1718E+01	2.0361E+00	6.8952E+03	1.0000E+00

Notes: atm=atmospheres, bar=bars, cm=centimeters, in=inches, mm=millimeters, Pa=Pascals, lbs=pounds; $1\ Pa = 1\ N/m^2$, m=meters.

APPENDIX C: GLOSSARY

Adsorption refers to the adherence of ions or molecules in solution to the surface of solids.

Air sparging refers to the injection of air below the water table to strip volatile contaminants from the saturated zone.

Advection is the process whereby solutes are transported by the bulk mass of flowing fluid.

Biodegradation, a subset of biotransformation, is the biologically mediated conversion of a compound to more simple products.

Biotransformation refers to chemical alteration of organic compounds brought about by microorganisms.

Bond number represents the ratio of gravitational forces to viscous forces that affect fluid trapping and mobilization. It can be given as a dimensionless number, such that $N_B = \Delta\rho \ g \ r^2 / \sigma$ where $\Delta\rho$ is the fluid-fluid density difference, g is gravitational acceleration, r is a representative grain radius, and σ is the fluid-fluid interfacial tension. For soils with a wide grain size distribution, r can be replaced by intrinsic permeability.

BTEX is an acronym for Benzene, Toluene, Ethylbenzene, and Xylenes, which are volatile, monocyclic aromatic compounds present in coal tar, petroleum products, and various organic chemical product formulations.

Bulk density is the oven-dried mass of a sample divided by its field volume.

Capillary forces are interfacial forces between immiscible fluid phases, resulting in pressure differences between the two phases.

Capillary fringe refers to the saturated zone overlying the water table where fluid is under tension.

Capillary hysteresis refers to variations in the capillary pressure versus saturation relationship that depend on whether the medium is undergoing imbibition or drainage. Capillary hysteresis results from nonwetting fluid entrapment and differences in contact angles during imbibition and drainage that cause different wetting and drying curves to be followed depending on the prior imbibition-drainage history.

Capillary number represents the ratio of viscous forces to capillary forces that affect fluid trapping and mobilization. It can be given as $N_C = k \ i_w \ / \ \sigma$ where k is the intrinsic permeability, i_w is the water phase pressure gradient, and σ is the fluid-fluid interfacial tension.

Capillary pressure causes porous media to draw in the wetting fluid and repel the nonwetting fluid due to the dominant adhesive force between the wetting fluid and the media solid surfaces. For a water-NAPL system with water being the wetting phase, capillary pressure equals the NAPL pressure minus the water pressure.

CERCLA is an acronym for the Comprehensive Environmental Response Compensation and Liability Act of 1980 which established a national program in the U.S. to respond to past releases of hazardous substances into the environment. CERCLA created the Superfund for financing remedial work not undertaken by responsible parties. Approximately 1200 sites are scheduled for cleanup under the CERCLA program.

CMS is an acronym for RCRA Corrective Measures Study.

Conservative solutes are chemicals that do not react with the soil and/or native groundwater or undergo biological, chemical, or radioactive decay.

Contact angle refers to the angle at a fluid-solid interface which provides a simple measure of wettability. An acute solid-water contact angle measured into the water in a DNAPL-water system indicates that water, rather than DNAPL, preferentially wets the medium (and vice versa).

Critical DNAPL height typically refers to the height of a DNAPL column required to exceed the threshold entry pressure of a medium.

Density is the mass per unit volume of a substance.

Desorption is the reverse of sorption.

Diffusion refers to mass transfer as a result of random motion of molecules; it is described by Fick's first law.

Dispersion is the spreading and mixing of chemical constituents in groundwater caused by diffusion and mixing due to microscopic variations in velocities within and between pores.

Dissolution is the process by which soluble organic components from DNAPL dissolve in groundwater or dissolve in infiltration water and form a groundwater contaminant plume. The duration of remediation measures (either clean-up or long-term containment) is determined by: (1) the rate of dissolution that can be achieved in the field, and (2) the mass of soluble components in the residual DNAPL trapped in the aquifer.

Distribution coefficient refers to the quantity of the solute sorbed by the solid per unit weight of solid divided by the quantity dissolved in the water per unit volume of water.

DNAPL is an acronym for *d*enser-than-water *n*on*a*queous *p*hase *l*iquid. It is synonymous with denser-than-water immiscible-phase liquid.

DNAPL body refers to a contiguous mass of DNAPL in the subsurface.

DNAPL entry location refers to the area where DNAPL has entered the subsurface, such as a spill location or waste pond.

DNAPL site is a site where DNAPL has been released and is now present in the subsurface as an immiscible phase.

Drainage refers to a process during which the saturation of the wetting fluid is decreasing and the saturation of the nonwetting fluid is increasing in a porous medium.

Effective porosity is the ratio, usually expressed as a percentage, of the total volume of voids available for fluid transmission to the total volume of the porous medium.

Effective solubility is the theoretical aqueous solubility of an organic constituent in groundwater that is in chemical equilibrium with a mixed DNAPL (a DNAPL containing several organic chemicals). The effective solubility of a particular organic chemical can be estimated by multiplying its mole fraction in the DNAPL mixture by its pure phase solubility.

Emulsion refers to a dispersion of very small drops of one liquid in an immiscible liquid, such as DNAPL in water.

EOR is an acronym for *e*nhanced *o*il *r*ecovery; EOR refers to processes (such as cosolvent or steam flooding) for recovering additional oil (or NAPL) from the subsurface.

Fingering refers to the formation of finger-shaped irregularities at the leading edge of a displacing fluid in a porous medium which move out ahead of the main body of fluid.

Free-phase NAPL refers to immiscible liquid existing in the subsurface with a positive pressure such that it can flow into a well. If not trapped in a pool, free-phase DNAPL will flow vertically through an aquifer or laterally down sloping fine-grained stratigraphic units. Also called mobile DNAPL or continuous-phase DNAPL.

Gravity drainage refers to the movement of DNAPL in an aquifer that results from the force of gravity.

Halogenated solvents are organic chemicals in which one or more hydrogen atoms in a hydrocarbon precursor such as methane, ethane, ethene, propane, or benzene, has been replaced by a halogen atom, such as chlorine, bromine, or fluorine. Chlorinated solvents (e.g., 1,1,1-trichloroethane, trichloroethene, and tetrachloroethene) have been widely utilized cleaning and degreasing operations. Halogenated solvents are DNAPLs.

Henry's Law Constant is the equilibrium ratio of the partial pressure of a compound in air to the concentration of the compound in water at a reference temperature. It is sometimes referred to as the air-water partition coefficient.

Heterogeneity refers to a lack of uniformity in porous media properties and conditions.

Hydraulic conductivity is a measure of the volume of water at the existing kinematic viscosity that will move in a unit time under a unit hydraulic gradient through a unit area of medium measured at right angles to the direction of flow.

Hydraulic containment refers to modification of hydraulic gradients, usually by pumping groundwater, injecting fluids, and/or using cut-off walls, to control (contain) the movement of contaminants in the saturated zone.

Hydraulic gradient is the change in head per unit distance in a given direction, typically in the principal flow direction.

Imbibition is a process during which the saturation of the wetting fluid is increasing and the saturation of the nonwetting fluid is decreasing in a porous medium.

Immiscible fluids do not have complete mutual solubility and co-exist as separate phases.

Immobile NAPL is at residual saturation, or contained by a stratigraphic (capillary) trap or hydraulic forces, and therefore, cannot migrate as a separate phase.

Interface refers to the thin surface area separating two immiscible fluids that are in contact with each other.

Interfacial tension is the strength of the film separating two immiscible fluids (e.g., oil and water) measured in dynes (force) per centimeter or millidynes per centimeter.

Interphase mass transfer is the net transfer of chemical compounds between two or more phases.

Interstitial velocity is the rate of discharge of groundwater per unit area of the geologic medium per percentage volume of the medium occupied by voids measured at right angles to the direction of flow.

Intrinsic permeability is a measure of the relative ease with which a porous medium can transmit a liquid under a potential gradient. Intrinsic permeability is a property of the medium alone that is dependent on the shape and size of the openings through which the liquid moves.

LNAPL is an acronym for less-dense-than-water nonaqueous phase liquid. It is synonymous with less-dense-than-water immiscible-phase liquid.

Linear soil partition coefficient refers to the ratio of the mass concentration of a solute phase to its mass concentration in the aqueous phase.

Macropores are relatively large pore spaces (e.g., fractures and worm tubes) that characteristically allow the enhanced movement of liquid and gas in the subsurface.

Mass exchange rate refers to the product of the mass exchange coefficient (dissolution rate) and some measure of NAPL-water contact area. It defines the strength of the dissolved contaminant source.

Miscible means able to be mixed.

Mobile NAPL refers to contiguous NAPL in the subsurface that is above residual saturation, not contained by a stratigraphic (capillary) trap or hydraulic forces, and therefore able to migrate as a separate phase is said to be mobile.

Mobility is a measure of the ease with which a fluid moves through reservoir rock; the ratio of rock permeability to apparent fluid viscosity.

NAPL wet refers to media which are preferentially wet by a particular NAPL rather than water.

PAH is an acronym for polycyclic aromatic hydrocarbons, a group of compounds composed of two or more fused aromatic rings (i.e., naphthalene, anthracene, chrysene, etc.). PAHs are introduced into the environment by natural and anthropogenic combustion processes (i.e., forest fires, volcanic eruptions, automobile exhaust, coking plants, and fossil fuel power plants). Creosote and coal tar are DNAPLs that contain a high PAH fraction.

Partitioning refers to a chemical equilibrium condition where a chemical's concentration is apportioned between two different phases according to the partition coefficient, which is the ratio of a chemical's concentration in one phase to its concentration in the other phase.

PCB is an acronym for Polychlorinated Biphenyl compounds. PCBs are extremely stable, nonflammable, dense, and viscous liquids that are formed by substituting chlorine atoms for hydrogen atoms on a biphenyl (double benzene ring) molecule. PCBs were manufactured primarily by Monsanto Chemical Company for use as dielectric fluids in electrical transformers and capacitors.

Phase refers to a separate fluid that co-exists with other fluids.

Plume refers to a zone of dissolved contaminants. A plume usually will originate from the DNAPL zone and extend downgradient for some distance depending on site hydrogeologic and chemical conditions. To avoid confusion, the term "DNAPL" plume should not be used to describe a DNAPL pool; "plume" should be used only to refer to dissolved-phase organics.

Pool refers to a zone of free-phase DNAPL at the bottom of an aquifer. A *lens* is a pool that rests on a fine-grained stratigraphic unit of limited areal extent. DNAPL can be recovered from a pool or lens if a well is placed in the right location.

Porosity is a measure of interstitial space contained in a rock (or soil) expressed as the percentage ratio of void space to the total (gross) volume of the rock.

Raoult's Law relates the ideal vapor pressure and relative concentration of a chemical in solution to its vapor pressure over the solution: $P_A = X_A P_A^\circ$ where P_A is the vapor pressure of the solution, X_A is the mole fraction of the solvent, and P_A° is the vapor pressure of the pure solvent. It can similarly be used to estimate the effective solubility of individual DNAPL components in a DNAPL mixture based on their mole fractions.

RCRA is an acronym for the Resources Conservation and Recovery Act which regulates monitoring, investigation, and corrective action activities at all hazardous treatment, storage, and disposal facilities. RCRA will provide the framework for environmental investigations and cleanup at an estimated 5000 operating and closed facilities.

Relative permeability is the permeability of the rock to gas, NAPL, or water, when any two or more are present, expressed as a fraction of the single phase permeability of the rock.

Residual saturation is the saturation below which fluid drainage will not occur.

Retardation is the movement of a solute through a geologic medium at a velocity less than that of the flowing groundwater due to sorption or other removal of the solute.

Retardation factors can be multiplied by the average linear velocity of groundwater to determine the rate of movement of dissolved chemicals. The retardation factor equals $[1 + (\rho_b/n)K_d]$ where ρ_b is the bulk density of the media, n is porosity, and K_d is the distribution coefficient between media and water.

RFI is an acronym for RCRA Facility Investigation.

Risk assessment involves evaluation of the potential for exposure to contaminants and the associated hazard.

Saturation is the ratio of the volume of a single fluid in the pores to pore volume expressed as a percent and applied to water, DNAPL, or gas separately. The sum of the saturations of each fluid in a pore volume is 100 percent.

Soil flushing refers to the forced circulation (e.g., by use of injection and extraction wells) of water, steam, cosolvents, surfactants, or other fluids to enhance the recovery of contaminants (i.e., immiscible, dissolved, or adsorbed) from soil.

Soil gas refers to vapors (gas) in soil above the saturated zone.

Soil gas surveys are used to collect and analyze samples of soil gas to investigate the distribution of volatile organic compounds in groundwater or soil.

Solidification/stabilization refers to several processes which utilize cementing agents to mechanically bind subsurface contaminants and thereby reduce their rate of release.

Solubility refers to the dissolution of a chemical in a fluid, usually water. *Aqueous solubility* refers to the maximum concentrations of a chemical that will dissolve in pure water at a reference temperature.

Sorption refers to processes that remove solutes from the fluid phase and concentrate them on the solid phase of a medium.

Source characterization involves investigating conditions in the areas of NAPL entry or release.

Specific gravity is the ratio of a substance's density to the density of some standard substance, usually water.

Surface tension refers to the interfacial tension between a liquid and its own vapor typically measured in dynes per centimeter.

Threshold entry pressure is the capillary pressure that must be overcome for a nonwetting NAPL to enter a water-saturated medium. It is also known as the displacement entry pressure.

Vacuum extraction refers to the forced extraction of gas (with volatile contaminants) from the vadose zone, typically to prevent uncontrolled migration of contaminated soil gas and augment a site cleanup.

Vadose zone is the subsurface zone that extends between ground surface and the water table and includes the capillary fringe overlying the water table.

Vapor plume refers to a zone of vapor in the vadose zone.

Vapor pressure is the partial pressure exerted by the vapor (gas) of a liquid or solid substance under equilibrium conditions. A relative measure of chemical volatility, vapor pressure is used to calculate air-water partition coefficients (i.e., Henry's Law constants) and volatilization rate constants.

Viscosity is the internal friction derived from internal cohesion within a fluid that causes it to resist flow. Absolute viscosity is typically given in centipoise. Kinematic viscosity is the absolute viscosity divided by the fluid density.

Volatilization refers to the transfer of a chemical from liquid to the gas phase.

Volumetric retention capacity is the capacity of the vadose zone to trap NAPL which is typically reported in liters of residual NAPL per cubic meter of media.

Water wet refers to media that are preferentially wetted by water relative to another immiscible fluid.

Weathering is a process whereby preferential and selective dissolution of relatively soluble and volatile NAPL components leaves behind a less soluble residue. Weathering causes the ratios of chemicals in the NAPL and dissolved plume to change with time and space.

Wettability refers to the relative degree to which a fluid will spread on (or coat) a solid surface in the presence of other immiscible fluids.

Wetting fluid refers to the immiscible fluid which spreads on (or coats) the solid surfaces of a porous medium preferentially relative to another immiscible fluid. In DNAPL-water systems, water is usually the wetting fluid.

Printed and bound by CPI Group (UK) Ltd, Croydon, CR0 4YY

23/10/2024

01778098-0001